Processes in Physical Geography

Processes in Physical Geography

R D Thompson
A M Mannion
C W Mitchell
M Parry
J R G Townshend

Longman
Scientific &
Technical

Copublished in the United States with
John Wiley & Sons, Inc., New York

Longman Scientific & Technical
Longman Group Limited
Longman House, Burnt Mill, Harlow
Essex CM20 2JE, England
and Associated Companies throughout the world

First published 1986
Reprinted 1992, 1993

British Library Cataloguing in Publication
Processes in physical geography.
  1. Physical geography.
  I. Thompson, R. D.
  910′.02   GB54.5

  ISBN 0-582-30136-X

Library of Congress Cataloging in Publication Data
Main entry under title:

Processes in physical geography.

  Bibliography: p.
  Includes index.
  1. Physical geography. I. Thompson, Russell D.
GB60.P74   1986   551   84-25005
ISBN 0-582-30136-X

Set in Linotron 202 9½/11 pt Times
Produced by Longman Singapore Publishers (Pte) Ltd.
Printed in Singapore

To Maurice Parry, our colleague and co-author, in appreciation of contributions to Reading Geography over 36 years

Contents

CONTENTS

Preface

This book aims to provide an up-to-date review of the various processes operating in the physical environment. The emphasis on processes is based on the authors' belief that only through a detailed study of their operation is it possible to move towards a full understanding of physical geography. Thus, within the book, emphasis is placed on the various processes which influence the mobility of the atmosphere, the erosion–transport–deposition of materials, the development of soils, the growth of plants and the behaviour of the oceans. Consequently, no detailed consideration is given to the results of these processes in terms of world climates, landforms, soil zones and vegetation types. It is believed that such topics should only be dealt with at a more advanced level after the fundamentals of operating processes have been considered in adequate detail.

To achieve this objective, and in view of the structure of senior school and undergraduate courses in physical geography, a traditional approach has been adopted which is manifested in the main section headings. We are aware that this subdivision into atmosphere, sediment transport, soils, vegetation and oceans can create a fragmentary approach to a discipline which relies on synthesis for its uniqueness. Consequently, we have attempted to overcome the problem of fragmentation, and to integrate the processes studied, by adopting a modern systems approach as a central unifying theme. Our emphasis, however, is not on General Systems Theory *per se* but is on a more liberal application in order to express the interrelationships between those processes which have moulded earth surface features. This approach is apparent both in the integrated text and the extensive cross-referencing between chapters.

A positive advantage of the traditional element in our approach to physical geography is that it has allowed five authors to provide synopses of their particular specialisms, namely: J. R. G. Townshend (Chapters 1, 15, 17–23); R. D. Thompson (Chapters 2–7, 33, 36 (section 36.2) and general editing); M. Parry (Chapters 8–14, 34, 36 (section 36.1)); C. W. Mitchell (Chapters 16, 24–27); A. M. Mannion (Chapters 29–32, 35, 36 (section 36.3)). This is unparalleled in physical geography texts, many of which have been written with considerable strength in some areas but are notably deficient in others. We have all contributed substantially to undergraduate courses in physical geography at Reading in which we recognise not only the significance of physical, chemical and biological processes in the environment but also their modification by human agencies.

This book is intended for first-year undergraduate courses in physical geography and environmental science and is also relevant to senior school students. A basic training in science and mathematics is assumed and is necessary to understand the equations and scientific laws which underpin environmental process. International system (SI) units are used except, as in Fig. 2.2, where conversion was considered unnecessary. Terms have been fully defined as far as possible where they occur in the text but reference to Whittow (1984) will provide more complete definitions. As an *aide-memoire*, key topics are given at the end of each chapter along with further reading material for those students who wish to pursue the subject-matter beyond this introductory level.

Acknowledgements

We are indebted to Keith Crabtree, Brian Knapp, Michael Turnbull, Phoebe Walder and John Whittow for advice on original material although we are responsible for any inaccuracies and deficiencies which remain in the text. The production of this book was greatly facilitated by the cooperation of Brian Goodall, head of the geography department at Reading University, by Pamela Dixon who provided the skill and patience of an invaluable typist and by the departmental drawing office, who readily offered cartographic advice. We also acknowledge all the help and encouragement from our respective families.

The authors and publishers wish to thank the following who have kindly given permission for the use of copyright material:

Academic Press Inc & the Author, for Table 2.6 from Table 8.4 p 130 (W. P. Lowry, 1967); American Association for the Advancement of Science for fig 31.2 from fig 1 (Likens et al, 1978) Copyright AAAS 1978, Table 32.1 from Table 1 (E. P. Odum, 1969) Copyright AAAS 1969; Edward Arnold Ltd for figs 21.4, 21.5· from figs 10.12, 10.15 (A. Warren, 1979), Table 30.1 from Table 2.2 (J. M. Anderson, 1981) in *Resource & Environmental Science Series*; Associated Book Publishers Ltd for Tables, 2.1, 2.2 from Tables 1.15, 1.14 pp 24–5 (R. G. Barry, 1969); Association of American Geographers for Table 31.6 from Table 1 p 229 (P. J. Gersmehl, 1976); Blackwell Scientific Publications Ltd for Tables 35.2, 35.4 from Tables 1.2, 1.3 pp 21, 32 (R. S. K. Barnes & R. N. Hughes, 1982); Butterworth & Co Ltd for fig 9.3 from fig 6 p 268 (H. H. Lamb, 1964); Cambridge University Press for fig 18.1 from fig 5.2 p 1000 (M. A. Carson & M. J. Kirkby, 1972), Table 32.3 from Table 2 p 124 (D. Walker, 1970); Gerald Duckworth & Co Ltd for fig 25.3 from fig 3.2 p 45 (I. M. Fenwick & B. J. Knapp, 1982); the Editor, Journal of Ecology for adapted Table 32.2 from Tables 1 & 2 pp 154, 156 (R. D. Roberts, Marrs, R. A. Skeffington & Bradshaw, 1981); Geological Association for fig 33.3 from figs 1 & 8 pp 301, 312 (E. R. Oxburgh, 1974); Gebruder Borntraeger for Table 17.1 from Table 6 p 45 (M. J. Selby, 1982); The Controller of Her Majesty's Stationery Office for figs 11.1a + b from figs 3 + 9 (Lamb et al, 1973), 11.5 a + b parts of 12 h surface charts + 15 h 500 mb contour charts from *DWR* and *DAR* 30.11.51, 9.2.56; Dr. W. Junk, BV Publishers for Table 29.2 from Table 2 p 97 (O. W.H. Heal & S. F. McLean, 1975); the Author, Professor K. King for fig 2.4 (K. King, 1961); Longman group Ltd for figs 15.1 from fig 3.3 p 29 (C. D. Ollier, 1981), 28.3 from fig 2.3 p 24–5 (G. Jones, 1979), Table 27.2 extracted from Table 2.1 p 24 (E. W. Russell, 1973); Macmillan Accounts & Administration Ltd for fig 6.2 from fig 3.9 & Table from Table 2 p 75 (K. S. Smith, 1972); Macmillan Publishing Co Inc for figs 27.3, 27.4 from figs 14.8, 16.10 Table 27.1 from Table 2.33, pp 388, 450, 24 (N. C. Brady, 1974) Copyright © 1974 by Macmillan Publishing Co Inc, Table 29.1 from Table 2 (R. H. Whittaker, 1970) Copyright © 1970 by Macmillan Publishing Co Inc; McGraw Hill Book Co for figs 3.5b from fig 4.20 (G. T. Trewartha, 1968), 4.1a from fig 4.3 (based on data from *Lapworth & S. T. Harding in Hickman*), 4.1b, 6.1, 7.2, 7.3 from (R. C. Ward, 1975), 26.1 adapted from plate IX (V. C. Finch & G. Trewartha 1942, Strahler 1969); Prentice Hall Inc for Table 33.1 from Table 2.1 (data from *J. Geophys. Res.* 71, 4505–25) in (J. P. Kennett, 1982); Royal Meteorological Society for figs 9.4–6, from figs 1–2, 10 (J. S. Sawyer, 1957); Springer Verlag (Heidelberg) for fig 29.2 from fig 2.1 p 7 (Leith & Whittaker, 1975); Springer Verlag (New York) Inc & the Authors for Tables 31.2–4 from Tables 26–7, 5.1 pp 68–9, 148 (F. H. Bormann & G. E. Likens, 1979); Thames Water Authority for fig 7.4 from *recorder charts for River Emborne & River Winterbourne discharge 25.11–2.12, 1975*; University of Chicago Press for figs 2.2, 2.5 from figs 8, 33 (W. D. Sellers, 1965).

Introduction

Chapter 1 Physical geography in perspective

Physical geography or 'natural' geography (Greek *physis* – nature) has always been concerned with process and form. An early book (Herbertson 1901) refers to 'movement' (processes) and the forms to which they give rise. In it the forms, whether of landscape, appearance of sky or vegetation cover, are carefully (even lovingly) described, while the 'movements' tend to be inferred from observation of the forms, an approach with a venerable tradition that can be traced back to the Greek philosophers. Later, as many then highly regarded textbooks testify, climatology concerned itself largely with description based on rapidly accumulating data (see for example Kendrew 1922) and geomorphology became fixated on the concept of the cycle of erosion as propounded by W. M. Davis (see Wooldridge and Morgan 1937). Currently, however, interest has focused more firmly on the processes themselves, with understanding depending less on inference, however perceptive, and more on careful observation, measurement and laboratory experiment. It is with this more scientific approach and with a rational linkage of process and form that this book is concerned.

1.1 The nature and position of physical geography

Physical geography is concerned with the description and explanation of features at the earth's surface and in the adjacent atmosphere and its subject-matter therefore includes climate, water in all its various forms, landforms, soils and vegetation. From such a definition, it is apparent that physical geography has strong linkages with such subjects as meteorology, hydrology, geology, pedology and botany. Given the existence of these other subject areas, it might appear that they overlap so strongly with physical geography as to negate the distinctive role of our subject. The first point to note in the defence of physical geography is that there are no sharp boundaries delimiting any subjects. For example, geology and physical geography may overlap, but geology itself has overlaps with chemistry in geochemistry, with physics in geophysics and with zoology in palaeontology. Moreover, studies near the boundaries between subject areas are often among the most exciting and fruitful and the position of physical geography, at the meeting of these subjects, is not therefore seen as a disadvantage. On the contrary, one of the most important contributions of the subject lies in its focus on the interrelationships between the varied components of the biophysical environment, as they affect the earth's surface features.

The approach in this book is to provide essential introductory material to the biophysical processes operating at the earth's surface. As such, emphasis is placed on ideas and methods developed in the physical and biological sciences and, as appropriate, mathematical and chemical equations have been used to describe relationships. The emphasis on processes, rather than on the results of these processes, deserves a word of explanation, since the reader may be disappointed not to find an explicit account of topics such as global climates and vegetation types. This stems from a belief that an understanding of processes must precede any attempt to explain the resultant spatial distributions of environmental types. The latter should form the subject-matter of a subsequent more advanced text. Furthermore, attempted explanation of the earth's surface features, without a proper comprehension of the biophysical processes operating, is likely to be naïve at best.

It should be clear from the previous introduction that strong linkages are recognised between physical geography and the physical and biological sciences. A second and vitally important set of linkages exists between physical geography and the human world, primarily through its relationships with human geography. The most profound role of humans in the physical environment derives from changes to the vegetation cover since, globally, few vegetation communities can now be found which have not been profoundly altered by humans. Such changes have impacts upon local energy and water balances and these in turn often affect rates of soil erosion and sediment deposition. Of course, human activities do not alter the physical principles underlying the operation of processes in physical geography, but they often strongly impact upon the magnitude and frequency of mass and energy fluxes at the earth's surface. On the other hand, it is important not to overstate the case. For example, the global distribution of atmospheric systems and winds is, to all intents and purposes, unaffected by human activities.

Within the discipline of geography, physical geography is significant because of the many ways in which the physical environment affects human societies. These effects should not be viewed as crude mechanistic forces directly shaping societies: human societies are much more complex and subtle systems than is implied by such explanations. Conversely, failure to recognise the importance of the many contemporary problems presented to humankind by the physical environment represents an almost criminally blinkered view.

In summary, physical geography is a subject area with the task of describing and explaining features at the earth's surface and in the surrounding atmosphere. Its distinctive role lies in the emphasis on the interrelationships between all the varied processes operating in the biophysical environment, rather than concentrating on any individual one. The distinctiveness of physical geography also stems from recognition of the integral role of human activities within the biophysical environment.

In the next section a discussion of the main ways in which scientific theories are developed and evaluated is presented. This is followed in the rest of the chapter by an outline of the main elements of the systems approach. The latter does not provide an alternative to the conventional scientific approach but provides a valuable extension of scientific methodology, especially appropriate to the investigation of complex natural phenomena.

1.2 Explanation in physical geography

In trying to explain the biophysical environment (or any other phenomena, for that matter) generalisations and ultimately laws can be derived or, alternatively, attention may be focused on explaining individual situations or unique events. The first is known as the *nomothetic approach*, which forms the bulk of conventional science, and the latter is called the *idiographic approach*, which has until recently been overwhelmingly dominant in history.

In order to highlight the difference between the two approaches, reference should be made to the ways in which a nomothetic geomorphologist, Nomos, and idiographic geomorphologist, Idios, might investigate a river system. Idios is concerned with the individual character of this particular river: how it is changing; how it has evolved in the past and how it differs from other rivers. Idios believes that the real world is so complicated that generalisations, especially quantitative ones, are almost impossible to derive. Moreover, even if generalisations can be derived about rivers in general, so much of the character of rivers is unique that the content of such generalisations will be very minor. Idios views the efforts of Nomos with disdain because they are so unrealistic and abstract.

On the other hand, Nomos views the work of Idios with ill-disguised contempt. Nomos is involved in making a set of quantitative observations of bank erosion on river bends in order to test Einstein's theory of meander development and to assess the rates of bank erosion with estimates made for a similar-sized river in Arctic Canada.

Nomos thinks it is quite impossible to study the varied aspects of the river as Idios is attempting to do and, in any event, Nomos believes that there is little point in studying rivers without placing such investigations in a broader context. A moment's consideration reveals that Idios and Nomos do not have such conflicting views as they think. Idios has to have some theories of river behaviour at the back of his mind, otherwise no explanation of any type could be given concerning the river: even pure description would be very difficult. Equally, the investigations of Nomos have a strong idiographic flavour. Practical problems prevent Nomos from studying several rivers or as many river bends as ideally would be measured. How representative are the observations of Nomos of river bends in general?

Nevertheless, the two approaches do represent very different ways of making explanations. In this book, the approach adopted is essentially nomothetic, since interest lies in the principles underlying the processes operating in the biophysical environment. This does not mean that the idiographic approach is completely rejected since, especially in applied physical geography, it is inevitable that these principles are applied to specific geographic locations. Derivation of generalisations or laws proceed by one of two basic routes, namely *inductivism* or *falsificationism* (Table 1.1). In inductivism, specific empirical observations are used to make generalisations. In what is termed naïve inductivism (Chalmers 1978) it is supposed that all theory stems from observations. Thus, by observing relationships between variables, generalisations are discovered and theories are derived.

In sophisticated inductivism, it is recognised that theories can arise in many different ways such as inspired guesswork, luck and, of course, following careful observations. However, the process of assessing the validity of the theory is essentially inductive. Thus, a theory is accepted if observations of sufficient reliability and regularity are made which agree with it, although the observations are expected to be numerous and made in a large range of circumstances. Moreover, the theory would not be expected to contradict other established theories. Once established, it is possible to give explanations and make predictions by deduction from the theory.

Now although this approach may seem to be a reasonable one, and is often used by earth scientists, it is usually regarded as an invalid approach by philosophers of science. There are two main objectives to inductivism. Firstly, there is no logical way by which induction can be proven since, just because X has always previously followed the occurrence of Y, it does not follow that this will invariably happen in the future. Nor can it be argued that because the principle of induction has worked numerous times in the past it will continue to work, since this itself is an inductive argument. Secondly, it is impossible to state how numerous the observations need to be and how wide the range of circumstances should be.

Naïve inductivism has the additional criticism levelled at it that theory-free observations are impossible. What is observed is intimately affected by existing knowledge. Deciding what to measure, in order to understand the operation of the global energy cascade, will be determined by current the-

Table 1.1 Principal approaches to law-making. (Based partly on Chalmers 1978)

Stages in making theories and laws	Naïve inductivism	Sophisticated inductivism	Falsificationism
Generation of theories	Use of observations alone	Use of observations but also insight, guesswork, luck, etc.	Deductions from other theories, but can also initially include use of observations, insight, guesswork, luck, etc.
Testing of theories	Use of observations alone, though major contradictions of other theories unacceptable	Use of observations alone, though major contradictions of other theories unacceptable	Design of critical experiments, which attempt to falsify theories. Fittest theory accepted
Use of laws	Used for explanation and prediction, until sufficient observations provide contradiction	Used for explanation and prediction, until sufficient observations provide contradiction	Used for explanation and prediction until another theory is shown to be fitter

ories of what processes will be operating. Measurements are hardly likely to be made of the concentration of carbon dioxide in the atmosphere unless they are related to its influence on the global energy balance.

The most common alternative to inductivism is falsificationism, which has received its most formal treatment by the philosopher Karl Popper. The falsificationist accepts that theory does indeed control the observations which one chooses to make. In traditional science, the origins of the theories are reached entirely by deduction from accepted theory or law. It is now usually acknowledged that factors such as inspiration, luck and chance observation also have a role to play in theory generation. But, it is expected that even if a theory was not derived by deduction initially, it should be possible to do this subsequently. Theories are initially regarded as tentative suggestions, which must be rigorously tested, and all but the fittest should then be discarded. Theories are thus never proven, but an accepted theory is one that has survived a series of attempted refutations better than any other. Compared with inductivism, falsificationism is logical in that singular observations can disprove generalisations although they can never prove them.

It follows that a theory must be falsifiable: that is, it must be testable. The statement that 'Erosion may occur during rainfall' is useless as a theory, since no observation can be made which will refute it. Conversely, the statement that 'The rate of mass movement is directly proportional to slope angle' is acceptable as a theory. The relative merit of theories and laws is related to how general and how precise are their predictions, as well as the facility with which they can be disproved. A theory which is extremely difficult to disprove may be of little value.

It is questionable whether or not falsificationism provides an ideal and complete framework for explanation in physical geography since, although it may be more satisfactory logically than inductivism, it has its difficulties. The first is that observations are not always accurate. Thus, in testing the theory that the rate of mass movements is directly proportional to slope angle, it would be necessary to be fully aware of the errors of measurement which might cause considerable 'scatter' in the expected relationship. Moreover, many other factors might disturb the relationship. Thus, just because one set of observations does not agree with an expected relationship, it might be difficult to refute the theory, especially if it had been derived from sound physical principles which were based on well-established laws. Thus, pure falsificationism is itself a model of how scientists and others make generalisations. Those wishing to explore the topic of explanation in physical geography are encouraged to look at the books by Chalmers (1978) and Harvey (1969). However, it is worth noting that it will be necessary for the reader to provide the appropriate examples from physical geography.

1.3 The systems approach to physical geography

This text is broadly based on a systems approach to physical geography. The advantages of this will become more apparent in the following chapters, but it is appropriate at this stage to give some definitions of systems terminology and to explain in outline the advantages of such an approach. Firstly, in defining a system, a rather abstract definition is that 'A system is a structured set of components of variables (i.e. phenomena which are free to assume variable magnitudes) that exhibit discernible relationships with one another and operate together as a complex whole, according to some observed pattern.' (Chorley and Kennedy 1971)

An example should help to clarify this definition. One of the most commonly described systems is the global hydrologic cycle, in which the movement and storage of water in its various forms is analysed (Ch. 2). Its various components, such as rainfall, evapotranspiration and runoff, are functionally interrelated at a global scale. However, this global system can be subdivided into many similar subsystems and, thus, it is feasible and often useful to examine the hydrological balance within individual drainage basins or even for individual slope segments within a drainage basin. At an even more detailed local scale, an individual raindrop might be regarded as a system, with water being added and lost throughout its brief life as a result of the inputs to and outputs from its local environment.

In the present context, therefore, a system is part of the physical world delimited in space and time, which is characterised not only by its objects and processes in isolation but by the nature of their interrelationships as a whole. The benefits of a systems approach can be justified on several counts. Firstly, because the world is so complex and its components so closely interrelated, the study of isolated parts is very artificial and is potentially misleading. A systems approach provides a rational set of procedures for subdividing the environment into manageable functional parts and then investigating the relationships both within and between terrestrial and atmospheric systems.

Secondly, a study of systems stresses the relationships between components of the environment. The nature of such relationships is as important as the magnitude of individual attributes in characterising and understanding the environment. Thirdly, the adoption of a systems approach is more likely to lead to the inclusion of all relevant variables and objects of importance. Thus, a systems approach in itself will facilitate the understanding of individual components of the environment, through a knowledge of their interdependence with other components.

Fourthly, through the objective definition of elements and variables within systems, a quantitative approach is thereby encouraged. This is an essential precursor for a full understanding of the environment and consequently should be welcomed. Fifthly, deciding how the environment is going to change in the future is often an extremely difficult task and, consequently, a systems approach assists the formulation of such predictions. Finally, for similar reasons to those for prediction, the chances of assessing the repercussions of human activities are enhanced by a systems approach. Consequently, in many circumstances, humans have enhanced possibilities of controlling the environment to improve its use and conserve its resources.

1.4 Morphologic, cascading, process-response and control systems and ecosystems

Conventionally, it is helpful in the understanding of systems to recognise five principal types of systems, namely morphologic, cascading, process-response and control systems and ecosystems. These do not describe different parts of the environment, but represent increasingly sophisticated views of the operation of systems. A *morphologic system* is simply composed of immediate physical properties, which together comprise a part of physical reality, and is characterised by the strength of the relationships between these morphologic properties. For example, a sand-dune system can be characterised by form variables such as slope angles and the height of dune crests, along with material properties (such as grain size) and resistance to wind shear.

As its name implies, a *cascading system* involves a flux (or cascade) of energy, matter or both. Specifically, it is a system having a definite geographic location and extent within which two or more subsystems are dynamically linked by a cascade of mass or energy. The most fundamental cascading system is the solar energy cascade (Ch. 2) in which energy enters the earth–atmosphere system and is transformed into a variety of forms, ultimately affecting every system of concern to physical geographers.

A cascading system thus defines the routes of throughput of energy or matter within a system. Regulations control the operation of cascading systems by determining the proportion of energy or matter which goes into storage and that which produces throughput to the next subsystem. An example of a regulator is infiltration capacity (Ch. 6) which describes the maximum rate at which water may pass into the surface materials of a slope. If it is exceeded, then runoff takes place (Ch. 7) leading to rapid removal of water from this part of the slope system.

Most physical geographers are interested not only in the immediate physical state of the environment, as described by morphologic systems or only in the cascading of energy or matter, but in the interrelationships between them. The latter are described in *process-response systems* which, on the basis of previous definitions, are characterised by the combination or intersection of at least one morphologic and one cascading system. Commonly, a process-response system operates in order to reach at least a short-term equilibrium between the morphological variables.

Such a balance is normally achieved by *negative feedback loops* i.e. the path whereby changes counteract the impact of the initial variable, which create *self-regulating mechanisms* (and are explained below). An example of adjustment of morphological variables may be found in the relationships commonly displayed between stream slope, bed material size and discharge as a result of transport of sediment by fluvial processes. On the other hand, changes in one of the morphologic variables may affect the operation of the cascading system. For example, changes in vegetation cover may affect the magnitude of the infiltration capacity regulator and hence substantially alter the hydrologic cascade of the slope. If some intelligence is used to modify a process-response system, then the term *control system* is used to describe it.

A rather special case of the process-response system becomes apparent when not only physical and chemical, but also biological processes and components are involved. This is known as an *ecosystem* (Ch. 28), which is obviously concerned with linkages between plants and animals and their physical environments. These living components are capable of influencing the physical processes and components, often by way of feedback loops and operate at scales from the microscopic to global. Furthermore, looking beyond the scope of this book, it is also clear that human beings organise themselves in *social systems* and the linkages between these and ecosystems produce a class of human ecosystems.

1.5 Open, closed and isolated systems

Most systems of interest to physical geographers are *open systems*, which means that both energy and matter can enter and leave a system. Clearly, nearly all recognisable parts of the environment are open systems. For example, in the case of soil profiles, water enters the profile chiefly from the surface and from the upslope margins and leaves at the base and downslope sides. It forms the prime cascade by which materials of the soil are altered, usually to produce an equilibrium profile in which separate horizons are produced (Ch. 24), creating a distinctive morphologic system. Additionally materials are added to and removed from the surface of the slope by geomorphological processes.

Closed systems have boundaries which prevent the import and export of matter, but which do allow the transfer of energy across them in both directions. The most obvious examples are present at the global scale, as in the case of the hydrological cascade. At more local scales, open systems are usually evident but examples of closed systems may be found. In the case of some isolated beaches, there may be movement of sand within the beach system but little transfer of sediment into or away from it. Some closed mountain basins have no outlet and thus there may be little transfer of sediment out of the system, until infilling of the basin and erosion of the margins eventually leads to integration with surrounding geomorphic systems. In both these examples, the systems are only closed with respect to sediment and not to water, which will enter and leave the systems.

In *isolated systems*, no transfer of either energy or matter takes place across their boundaries. Although, in practice, such systems do not occur in the biophysical environment, there are occasions when it is useful to assume their existence for the purposes of modelling systems behaviour.

1.6 Positive and negative feedback in systems

One of the most important insights of the systems approach is provided by the idea of feedback. This refers to the process by which a change in one of the system's variables is transmitted through the system structure, so that eventually the effects are returned to the original variable (section 28.4(a)). In the case of *negative feedback*, these effects tend to remove or dampen the original variable and such systems are consequently self-regulating. For example, an animal population may increase temporarily for some reason but, on maturing, the animals will subsequently deplete the food supply and thus ultimately reduce the population to its previous levels.

In *positive feedback* loops, reinforcement of trends is common. For example, removal of vegetation cover exposes the soil below and will often reduce the infiltration capacity, causing an increase in surface runoff which in turn causes increased slope erosion. If this erosion exposes less permeable soil, as is often the case, then the infiltration capacity will be reduced yet further and will lead to even more slope erosion. Key regulators may cause positive or negative feedback mechanisms to be present and hence lead to completely different end results.

1.7 Thresholds, lag times and relaxation in systems

A common characteristic of many biophysical systems is that they may operate at a stable rate through the operation of negative feedback loops for a considerable time. However, a small change in one particular variable may occur and cause a regulator to switch output into a different part of the system structure. If this part of the system contains a positive feedback loop, then the transition may well be irreversible at least in the short term. The transition from one state of system operation to another is called a *threshold*. Furthermore, following the crossing of a threshold, a different state of stability may be reached with its own self-regulating feedback loops (section 28.4(6)).

In many semi-arid areas, sheetwash is a common process of sediment transport. In this process, a thin layer of water travels across a high proportion of the slope surface causing significant (but relatively low) rates of erosion per unit area. However, with a reduction in vegetation cover or especially high rainfall intensity, channelled flow may be initiated forming rills and with enlargement the creation of gullies. This increase in the density of drainage lines frequently results in a permanent shift to an erosional system dominated by fluvial transport.

It should be apparent that terrestrial systems are often exceedingly complex, being comprised of numerous inputs, outputs, regulators and stores. One consequence of this is that the impact of changed inputs of energy or mass may take a considerable time to pass through a system. The length period between the initial input and subsequent impact is called the *lag time*. A well-known example is the time between the maximum rainfall in a storm and the peak discharge of streams (see Fig. 7.1), caused by the travel time of water from various parts of a drainage basin to the stream cross-section, where discharge is measured.

It may be difficult to assess how long lag times are because of the confusing influence of many other variables. For example, it is now confidently believed that major volcanic eruptions produce 'dust veils' which can affect the global energy cascade. However, it is far from certain whether these effects are first detected within a few months or take a year or more to have an impact. Particularly long lag times are found in the readjustment of the ground level following the disappearance of ice sheets and such adjustments are still taking place in Sweden, some 10 000 years after ice retreat began. This example also illustrates the complexities of the relaxation of systems. The land surface does not simply rise to its final equilibrium level but overshoots before reaching this level.

1.8 Retrospect

The list of contents of this book will show that although a systems approach has been adopted (to provide an essential interaction between processes), the material has been arranged fairly conventionally in terms of atmospheric mass, energy and circulation, sediment transport and so on. One problem inevitably faced, both in teaching this material and in writing this book, is that the subject-matter has to be arranged in a linear fashion, whereas one of the recurrent themes of the book is the interrelationships of all these systems. On the basis of teaching experience, the present order has been found to lead to the fewest problems of introducing material in outline to explain one part of the subject, before fully dealing with it later on. On the other hand, it is hoped that the reader will find sufficient cross-references in every section so that alternative orders of reading and course presentation are readily feasible

Key topics

1. Explanation of the relationships between physical geography, the physical/biological sciences and human ecology.
2. Routeways for the derivation of laws for explanation in physical geography.
3. The benefits of a fivefold systems approach to physical geography.
4. Regulation of systems via positive and negative feedback and changes effected by crossing thresholds.

Further reading

Bennett, R. J. and Chorley, R. J. (1978) *Environmental Systems*. Methuen; London and New York.

Dury, G. H. (1981) *An Introduction to Environmental Systems*. Heinemann; London and Exeter, New Hampshire.

Trudgill, S. T. (1977) *Soil and Vegetation Systems*. Clarendon; Oxford.

White, I D., Mottershead, D. N. and Harrison, S. J. (1984) *Environmental Systems*. Allen and Unwin; London and Boston.

Part I
Atmospheric energy and mass systems
Chapters 2–7

The following six chapters introduce the mass and energy cascades which, through the hydrologic cycle, initiate and sustain all atmospheric systems and influence virtually all earth surface processes. The characteristics of the inputs of mass and energy are discussed, particularly water balance and radiative/turbulent energy cascades at every scale, and reference is made to the role of vertical motion in mass/energy changes (especially stable and unstable atmospheric conditions). However, emphasis is placed on the physical processes (both meteorological and environmental) which are involved in the major components of the hydrologic cycle, namely evaporation/transpiration, condensation/precipitation, subsurface water and surface runoff.

Mass and energy transfers along the Hong Kong – Chinese border.

Chapter 2 Characteristics of mass and energy inputs

Atmospheric systems are initiated and sustained by the supply of mass and energy, where the respective channelling or flux of water and heat represents two fundamental cascades in the earth–atmosphere system. Furthermore, it should be noted that whereas mass and energy are conserved at all times, they may be represented by different, continually transformed states which are transported by a variety of subsystems or modes. For example, water may be present in a vaporised, liquid or solid state and is transported by convection, precipitation, percolation and runoff. Similarly, energy exists in four different forms (i.e. radiant, thermal, kinetic or potential energy), with continuous transformation from one form to another, and is transmitted by radiation, convection and conduction.

It is also worth noting that the flow of mass and energy through an atmospheric system over the long term (i.e. a year) may not change the storage of water and heat within the system. In this case, input balances output although the mass and energy leaving the system are not necessarily in the same state or mode of transmission as when they entered. Conversely, over shorter periods (i.e. diurnal patterns), mass and energy storage changes can occur. These are due to the accumulation or depletion of water and heat within the system, which destroy the prevailing mass/energy balance and introduce respective positive and negative budgets.

This chapter is concerned with the properties and characteristics of mass and energy cascades. In terms of atmospheric mass, it examines the *hydrologic cycle* and emphasises the annual water balance of the continents and oceans. Global energy is discussed separately as radiative and turbulent cascades. The former cascade concerns the input of *solar radiation* which indeed represents a great 'engine' which drives the earth's atmospheric and oceanic circulations and generates the weather systems. It also includes the loss of *terrestrial radiation* to the atmosphere, which is mostly re-radiated back to the earth's surface. *Turbulent energy cascades* involve the transfer of radiant energy into the atmosphere by conduction and convection, where the most important convective transfer is associated with evaporation–transpiration/condensation processes and related latent heat fluxes.

2.1 Mass cascades

Ocean storage contains some 97% of all global free water which represents an estimated volume of 1.3×10^{18} m³ of water. The remaining 3% is contained in atmospheric and, especially, continental fresh water of which some 74.61% is in the form of inaccessible ice sheets and glaciers. About

25% of the readily available fresh water is stored in porous rocks as groundwater although less than half has accumulated at accessible depths within 750 m of the surface. Lakes (0.30%), soil moisture (0.06%) and rivers (0.03%) complete the continental storage while the atmosphere contains only 0.03% of all the fresh water available for the mass cascade (Barry 1969). However, even though the amount of water stored in the air is very small in relative terms, it would yield an average 25 mm of rainfall over the entire globe when released as precipitation (Ch. 5). Furthermore, active exchanges with the land and oceans are very considerable and give atmospheric water a key role in the global cascade of water. This cascade is represented by the hydrologic cycle (Fig. 2.1) with its constant state of transformation between the three dominant modes (i.e. *evapotranspiration*, *condensation/precipitation* and surface *runoff*).

Figure 2.1 also illustrates the average annual amounts of water involved in each mode within the global cascade. The cycle is initiated by the energy cascade (see Fig. 2.2) which is used in evapotranspiration (E) from water surfaces, soil and vegetation. This water vapour is transported into the atmosphere by free or forced convection and the rising air cools adiabatically to its *dew point*, when condensation takes place. The continued growth of these droplets and ice crystals eventually exceeds the size where they can be held in suspension by the rising air parcels and these *hydrometeors* fall to earth as precipitation (P). These complex states and modes involved in the workings of the hydrologic cycle will be analysed

fully in the remaining chapters of this section. Over the oceans, E is greater than P and the 13 cm excess water vapour is transported towards the land masses by atmospheric advection (A in Fig. 2.1), where it supplements P which exceeds E by 13 cm. Finally, this surplus land water is transported back to the oceans in the form of surface runoff (R) from streams and rivers.

For the long-term and at the global scale, the water input will equal the water output for the land and ocean subsystems and the annual *water balance* for the continents may be written:

$$P = E + \triangle R \qquad [2.1]$$

where $\triangle R$ is the net runoff, as illustrated in Table 2.1. Table 2.1 indicates that the water balance of the arid lands of Africa and Australia is dominated by E with minimal contributions from $\triangle R$ (i.e. less than 24%). This contrasts with the greater contribution from $\triangle R$ in the more temperate, pluvial continents of Europe and North America. However, the annual water balance for the oceans (Table 2.2), must include the role of ocean currents which modify equation [2.1] as follows:

$$P + R = E \pm \triangle W \qquad [2.2]$$

where $\triangle W$ is the water exchange with adjacent oceans. Table 2.2 data show that the water balance of the Atlantic and Indian Oceans (where E exceeds both P and R) is maintained by the inflow of 'surplus' water from the adjacent Arctic and Pacific Oceans respectively where E is considerably less than P and R. However, it should be emphasised at this global scale that despite the above

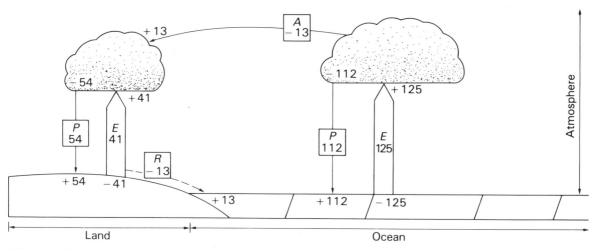

Fig. 2.1 Schematic representation of the average annual hydrologic cycle. Modes and symbols are explained in the text. Values are expressed in cm yr⁻¹. (Data from Budyko 1962)

Table 2.1 Annual water balance of the continents (cm yr⁻¹; E and ΔR values are also expressed as a % of P. (After Barry 1969)

Continent	Precipitation (P) =	Evapotranspiration (E) +	Net runoff (ΔR)
Africa	67	51 (76%)	16 (24%)
Asia	61	39 (64%)	22 (36%)
Australia	47	41 (87%)	6 (13%)
Europe	60	36 (60%)	24 (40%)
N. America	67	40 (60%)	27 (40%)
S. America	135	86 (64%)	49 (36%)

$P = E + \Delta R$ (handwritten)

Table 2.2 Annual water balance of the oceans (cm yr⁻¹). (After Barry 1969)

Ocean	Precipitation + (P)	Runoff from adjacent land (R)	= Evaporation + (E)	Water exchange (ΔW) with adjacent oceans
Atlantic	78	20	104	−6
Arctic	24	23	12	35
Indian	101	7	138	−30
Pacific	121	6	114	13

$P + R = E \pm \Delta W$ (handwritten)

continental and oceanic transfers, the total earth–atmosphere system is closed to the import and export of mass. Therefore the overall annual water balance may be simplified to:

$$P = E \qquad [2.3]$$

where all *advective transfers* (including airstreams, surface runoff and ocean currents) are internal and the net water storage for the global system is zero.

Whereas continental water balances are controlled by P, E and ΔR, *soil moisture* (Ch. 6) must be included in the mass cascade of small-scale surface –atmosphere interactions, as follows:

$$P = E + \Delta R + \Delta S \qquad [2.4]$$

where ΔS represents the net change in soil moisture content. At the same time, it is apparent that P is not a continuous daily function like E and,

during drought conditions, the soil moisture store becomes the dominant input to vapour flux. In these circumstances, equation [2.4] reduces to:

$$E = \Delta S \qquad [2.5]$$

particularly on level terrain where ΔR is negligible. Consequently, on the short time-scale, ΔS is well above zero and becomes a significant factor in the water cascade. This contrasts with the annual situation described above where the net water storage in the earth–atmosphere system was zero . The short-term water balance variations are evident in Table 2.3 where the use of stored soil water, i.e. 102 mm which represents the water available to most crops according to Thornthwaite (1948), postpones the summer water deficiency. Conversely, the winter water surplus is delayed by some two months until the 102 mm soil water has been recharged.

Table 2.3 Water balance at Harrogate, England (54 °N), according to the Thornthwaite method (mm). (After Smith 1972)

	J	F	M	A	M	J	J	A	S	O	N	D	Year
PE*	8	10	25	42	70	95	110	97	68	42	21	12	600
Rainfall	80	62	49	53	62	51	72	74	63	75	79	71	791
Storage change	0	0	0	0	−8	−44	−38	−12	0	+33	+58	+11	
Storage	102	102	102	102	−94	50	12	0	0	33	91	102	
Soil moisture deficiency	0	0	0	0	8	52	90	102	102	69	11	0	
Water deficiency	0	0	0	0	0	0	0	12	5	0	0	0	
Water surplus	72	52	24	11	0	0	0	0	0	0	0	49	208

* PE = *potential evapotranspiration*, as defined in Chapter 4.

It should be reiterated that the hydrologic cycle and all water balances are initiated and sustained by the supply of energy which is required for evapotranspiration and the transfer of water vapour by convection. This energy is supplied by radiative and turbulent cascades over the surface of the globe as described in the next two sections.

2.2 Global radiative energy cascades

The total earth–atmosphere system is closed to the export or import of mass but it does allow the exchange of energy with space. The sole input to the energy flow through this system is radiation emitted by the sun, some 150 million km away from the earth, with a surface temperature of 6000 K. This solar radiation represents the 'engine' which drives the energy cascade of the system and is radiant energy in the form of *electromagnetic waves* (with increasing wavelength and decreasing frequency from gamma radiation through to radio waves) travelling at the speed of light (3×10^8 m s^{-1}). All bodies with a temperature above *absolute zero* (0 K or –273.2 °C) emit radiant energy and a *black body* is one which absorbs and emits all the radiation falling upon its surface. Furthermore, the rate of flow (flux) of radiation from such a body is obtained from the *Stefan–Boltzmann law* which states that this flux is directly proportional to the fourth power of its *absolute temperature* as follows:

$$\text{Radiation flux} = \sigma T^4 \qquad \textbf{[2.6]}$$

where σ is the Stefan–Boltzmann constant (5.67×10^{-8} W m^{-2} K^{-4}) and T is the surface temperature of the body in degrees Kelvin (K).

The sun is assumed to be a black body and radiates about 99.97% of the energy required by the earth–atmosphere system. The *solar flux density* as obtained from equation [2.6] with a surface temperature of 6000 K, is equal to 1353 ± 10 W m^{-2}. This value, known as the *solar constant*, is that received outside the atmosphere on a plane surface placed normal to the solar beam. According to the *Wien Displacement Law*, the wavelength of maximum intensity of emission

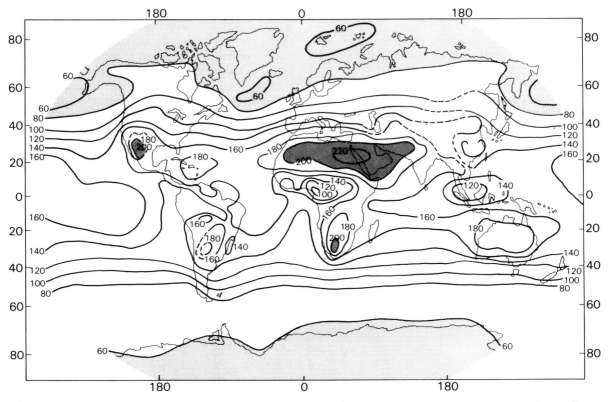

Fig. 2.2 The average annual solar radiation on a horizontal surface at the ground. The units are kilolangleys per year. (After Sellers 1965)

(λ max) from a black body is inversely proportional to the absolute temperature (T) of the body:

$$\lambda \text{ max } (\mu m) = 2897T^{-1} \qquad \textbf{[2.7]}$$

For the sun (at 6000 K), the wavelength of maximum emission is 0.48 μm which is in the green of the visible portion of the spectrum. Furthermore, almost 99% of the solar radiation is contained in the so-called short wavelength of the spectrum between 0.15 and 4.0 μm. Of this, 9% is in the ultraviolet (u.v.) ($\lambda < 0.4$ μm), 45% is in the visible ($\lambda = 0.4$–0.74 μm) and 46% is in the near-infrared (up to 4.0 μm).

The amount of solar radiation actually incident on the top of the atmosphere depends on the time of year, time of day and the latitude. However, on an annual basis and spread uniformly over the outer edge of the atmosphere, the amount of solar radiation received per unit area and time is approximately 338 W m^{-2}. This average value is equal to about 25% of the solar constant which represents the maximum possible short-wave receipt at any point in the earth–atmosphere system. Figure 2.2 illustrates that the distribution of solar radiation reaching the earth's surface is not uniform throughout this system. Due to the greater obliquity of the solar beam with increasing latitude, the annual value in the subtropics (Fig. 2.2) is about three times that in polar areas (Sellers 1965). Figure 2.2 also reveals the relatively low values of solar radiation arriving in equatorial areas associated with the persistent cloud cover and high reflection rates (see Fig. 2.3). However, Fig. 2.2 does not illustrate the fact that the actual solar constant can be approached at some subtropical locations at high

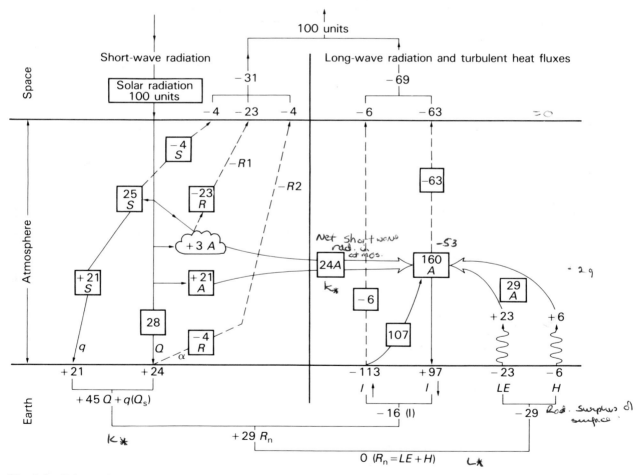

Fig. 2.3 Schematic representation of the atmospheric energy cascade. Transfers and symbols are explained in the text. Values are expressed as percentage of incoming solar radiation at the outer edge of the atmosphere. (Adapted from Barry and Chorley 1982)

elevations when the sun is directly overhead (at zenith) in cloudless, impurity-free atmospheric conditions.

During its passage from the top of the atmosphere to the earth's surface, the solar radiation has to penetrate and diffuse through clouds and a variety of atmospheric constituents like water vapour, aerosols and numerous gases. Consequently, part of the spectrum is scattered or reflected, part is absorbed and the remainder transmitted to the earth's surface as direct radiation. Figure 2.3 illustrates this atmospheric energy cascade and all fluxes are represented as a percentage of the 338 W m^{-2} solar energy arriving at the top of the atmosphere. Some 25% of the beam is scattered (S) by air molecules and minute impurities, a selective process in that the shorter (blue) wavelengths at about 0.4 μm are scattered more readily than the longer (red) wavelengths. This is frequently referred to as *Rayleigh scattering* and accounts for the blue colour of the sky. However, only a relatively small portion (4%) of the scattered short-wave radiation is lost to space since 21% is scattered downwards and reaches the earth as indirect *diffuse radiation* (q) or *skylight*.

Reflection (R) is mainly from cloud tops, where the proportion of incident radiation reflected or the *albedo* (α) averages 55%, and 23% of the incident radiation is reflected back to space. Absorption (A) by clouds (3%) and atmospheric gases (21%) occurs primarily at the shorter wavelengths (below 0.3 μm) where oxygen and ozone absorb the bulk of the X-rays, gamma rays and lethal u.v. radiation. Water vapour and carbon dioxide are more effective absorbers with wavelengths beyond 0.8 μm. Finally, 28% of the incident radiation reaches the earth's surface as direct-beam solar radiation (Q). However, some 4 units are reflected (R) back to space immediately due to the earth's albedo (α) which is a variable factor ranging from 5% for dark wet sand to 20% for deciduous forest and to 95% for fresh snow.

Figure 2.3 indicates that, on average, 31 units of solar radiation are transmitted back to space by scattering and reflection and a further 24 units are absorbed in the atmosphere. Therefore, slightly less than half of the solar radiation received at the outer edge of the atmosphere (45% or 152 W m^{-2}) actually reaches and is absorbed by the earth's surface as Q_s (24% direct beam Q and 21% diffuse skylight q). The energy is converted from radiation into thermal energy which heats the earth's surface which, in turn, becomes a source of long-wave radiation ($I\uparrow$) together with a considerable

amount of energy released by the earth's materials (Fig. 2.3). At 293 K, this terrestrial radiation is mostly emitted in the *infrared spectral range* from 5 to 50 μm with a peak emission at 10.2 μm (equation [2.7]) and, using equation [2.6], this radiation transfer equals 113 units.

Although the atmosphere is semi-transparent to short-wave radiation (absorbing only 24% of the incident solar beam), it readily absorbs some 90% of the emitted long-wave terrestrial radiation. The principal absorbers are water vapour (5.3–7.7 μm and beyond 20 μm), ozone (9.4–9.8 μm), carbon dioxide (13.1–16.9 μm) and clouds (all wavelengths). Consequently, only 6 units of the terrestrial radiation escape directly to space mainly through the *atmospheric 'window'* between 8.5 and 14.0 μm, where water vapour and carbon dioxide are weak absorbers. The rest of the radiation is absorbed (A) by the atmosphere (107 units) which, in turn, re-radiates the absorbed energy partly to space (10 units) and mainly (97 units) back to the earth's surface as *counter-radiation* ($I\downarrow$) or the so-called '*greenhouse effect*' (sections 30.4 and 36.1). Thus, the net or effective outgoing long-wave radiation loss from the earth's surface is reduced considerably from what would be observed with a perfectly translucent atmosphere. Furthermore, the absence of this counter-radiation would reduce the temperature at the earth's surface by 30–40 °C (Sellers 1965).

Table 2.4 compares the outgoing terrestrial radiation ($I\uparrow$) with the incoming solar radiation (Q_s) discussed earlier. The latitudinal variation in absorbed Q_s is again apparent, where the equatorial values are approximately two and a half times the polar values. However, the $I\uparrow$ values

Table 2.4 Distribution of incoming solar radiation (Q_s) and outgoing terrestrial radiation ($I\uparrow$), northern hemisphere (W m^{-2}). (Adapted from Trewartha 1968)

Latitude (°)	Q_s Absorbed	$I\uparrow$ lost	Difference
0	237	189	48
10	233	197	36
20	223	198	25
30	207	198	9
40	186	197	−11
50	162	193	−31
60	135	190	−55
70	112	181	−69
80	100	176	−76
90	98	176	−78

are all within 22 W m^{-2}, with the greatest losses occurring in the dry cloud-free subtropics where counter-radiation is weakest. The difference (or *net radiation*, as described in the next section) reveals a radiative energy surplus or heat source south of 35 °N while negative values and a heat sink characterise latitudes to the north, with peak deficiencies occurring in the High Arctic where they exceed 69 W m^{-2}. It is worth noting here that these latitudinal energy surpluses and deficiencies are responsible for the primary or global circulation systems in the atmosphere which are discussed in Chapter 11.

The total long-wave radiation loss to space (69 units) in Fig. 2.3 also includes 24 units radiated from the absorbed short-wave radiation described earlier and 29 units transferred by the turbulent energy cascade. This turbulent transfer is necessary to balance the atmospheric energy cascade, where the absorbed solar radiation is 'consumed' by the outgoing long-wave radiation ($I\uparrow$) and convective/conductive heat transfers explained in the following section.

2.3 Global turbulent energy cascades

The previous section indicates that the globe has an annual radiant energy surplus of 29 units which would increase surface temperatures at a rate of about 250 °C day^{-1} in the upper centimetre (with a corresponding atmospheric cooling of 1 °C day^{-1}) without an effective atmospheric transfer by non-radiative fluxes (Sellers 1965). The required transfer is accomplished by the turbulent cascade of energy away from the earth's surface associated with conduction and convection. *Thermal conduction* is the process whereby heat is transmitted within a substance by the collision of rapidly moving molecules. Naturally, this transfer is most effective in solids and least effective in gases. Consequently, molecular conduction is only significant in the atmosphere within a few millimetres of the earth's surface (in the so-called non-turbulent *laminar boundary layer*) and is quite negligible on the atmospheric scale represented in Fig. 2.3.

On the other hand, convection involves the vertical turbulent movement of energy which can only occur in liquids and gases. In the atmosphere, local air movements or *eddies* transport energy and mass upwards through turbulent motions initiated by free or forced convection. *Free convec-*

tion relates to parcels of air which rise due to a higher temperature and lower density inside the parcel compared with the surrounding environment. On the other hand, *forced or mechanical convection* occurs when air parcels are moved aloft over surface obstacles such as mountain ranges (*orographic displacement*) or when contrasting air masses are in juxtaposition (*frontal uplift*). The actual mechanics of forced and free convection will be discussed in section 3.1 but it is apparent that these processes transport energy into the atmosphere in the form of both *sensible heat* (H) and *latent heat* (LE). However, of the 29 units transferred this way (Fig. 2.3), only 6 are accomplished by sensible heat transfers (or the direct transfer of atmospheric heat energy) and this occurs when the surface is warmer than the lower atmosphere and the sensible heat is released upon mixing with the cooler air.

The most important convective transfer (involving 23 units) is associated with the evapotranspiration/condensation processes and the related latent heat fluxes. On average, evapotranspiration from water bodies, soil and plants consumes 23 units of available radiative energy, since 2.45 MJ kg^{-1} is required for the latent heat of vaporisation at 20 °C. Furthermore, this energy is stored in the vapour and is transported into the atmosphere by the free or forced convective systems described earlier. The rising air parcels will eventually cool beyond their dew point levels and the vapour then condenses to form clouds and possibly precipitation. At this time, the stored energy is released into the atmosphere as the latent heat of condensation (i.e. 2.45 MJ kg^{-1} at 20 °C) which can influence the equilibrium or *stability tendency* of the air (section 3.4).

It should be noted that these 29 units of latent and sensible heat are stored temporarily in the atmosphere until they are finally transferred into space by *atmospheric radiation* (Fig. 2.3). They combine with the radiative transfers described in section 2.2, to balance the earth–atmosphere energy cascade. Furthermore, the radiative energy available for the two turbulent heat fluxes represents the net exchange between all-wave incoming and outgoing radiation. This exchange is termed net radiation (R_n) and is expressed in equation [2.8], with the terms indicated in Fig. 2.3.

$$R_n = \underbrace{\frac{Q_s}{(Q+q)(1-\propto)}}^{+} \quad \underbrace{\frac{I}{I\downarrow - I\uparrow}} \qquad [2.8]$$

In terms of a global *energy balance*, R_n is partitioned between sensible heat (H) and latent heat (LE) fluxes as follows:

$$R_n = H + LE \qquad \textbf{[2.9]}$$

although the actual partitioning varies between land and ocean subsystems. For example, the world's oceans use about 90% of R_n to evaporate water (LE) whereas over land surfaces, the two turbulent heat fluxes are almost equally important (LE 51%, H 49%). However, there is considerable variation in the actual energy balance recorded by individual continents. This variability is illustrated in Table 2.5 and simply relates to the availability (or otherwise) of surface water vapour. The end column in the table indicates the ratio of sensible heat to latent heat flux (H/LE) commonly known as the *Bowen ratio* (β).

Table 2.5 Annual energy balance of the continents (H and LE values are expressed as a % of R_n). (Adapted from Sellers 1965)

Area	$R_n(\%)$	LE	H	$H/LE(\beta)$
Europe	100	62	38	0.62
N. America	100	57	43	0.74
S. America	100	64	36	0.56
Australia	100	31	69	2.18
Asia	100	47	53	1.14
Africa	100	38	62	1.61

Table 2.5 indicates that the continents can be divided into two main energy balance groups. Firstly, Europe and the Americas have energy balances dominated by the LE flux, with a β less than 1.0 since these continents are pluvial with freely evaporating and transpiring surfaces. Secondly, in Australia, Asia and Africa (which contain about 90% of the world's deserts), vapour flux is reduced and H becomes the main energy sink which is used to warm the air. The Bowen ratio now exceeds 1.0 which confirms the dominance of the H flux.

The global energy balance described above represents heat transfers over the long term, when net radiation storage does not occur and energy input equals the output of heat. However, with short-term diurnal cascades described in the next section, heat can be stored temporarily in the atmosphere and/or the ground, which is responsible for negative or positive energy balances.

2.4 Diurnal turbulent energy cascades at the earth's surface

This diurnal energy cascade is associated with the H and LE fluxes observed above on the global scale together with the conduction of heat to or

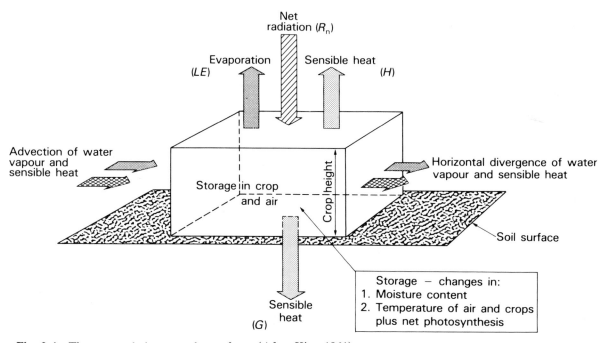

Fig. 2.4 The energy balance at the surface. (After King 1961)

from the underlying ground (G), which completes the surface energy balance as follows:

$$R_n = H + LE + G \qquad [2.10]$$

During the night, however, Q_s is zero and $I\uparrow$ maintains a negative R_n with the three surface fluxes acting as heat sources for the long-wave emissions. It should be mentioned that other important environmental responses, especially the energy requirements of *photosynthesis* (section 29.1) and respiration, are rather insignificant regarding the use of available net radiation (i.e. $< 1\% \, R_n$) and are generally omitted from energy balance considerations (Fig. 2.4). Figure 2.4 also shows the importance of advective energy in the diurnal cascade. For example, energy can be advected into or out of the system in the form of water vapour and sensible heat transfers. Under the former condition, the energy requirements at the site can exceed the available R_n and supplementary energy is provided by H.

It is apparent that the diurnal energy cascade is controlled by the nature of the surface and the relative abilities of the atmosphere and soil to transport heat. The major surface variables concerned are the resistance (or otherwise) to evapotranspiration and the insulation–thermal conductivity properties of the ground material. These environmental constraints are evident in Table 2.6 which makes qualitative assessments of energy balance differences over a variety of contrasting surfaces (Lowry 1967). For example, a cornfield or other 'lush' growing crop is dominated by LE, whereas H is the main energy sink over a pavement and desert surface. Also, leaf litter and the snow pack are excellent insulators and prohibit G, whereas the transparency and turbulent mixing of open water are conducive to large amounts of G. However, it must be remembered that no matter which flux dominates the energy cascade over a particular surface, it is clear that convection is the principal means of daytime heat transport away from the interface, as discussed earlier.

More specific data are provided by Fig. 2.5, which illustrates the diurnal energy balance recorded at three different sites in the USA. The diurnal patterns were similar at each site with negative values of net radiation and heat fluxes

Table 2.6 Some qualitative generalisations about the energy balance of a variety of environments. (After Lowry 1967)

Type of site	Remarks on the energy balance of the site
Mid-latitude vegetated areas generally	Dissipation modes in order of decreasing magnitude are LE, H, and G
Cornfield or other growing crop	LE as much as 80–85% of R_n
Meadow	LE largest dissipation mode during growth; H dominates when the grass is cut, and after it cures and mats to form an insulator, and a vapour barrier
Pavement or stonefield	LE is almost nil; H and G are about equal
Desert	Similar to a stonefield, except H larger than G
Ice cap (high latitude)	LE and G almost nil; H and R_n are about equal and opposite in sign – H positive and R_n negative
Open water	LE and G completely dominate H; LE is at the maximum possible given water and air temperatures; G large because of water's transparency and convective mixing downward below the surface
Still water	G is reduced as compared with open water, since convective mixing is reduced. As a result, the water surface becomes warmer, increasing both LE and H
Snow (mid-latitude)	Although snow is transparent to the incident sunlight, its large albedo keeps R_n comparatively small; as long as snow and air are below freezing, the excellent insulation of snow keeps G small, low vapour pressures keep LE small, and H becomes a large fraction of R_n. Snow at the freezing temperature results in melting and mass transfer within the G term
Leaf litter	The litter acts as a thermal insulator, keeping G very small; acts as a vapour barrier when dry; H is largest mode of dissipation

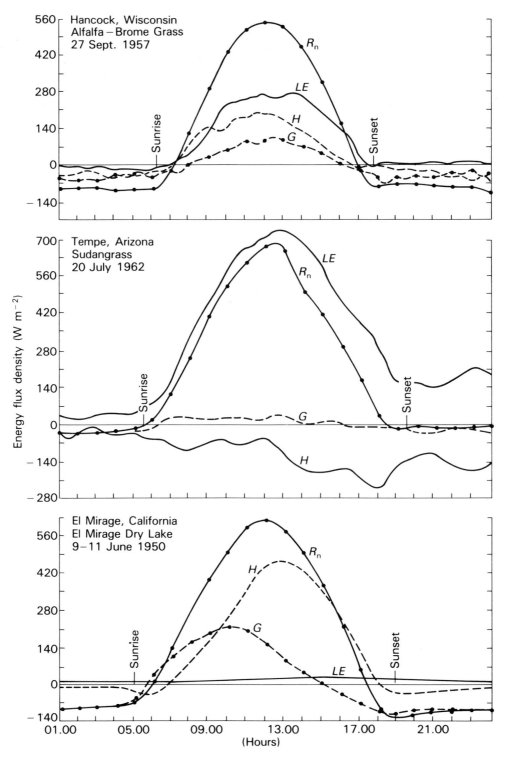

Fig. 2.5 Average diurnal variation of the components of the surface energy balance over grass at Hancock, Wisconsin and Tempe, Arizona and over bare soil at El Mirage, California. (After Sellers 1965)

before sunrise/after sunset and with obvious peak values around solar noon. However, the individual R_n partitioning at each site displayed quite considerable variation associated with the availability of surface water. For example, the data from the Wisconsin grasslands were typical of any mid-latitude vegetated site with R_n dissipated as LE, H and G in decreasing magnitude. Furthermore, the modest daytime demands made by evapotranspiration were mainly met by the available R_n apart from the hour or so before sunset. This was not the case over the irrigated Sudangrass in Arizona where the driving force for evapotranspiration (i.e. the vapour pressure difference between the air and grass surface) was much greater. Consequently, LE exceeded R_n at all times and the necessary energy requirements had to be supplemented by H, which was advected from the surrounding desert. Finally, at the barren dry lake site in California, evapotranspiration rates and the related LE flux were close to zero which represented an extreme case of resistance to vapour flux. The major energy sink was now H although G dominated the cascade between sunrise and 09.30 hours, when the soil heat capacity reached its maximum.

2.5 Retrospect

It is obvious that the input of mass and energy are fundamental cascades in atmospheric systems at every scale. For example, the hydrologic cycle and water balance of the continents and oceans represent significant closed mass cascades in the earth–atmosphere system. They combine with global energy cascades to develop major climatic patterns and prevailing circulation systems (Ch. 11). Furthermore, the complex cascades of global radiative energy (including both short-wave solar and long-wave terrestrial radiation) are balanced by the turbulent energy cascade. Here convective transfers (particularly the latent heat flux) are responsible for 42% of the total atmospheric radiation into space (Fig. 2.3). Diurnal energy cascades are influenced by short-term heat storage changes in the air and ground which lead to temporary negative or positive energy balances. These changes represent significant heat sources and sinks over the short-term which are associated more with synoptic weather systems (Chs 13 and 14).

Key topics

1. The global cascade of water effected through the hydrologic cycle.
2. The nature of inputs and outputs of radiation in the global radiative energy cascade.
3. The role of sensible and latent heat fluxes in the annual energy balance of the global turbulent energy cascade.
4. The contrasting energy balances as part of the diurnal turbulent energy cascade.

Further reading

Gates, D. M. (1962) *Energy Exchange in the Biosphere*. Harper and Row; London and New York, pp. 1–112.
Miller, D. H. (1977) *Water at the Surface of the Earth*. Academic Press; London and New York.
Miller, D. H. (1981) *Energy at the Surface of the Earth*. Academic Press; London and New York.
Ransom, W. H, (1963) Solar radiation and temperature, *Weather*, **8**, 18–23.

Chapter 3   Vertical motion and mass/energy changes

The previous chapter has emphasised that the transfers of mass and energy by turbulent fluxes are important subsystems in the global energy cascade. Without these transfers the atmosphere would be heated to an intolerable degree within the boundary layer, which would accentuate the latitudinal imbalance of insolation observed between the equator and the North and South Poles (see Fig. 2.2). It should be noted that advective transfers of sensible and latent heat also help to offset the excessive surface heat accumulation in the tropics and represent an important negative feedback mechanism. These horizontal fluxes are associated with the movement of mass and energy in ocean currents (e.g. the Gulf Stream/North Atlantic Drift) and airstreams, such as the prevailing southwesterly winds in the temperate areas. It is worth noting that subtropical ocean currents alone account for about 25% of the energy moved out of low latitudes.

This chapter examines the turbulent cascade of energy and mass within the troposphere by free and forced convection, and considers the temperature changes which occur in these rising parcels of air *vis-à-vis* the surrounding atmospheric environment. This leads on to a discussion of the thermal and density relationships between the rising parcels and the surrounding air in terms of basic stability tendencies, which control the degree of free convection in particular.

3.1 Types of vertical motion

Convection represents the most effective part of the turbulent cascade of energy and mass within the atmosphere and the associated vertical motion is accomplished by the free and forced transport of air aloft. Free convection refers to air parcels which are heated initially through contact with surface heat accumulation, associated with local concentrations of absorbed solar radiation. For example, surface heating irregularities are related to albedo differences which characterise urban and rural areas. The lower albedo of a city's bitumen (2%) contrasts markedly with that over rural grasslands (15–30%) and partly accounts for the *urban heat island*, which results in the greater free convection, cloud cover and precipitation over urban areas.

When an air parcel is heated from below, it will eventually develop a higher temperature and lower density (with expansion) compared with the surrounding air. The warmer, less dense air parcel now rises from the surface and is replaced by cooler, more dense air converging across the surface towards the area of excessive heating. The more buoyant air parcel will continue to rise until its density is in equilibrium with the surrounding air, at which level it will diverge horizontally.

Eventually, the cooling processes associated with its movement away from the heating source will make the parcel cold and dense enough to descend through the atmosphere and replenish the supply of air converging towards the overheated surface. This process generates the formation of thermal low pressure systems and hurricanes which are discussed in detail in sections 10.3 and 13.5.

Forced or mechanical convection represents the vertical transfer of energy and mass by *eddy currents* which are associated with orographic obstruction to smooth or *laminar air movement,* *frontal uplift* along the boundaries of contrasting air masses and the *mass convergence* of low pressure systems (section 10.3). Individual turbulent systems that develop from this deflection of surface air are known as eddies. They form within a zone of variable depth and height in the so-called *friction layer* (generally below 1000 m) and depend largely on the *surface roughness* (or degree of irregularity) and the prevailing wind velocity. Strong orographic and frontal displacement promotes a steady updraught of air and is responsible for mixing air with contrasting temperature, humidity and density from upper and lower levels. The resultant circulation can lead to distinctive cyclogenesis and the formation of *orographic lows* and *wave depressions* of middle latitudes which are discussed in sections 10.3 and 13.2.

3.2 Vertical distribution of temperature in the troposphere

Figure 3.1 illustrates the temperature distribution within the atmosphere where warming trends characterise the stratosphere and thermosphere associated with the absorption of u.v. solar radiation in ozone and atomic oxygen respectively. As a general rule, temperature decreases with increasing elevation throughout the mesosphere and troposphere up to the tropopause around 11 km. This is known as the *environmental lapse rate* (ELR) or vertical temperature gradient, which averages 0.65 °C per 100 m or 6.5 °C per 1000 m in the troposphere and refers to the temperature conditions existing in a large mass of stationary air at a given time and place. It should be remembered that this air has no vertical motion and represents the total atmospheric environment through which parcels of air are moved by free and forced convection. This lapse rate is approximately 1000 times greater than the average horizontal rate of

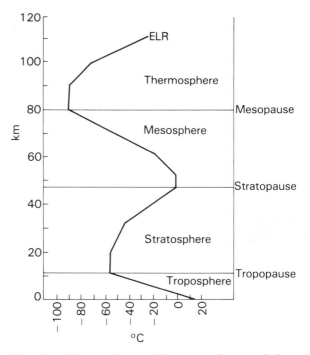

Fig. 3.1 The structure of the atmosphere and the distribution of temperature (ELR).

temperature change with latitude (Trewartha 1968) and is associated with a number of atmospheric factors. Firstly, the highest air temperatures occur at low elevations next to the earth's surface, which indicates that the air is heated from below by conduction and convection. Furthermore, an increase in elevation will result in increasing remoteness from this source of heating and a progressive air cooling. Secondly, the decreasing amount of water vapour recorded aloft will lead to a greatly reduced 'greenhouse effect' (section 2.2) and marked reduction in the absorption of infra-red long-wave radiation emitted from the earth's surface. Finally, the air becomes thinner and less dense aloft which means a greatly reduced heat retention and absorption of terrestrial radiation.

It should be emphasised that although the air temperature decreases progressively with height, there is nothing constant about this rate of change. In fact, the ELR is highly variable and this characteristic is associated with dramatic short-term changes in the receipt of insolation (and degree of cooling) at the earth's surface and the concentration of water vapour in the troposphere. Consequently, the average ELR does conceal the variability of the distribution of temperature aloft as illustrated in Fig. 3.2. It is obvious that the slope of the ELR varies considerably from a constant decrease with height (condition *A* in Fig. 3.2(a)) to a pronounced increased (condition *C*) or decreased (condition *B*) rate of change. However, the most extreme deviations from the average trend are associated with *lapse-rate steepening* (condition *E*) and a *temperature inversion* (condition *D*). It is worth noting that conditions *D* and *E* represent the diurnal variation of the ELR under clear skies, following terrestrial radiation loss by night and solar radiation accumulation by day. Furthermore, steepened lapse rates or *super-adiabatic conditions* are characterised by very rapid temperature decreases in the layer of air close to the earth's surface, when the ELR greatly exceeds the average rate to approach 1.4 °C per 100 m. They develop on hot, sunny, summer days when excessive heat accumulation occurs at the earth's surface which is transferred slowly by turbulent fluxes into the lowest air layers. As will be seen later, these conditions can develop into extreme *instability* and the formation of severe thunderstorms.

However, the most dramatic ELR deviations are associated with temperature inversions which reverse the lapse rate trend in the atmosphere when, occasionally, temperature increases with increasing altitude. Figure 3.2(b) clearly illustrates this reversal associated with temperature inversions at surface layers and in the upper air, which develop in markedly different ways. Ideal conditions for *surface inversions* are those which favour the rapid cooling of the earth's surface and adjacent air by nocturnal terrestrial radiation. For example, a long night is highly effective in maximising the loss of long-wave infra-red radiation. Furthermore, this loss is accentuated by clear skies and dry air (with weak counter-radiation and surface warming) and calm conditions, which allow the air to stagnate and be chilled by conduction from below. Therefore, the best conditions are associated with winter anticyclones over a snow-covered surface particularly in areas of diverse relief where *cold air drainage* (sections 5.2 and 14.2) into valleys intensifies the radiative chill-

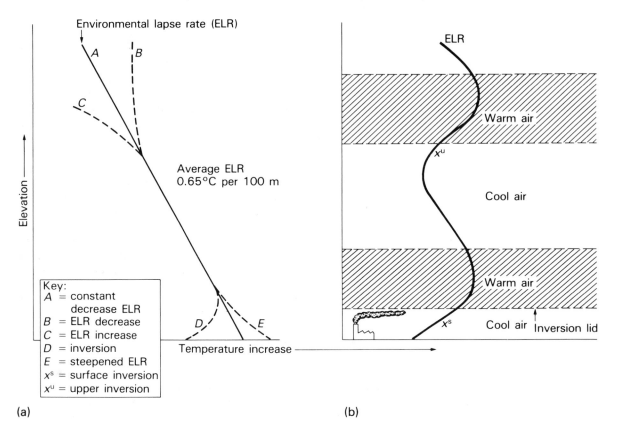

(a)   (b)

Fig. 3.2 Lapse rates (ELR) in the troposphere showing (a) varying rates and (b) surface and upper air inversions.

ing and leads to quite severe inversions.

The weather phenomena resulting from strong inversions are very localised forms of condensation, at or close to the earth's surface, which will be examined in section 5.2. Also, when the top of the cold air zone (i.e. the so-called *inversion lid*) is located at chimney height, then fumigation-type pollution episodes develop which lead to human discomfort and an increased death rate (e.g. the London *smog* of December 1952, which is described in section 5.2). As will be seen later, surface inversions make the air strongly *stable* which eliminates free convection from the earth's surface and aggravates smog accumulation. Besides being produced by radiative cooling, surface inversions are also caused by the advection of warm moist airstreams over cold surface air. Perhaps the best example occurs in central California when, particularly in summer, cool maritime air from the offshore cold current wedges under warmer continental air. This causes a strong surface inversion and a very stable, shallow air layer forms, with widespread *marine fog* (section 5.2).

Upper-air inversions also develop at various levels in the atmosphere and are mainly associated with warm-front advection in wave depressions or *subsidence warming* in strong anticyclones. The former type develops ahead of a warm front (Ch. 12) when warm moist air overrides a 'wedge' of cold air to the east. Some 300 km ahead of the front, the warm air is forced up to some 6000 m and represents an area of marked tropospheric warming compared with the colder air below the advection zone. However, the most widespread and persistent upper-air inversion is located in the semi-permanent *subtropical anticyclones* (section 10.4) where subsidence over 15 000 m feeds strong surface divergence. During this prolonged descent, the air is heated by compression and a layer of warm, dry air accumulates around 2000 m elevation. It should be noted that this warming does not extend to the earth's surface since nocturnal and, especially, winter chilling of the desert surface effectively counteracts the heat accumulation in the lowest air layers. Furthermore, friction associated with land-surface features creates strong eddy currents and this turbulence prevents the lower air from participating in the general subsidence, and collecting its 'share' of the general rise of temperature. In the subtropics, this *trade wind inversion* is particularly well developed and lowest (1–2000 m) over the eastern side of the oceanic trades.

Later in this chapter, it will become evident that upper-air inversions (like those at the surface) are responsible for strong atmospheric stability which prohibits the formation of deep turbulent eddies and causes serious pollution episodes over subtropical cities like Los Angeles and Sydney. Consequently, *tropical cyclones* do not develop under low-lying trade wind inversions along the eastern sides of the subtropical oceans like the Azores region. Instead, they are more common over western areas (e.g. the Caribbean) where the upper inversion is higher and weaker and more easily 'broken down' by the latent heat release of 'parent' *easterly waves* (section 10.3 and 13.5). *Subsidence inversions* dominate the upper air of all intensive anticyclones, including the *blocking highs* (section 11.4) which regularly influence midlatitudes. However, they are especially well-developed in the slow-moving high pressure systems of subtropical latitudes (section 10.4).

3.3 Adiabatic temperature changes in turbulent eddies

When a parcel of air is moved upward from the earth's surface by free or forced convection, it is subjected to very important temperature and density changes during its ascent through the surrounding stationary atmospheric environment. As the parcel rises and moves into a region of lower pressure, a net unbalanced force arises between the pressure force in the parcel and the environmental pressure force outside it. The volume increases just enough to balance these forces and causes the air parcel to expand, as is witnessed with a rising gas-filled balloon which has decreasing pressure exerted upon it during its ascent and eventually expands until it bursts. At the same time, the expanding gas uses up internal *potential heat* to supply the energy for the expansion process and, since air is a poor conductor of heat, it restricts the mixing of the parcel into the surrounding atmosphere. Consequently, no heat can be added to or extracted from the parcel and the expanding gas must cool gradually at a self-contained rate which is independent of the surrounding environmental air.

Conversely, descending air parcels will encounter increasing pressure from the surrounding air and will experience compression, with a marked rise in temperature (analogous of the warming air of a tyre when pumping takes place). As before, the compressional warming of the parcel takes place at a self-contained rate and such a change of state, in which no energy (or mass) enters or leaves a given system, is called *adiabatic*. Essentially all large-scale vertical motion in the atmosphere involves adiabatic expansion or compression. Also, it is worth noting that all subsiding air parcels will always grow warmer at the same rate at which rising air cools when ascending to an equivalent height, assuming of course, as will be soon observed, that saturation does not take place. Furthermore, since no energy crosses over the boundaries of the parcel, it recovers its initial state conditions upon returning to its original elevation. The rate at which the temperature of an air parcel changes as it rises/expands and falls/compresses is called the *adiabatic rate*, which is constant for dry, unsaturated air parcels at any temperature.

If the original temperature of the 'dry' parcel is known, together with the altitude change, then the temperature changes in rising or falling air parcels can be calculated by means of the constant *dry adiabatic rate* (DAR) of 1 °C per 100 m. It is apparent, therefore, that the rate of cooling of an ascending 'dry' air parcel is considerably more rapid than the normal vertical decrease in the atmospheric environment, where the ELR averages 0.65 °C per 100 m. However, it must be reiterated that the ELR displays pronounced temporal and spatial variability and is simply an expression of the prevailing temperature distribution in the atmospheric environment. Furthermore, it is completely isolated from the adiabatic temperature changes associated with air parcels ascending or descending through the air mass and this essential difference determines the stability tendency of the atmosphere, which is discussed in the following section.

When the unsaturated air parcel rises to its dew point level (section 5.1), the water vapour is forced to condense and the resultant air saturation will change the adiabatic rate of cooling (Fig. 3.3). The rising air will now cool more slowly than the DAR because, with condensation, latent heat is released into the parcel which opposes the primary cooling by expansion. The reduced rate of cooling, which occurs when condensation takes place, is known as the wet or *saturated adiabatic rate* (SAR). The SAR is most variable (unlike the constant DAR), since it depends on the amount of water vapour present in the parcel and the temperature, which controls the degree of saturation (section 5.1). As a result, the amount of liberated heat of condensation is much greater at high temperatures, so that the SAR cooling is slower in 40 °C air compared with 0 °C air (Trewartha 1968). Conversely, at very low temperatures (−40 °C) found in polar regions and very high elevations, the SAR approximates the DAR since

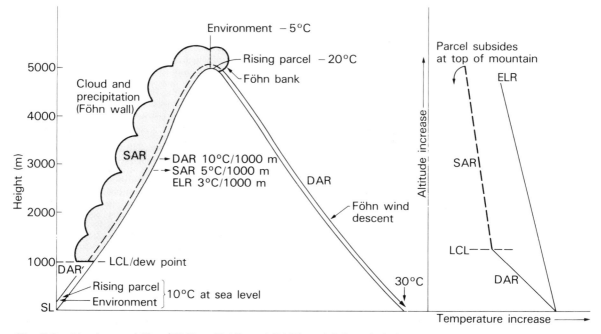

Fig. 3.3 Absolute stability (ELR < DAR and SAR) and föhn wind characteristics.

only minute amounts of water vapour are being converted into ice crystals (through *sublimation* with no intervening liquid state) and negligible amounts of latent heat are liberated. A 'rule of thumb' is that for temperatures above 30 °C, the SAR is about 5 °C per 1000 m rise and below 0 °C, the rate increases to 7 °C per 1000 m.

Even though the amount of latent heat release depends on the moisture content of the air, the average release in a 20 °C air parcel (2.45 MJ kg⁻¹) warms the air by 0.5 °C per 100 m. The net cooling effect (i.e. the DAR less the latent heat release) equals 0.5 °C per 100 m, which represents the average SAR in the lower troposphere. When moist air subsides, the resultant adiabatic warming may follow either the dry or saturated rates. If the water droplets are carried downwind with the air without evaporating, then the air warms at the DAR. However, with evaporation, latent heat is utilised to effect the change from liquid to vapour, and the rate of warming occurs at the SAR, which decreases by an amount that is related to the degree of evaporation (Donn 1965). This principle is best explained with reference to the *Föhn* or *Chinook wind* (Fig. 3.3), which is discussed further in section 14.7. The parcel ascent on the windward side of the mountain is dominated by the SAR, and the net gain of latent heat equals 20 °C. Consequently, at the top of the mountain (5000 m), the parcel's temperature equals −20 °C compared with −40 °C if the DAR had operated over the entire ascent. The descent on the leeward side is mainly without evaporation at the DAR, with only a minimum amount of latent heat used up near the crest to maintain the Föhn bank (see Plate 14.2). This, in effect, transports the 20 °C net heating to the foot of the mountain as a very warm (30 °C) dry wind which accelerates the ablation of the winter snow pack. This process is termed *pseudo-adiabatic* (Cole 1970) since there is an increase in the parcel's *potential temperature* (which is the temperature a dry parcel would have if brought adiabatically to a standard pressure of 100 kPa). It is a relatively rare phenomenon only occurring to the lee of the highest mountain ranges in the Rockies or Alps.

3.4 The relationship between the environmental lapse rate and adiabatic rates: stable and unstable atmospheric conditions

Air parcels are given vertical motion by the free and forced convective processes described earlier and ascend through the atmospheric environment with distinctive adiabatic cooling. The rate and amount of vertical motion that develops depends to a great extent on the type of equilibrium or mechanical balance existing in the atmosphere. The nature of this balance is determined by the prevailing thermal/density conditions, in particular the relationship between the ELR and adiabatic rates. For example, when the parcel of air resists the initial displacement and becomes non-buoyant, it comes to rest and eventually must sink downward. Conversely, if displacement results in air parcel buoyancy, it moves further away from the source of displacement up to considerable altitudes. The actual equilibrium tendency can be readily determined by comparing the actual ELR (as recorded from upper-air *radio-sonde* ascents) with the dry and saturated adiabatic rates. This relationship is of vital importance to weather forecasters, when the three prevailing rates are plotted on a temperature–height graph or *tephigram*, to facilitate the analysis of atmospheric equilibrium.

Stability occurs when the current ELR is less than the DAR, which means that the rising air parcel will cool initially at a faster rate than the air around it. This would occur with an ELR of 3 °C per 1000 m (Fig. 3.3) in the lower troposphere which is approximately one-third of the rate of cooling of the unsaturated parcel. As it rises, the parcel soon becomes colder and heavier than the surrounding atmospheric environment. The resultant non-buoyancy is opposed to vertical movement and the dense parcel will tend to sink back to its former position until its temperature and density equal that of the air at some lower level. Indeed, such an air parcel only rises in the first place when a forcing process (normally a mountain slope or a front) initiates the displacement since free convection is prohibited by the unfavourable density relationships at ground level. However, if equilibrium is restored just above the dew point level, then free convection is allowed over a limited vertical distance. Under these conditions, the ascent is soon terminated and the clouds can only develop into very shallow *cumulus* with distinctive flattened tops (section 5.3) which do not favour precipitation (Plate 3.1).

When the ELR is less than both DAR and SAR, then *absolute stability* exists, restricting upward movement of the parcel even when condensation and latent heat release occur. This is illustrated in Fig. 3.3 when non-buoyant dense air is forced to rise over a mountain range, where the

Plate 3.1 Absolute stability – cumulus humilis clouds over Alsace, France.

dew point and lifting condensation level (LCL) occur at 1000 m. At 5000 m, the air parcel is 15 °C colder than the atmospheric environment and the greater density forces the air parcel to sink down the leeward slopes (as a föhn wind described earlier). The cloud cover above the LCL on the windward side is shallow, extensive *stratus* or *stratocumulus* (see Fig. 5.1) with light, steady rain or drizzle. At the crest, the cloud soon dissipates at the distinctive föhn bank below which the leeward slopes are sunny and dry (see Plate 14.2). With frontal displacement, stability restricts cumuliform cloud development leading to thick layers of stratus cloud and drizzle at the warm front and shallow stratocumulus with only very light rain at the cold front (section 5.3).

As was mentioned in section 3.2, inversions in the ELR accentuate stability in the troposphere and tend to inhibit vertical motion. Because of the presence of a surface inversion, the adiabatic curve intersects the ELR at a much lower elevation, and this level of intersection marks the height at which the rising air will come to rest, and eventually sink back to earth. The fumigation-type

pollution episodes described earlier occur when instability occurs near the ground (ELR > DAR) and the inversion lid is located just above the chimney stack. This stability and sinking air prevent the smoke and gases from rising into the atmosphere. Instead, they are transferred horizontally until they are 'pulled' towards the earth's surface by turbulent eddies in the unstable air between the ground and chimney orifice. This results in pulses of effluent reaching the lowest levels, producing a chronic ground-level concentration and causing serious respiratory diseases.

Absolute instability occurs when the prevailing ELR is greater than both the DAR and SAR in the lower and middle troposphere. This means that the parcel will cool more slowly during its ascent to 8000 m or so and characterises an air parcel when the ELR exceeds the DAR in the surface layers (i.e. approaching 14 °C per 1000 m), following lapse-rate steepening or the superadiabatic conditions described in the last section. As the parcel moves upward by free or forced convection, it rapidly becomes warmer and less dense than the surrounding atmospheric environment.

Plate 3.2 Absolute instability – cumulus congestus clouds over the New South Wales Tablelands, Australia.

The resultant buoyancy exaggerates the vertical motion and the rising air accelerates freely up to the level where its temperature eventually reaches that of its surroundings at the cloud top (Plate 3.2).

This is illustrated in Fig. 3.4 where free convection is associated with surface heating. At the LCL, the parcel is already about 4 °C warmer than the surrounding atmosphere and this difference increases as the SAR takes over (which is approximately one-half the ELR). Consequently, at 3000 m the parcel is some 10 °C warmer (and more buoyant) than the surrounding air and continues to rise freely until about 7500 m, forming cumulus clouds of great vertical thickness. At this altitude, the ELR has decreased significantly with a slight upper inversion and isothermal trend (which characterises the tropopause) so that the SAR now exceeds the ELR and stability now operates, eliminating the buoyancy and producing the cloud top. Section 5.3 will reveal that deep cumuliform clouds form in these unstable conditions with torrential rain and thunderstorms (see Fig. 5.3). Furthermore, these cumuliform clouds characterise both warm and cold fronts with the displacement of warm, buoyant air.

Conditional instability occurs when the current ELR lies *between* the DAR and SAR throughout

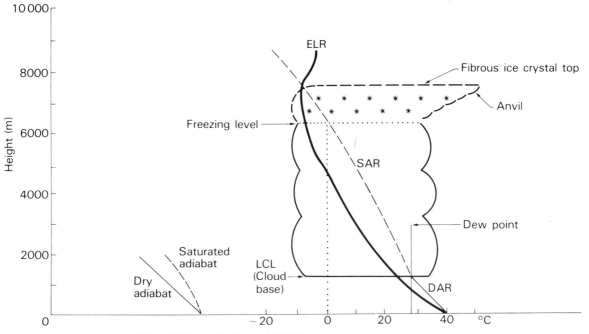

Fig. 3.4 Absolute instability (ELR > DAR and SAR).

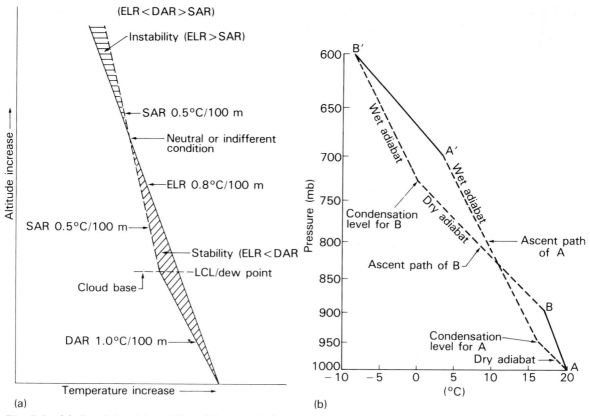

Fig. 3.5 (a) Conditional instability; (b) potential/convective instability. (After Trewartha 1968)

the ascent of a parcel and the resultant instability is conditional upon the saturation of air at all levels (Trewartha 1968). This situation is illustrated in Fig. 3.5(a) and it is apparent that in the early part of the ascent when the air is unsaturated, the DAR exceeds the ELR (viz. 1.0 °C per 100 m compared with 0.8 °C per 100 m). This makes the parcel colder than the environment and, at this level, the parcel becomes denser and resists displacement. Consequently, any further rise normally takes place with mechanical or forced ascent, especially when a parcel is forced over an active front (section 12.2) or a high mountain range. However, such orographic uplift will eventually force the parcel up to the condensation level (LCL) where latent heat is released and the SAR (0.5 °C per 100 m in Fig. 3.5(a)) takes over, which is less than the ELR. At this point, the ELR has become greater than the SAR and ultimately, beyond the *neutral equilibrium level*, the rising parcel becomes increasingly warmer than the surrounding air with a dramatic change to buoyancy. The parcel now rises freely on its own accord and the cloud now 'mushrooms' into a heaped cumulus of great vertical extent (Plate 3.3)

with heavy showers (section 5.3). Obviously, this dramatic change in equilibrium from stability to instability (and the associated weather phenomena) is conditional upon the release of latent heat of condensation, which reduces the adiabatic cooling to a rate much less than the ELR.

Whereas conditional and absolute instability are related to parcels of air rising through a stationary air mass, *potential* or *convective instability* occurs when the entire air-mass layer is forced to rise. Furthermore, when the lower layers of the air mass have a much higher relative humidity than the layers aloft, then the uplift of the air mass results in the early condensation of the lower more humid layers. The associated latent heat release causes these layers to cool more slowly (at the SAR) than the drier upper layers which cool at the DAR. A greatly steepened lapse rate thus develops within the air mass which converts the initially stable air to a convectively unstable condition.

Figure 3.5(b) illustrates convective instability where a layer of air (lying between 1000 and 900 mb) has an ELR indicated by line AB. This rate is obviously less than the DAR which leads to a condition of absolute stability. However,

Plate 3.3 Conditional instability – cumulus congestus clouds over the Colorado Rockies; USA

since the base of the layer at point A has a much higher relative humidity than at B, condensation will take place more quickly for A at the 950 mb level (compared with point B at 730 mb) when air-mass ascent occurs. As a result, the SAR takes over earlier from the DAR and this retarded rate of cooling characterises the rising air mass to point A' at the 700 mb level. At the same time, the drier air layer represented by point B is rising and cooling at the DAR to its condensation level (730 mb) before the SAR takes over up to 600 mb (B'). Therefore, following uplift, the original absolutely stable lapse rate (AB) has been steepened considerably to the new A'B' slope, which represents a condition of greater instability.

This conversion from stability to instability can take place with only modest forced ascent (as at a warm front), as long as the surface layers are more humid than those aloft. Such forced vertical displacement of large air masses characterises the windward sides of mountain ranges and also along the warm fronts of mid-latitude wave depressions, when a layer of warm tropical air is lifted over a cold stable 'wedge' of polar air. The associated clouds are usually a mixture of layer and cumuliform types (section 5.3) with *cumulonimbus* related to the buoyant instability of the later stages of the ascent, with heavy localised showers (see Fig. 5.1).

3.5 Retrospect

The turbulent cascade of mass and energy within the troposphere is accomplished by the free and forced convection of air parcels through a large stationary air mass. Normally, temperature decreases with increasing altitude but the resultant ELR is most variable and depends on short-term changes in surface heating or cooling and the concentration of water vapour in the troposphere. Indeed, extreme variations are evident with lapse rate steepening and temperature inversions. In the former case, very rapid temperature decreases in the layer of air close to the earth's surface (greater than the DAR) are associated with excessive heat accumulation on sunny, hot summer days. Conversely, inversions represent a complete reversal of the ELR when, occasionally, temperature actually increases with increasing altitude.

When parcels of air are forced aloft by convective mechanisms, they are subjected to rapid pressure changes which modify their temperature and density characteristics during their ascent through the surrounding atmosphere. For example, a parcel of 'dry' air moving upward from near sea-level (pressure ~ 1013 mb) to an altitude of 7000 m (pressure ~ 410 mb) would almost double its volume and the resultant expansion would lead to a temperature decrease of about 70 °C. If the same 'dry' parcel of air is returned to its original level, compressional warming would result in its temperature and volume returning to their initial values. Furthermore, these temperature changes are self-contained within the parcel, since there is no heat or mass exchange with the surrounding atmospheric environment. Whereas these adiabatic temperature changes are constant in dry parcels (DAR 1 °C per 100 m), saturation above the dew point liberates latent heat which opposes the primary cooling by expansion. The resultant cooling is at the saturated adiabatic rate (SAR) which varies according to the amount of water vapour present in the parcel (averaging 0.5 °C per 100 m) in the lower troposphere.

Therefore, the situation exists where air parcels are displaced upwards through the atmospheric environment with distinctive adiabatic cooling. The relationship between the prevailing ELR and the adiabatic rates leads to significant density contrasts which results in parcel buoyancy or nonbuoyancy. A criterion then for the nature of vertical stability may be expressed as follows:

ELR:	Equilibrium tendency:
Less than DAR and SAR	Absolutely stable
Greater than DAR and SAR	Absolutely unstable
Between the DAR and SAR	Conditionally unstable

Finally, potential or convective instability refers to an entire air mass where the lower layers have a much higher humidity (cooling at the SAR) than those aloft which cool at the DAR. With uplift, a greatly steepened lapse rate develops within the air mass and converts initially stable air to an unstable state.

Key topics

1. Free and forced convection as the major mechanisms of vertical air movement.
2. The factors involved in changing the environmental lapse rate.
3. The differences between the dry and saturated adiabatic rates.
4. The relationships between the environmental lapse rate and adiabatic rates in the troposphere.

Further reading

Barry, R. G. and Chorley, R. J. (1982) *Atmosphere, Weather and Climate*. Methuen; London and New York, pp. 68–74.

Donn, W. L. (1965) *Meteorology*. McGraw Hill; London and New York, pp. 62–77.

Miller, A. and Thompson, J. C. (1970) *Elements of Meteorology*. Merrill; Columbus, Ohio, pp. 111–21.

Pedgley, D. E. (1962) *A Course in Elementary Meteorology*. HMSO; London, pp. 9–19.

Chapter 4 Evaporation and evapotranspiration

Chapter 2 reveals that the combined effects of *evaporation* and transpiration represent a subsystem of the mass cascade which is responsible for a considerable part of the global energy cascade. Furthermore, the acquisition of vapour by the latent heat exchange represents the throughput of energy (section 1.3) which initiates the hydrologic cycle. Evaporation is simply defined as the process by which a liquid or solid is converted into a gas, which is transferred into the troposphere. This vapour flux from the earth's surface is primarily from the seas and oceans, particularly in low latitudes where radiant energy surpluses promote excessive amounts of evaporation (e.g. 200 cm yr^{-1} in the western Pacific and central Indian Oceans, around 15 °S). Evaporation also takes place from terrestrial water bodies, exposed wet soil and from snow and ice surfaces, although the latter transfer is referred to as sublimation. Another important vapour source is the transpiration of moisture from plant surfaces, partly acquired from adjacent soil in the root zone. This flux is termed evapotranspiration and has been further subdivided into actual or potential rates. This chapter examines the evaporation and evapotranspiration processes and emphasises the role of meteorological, geographical and botanical factors which control the rate of vapour flux from surface water.

4.1 Evaporation and transpiration processes

It should be noted that water molecules are in constant motion, whether in a large water body or a thin film covering soil or leaves. Furthermore, when heat is added to the water, the molecules become increasingly energised or excited and move more rapidly. Eventually, the molecules possess sufficient energy to break through the restrictive membrane of the water surface and diffuse into the atmosphere in a gaseous form. Similarly, some of the water molecules contained in this water vapour in the lowest atmosphere, that are also in motion, may penetrate the water surface and enter the liquid. Therefore, the rate of evaporation at a given time will depend on the number of 'fast' molecules leaving the water surface less the number of returning molecules. A positive exchange (i.e. more molecules leaving the water surface than are returning to it) will result in evaporation whereas the reverse exchange will terminate vapour flux and initiate condensation (as discussed in Ch. 5).

The molecules making up a volume of liquid water are in close proximity, with a separation of just over one molecular diameter. At such distances, the molecules interact with a strong bondage or attraction and a short-range force develops which weakens with increased separation. In water vapour, the molecules are much further apart (typically 10 or more molecular diameters), although the distance involved depends on the actual weight of the water molecules (or vapour pressure). Furthermore, with this increased separation, the *intermolecular bondage* (or molecule interaction) is very small indeed. Therefore, in order to create water vapour from liquid water, it is necessary to increase the separation between all the molecules by supplying kinetic energy (from solar radiation via sensible heat transfers) to work against the holding force. The amount of energy required is directly related to the number of molecules, which is directly proportional to the mass of water involved. The amount of energy per unit mass of liquid water is called the latent heat of vaporisation, which is supplied by the kinetic energy of the molecules within the liquid. The actual amount of energy required varies slightly with temperature, which will be discussed in the following section.

Evapotranspiration represents a more complicated use of water molecules by *transpiration* (the process by which they escape from the leaves to the atmosphere) which causes the plant nutrients in the soil to rise and aid the building of plant tissues. Indeed, Ward (1975) refers to the process as a 'consumptive use' since the water molecules represent a substantial part of the food-fibre production in the plant system. The process also represents the most important movement of water in the hydrological cycle since it accounts for the disposal of nearly 100% of the annual precipitation in arid regions and about 75% in humid areas (Harrold 1969). Transpiration takes place through the leaf openings or *stomata*, the lenticles or occasional holes in the bark, the surface of spongy leaf cells and the cuticles of leaves. Water molecules are also lost in the soil zone adjacent to the plant since the root system obtains water by absorption at surfaces in contact with soil moisture.

The transpiration process depends primarily upon the supply of radiant and sensible heat energy to the plants, which stimulates evaporation within the leaf. When the resultant water vapour has diffused through the stomata and has been carried into the air by turbulent mixing (section 3.1), the resulting water molecule loss produces a water deficiency within the leaf cells. This deficit represents a suction force or *water potential* which is transmitted by the cells from the roots to the leaf surface against the force of gravity. This water potential (which is negative in a transpiring plant cell) is expressed as:

$$\psi = O + P + Z \qquad [4.1]$$

where ψ is the total water potential/suction force, O is the osmotic potential of the cell sap in root and leaf (strongly negative if plant is actively transpiring), P is the cell wall pressure potential (approximately zero with active transpiration) and Z is the gravity potential (positive downwards with active transpiration).

As was mentioned in the introduction to this chapter, the evapotranspiration process takes place as either an actual or potential water loss. *Actual evapotranspiration* (AE) is the process by which water vapour escapes from the living plant (principally through the leaves) and enters the atmosphere, plus the water evaporated from the adjacent soil or snow cover. It is controlled by meteorological, soil and plant factors which will be discussed in the next section. Conversely, potential evapotranspiration (PE) represents the water need of a plant, which is the supply of soil moisture or precipitation stored on the surface which is at all times sufficient to meet the demands of transpiring plants. Thornthwaite (1944) aptly defined PE as the water loss which will occur if at no time there is a deficiency of water in the soil for the use of vegetation. Since this condition can only be maintained permanently by supplementing natural rainfall with irrigation, PE is almost entirely controlled by meteorological factors with stomata size and plant type the only other variables concerned (section 4.3).

4.2 Meteorological factors affecting evaporation and evapotranspiration

Meteorological factors combine with environmental characteristics to control the rate of vapour flux from free-water surfaces, soil and plants. However, it is difficult to determine the relative importance of all factors involved and, furthermore, this significance will change markedly with the actual process concerned, as is illustrated in Table 4.1. Table 4.1 reveals that meteorological factors control all three processes but, as will be explained shortly, the relative importance of these factors (particularly wind speed) will vary quite considerably. It is obvious that plant factors will have no control on the vapour flux from free-water surfaces and bare soil. Similarly, transpiration from plants is isolated from all the physical and chemical water properties. It is also apparent that PE

Table 4.1 The contribution of meteorological and environmental factors to the evaporation and evapotranspiration processes

Process	Atmosphere	Water	Soil	Plant
Evaporation	X	X	X	–
Actual evapotranspiration (AE)	X	–	X	X
Potential evapotranspiration (PE)	X	–	–	X*

* Stomata size and plant type only significant.

proceeds independently of soil factors and most plant characteristics since it involves the supply of moisture which is at all times sufficient to meet the demands of the transpiring vegetation cover.

Meteorological factors interact in the boundary layer to provide the energy and diffusion mechanisms which are essential for water loss from all surfaces. It was noted in the previous section that radiant and sensible heat energy provides water molecules with the kinetic energy that is required to work against the holding force of the molecules in a large water body or in a thin film covering soil grains and leaves. Consequently, solar radiation (or more correctly net radiation) provides the energy necessary for the latent heat of vaporisation which is used to effect the change from liquid to vapour. The actual amount of energy required decreases as the temperature increases by about 0.1% per degree Celsius since, at higher temperatures, the intermolecular spacing in the liquid is greater and the phase change takes place more easily. For example, the latent heat of vaporisation at 0 °C is 2.501 MJ kg^{-1}, at 20 °C is 2.450 MJ kg^{-1} and at 30 °C is 2.430 MJ kg^{-1}. Sublimation requires even more energy, since the latent heat of fusion (0.33 MJ kg^{-1}) is needed to melt the ice crystals before the latent heat of vaporisation is effective. At 0 °C, the latent heat of sublimation equals 2.83 MJ kg^{-1}.

Apart from providing the necessary thermal energy for the initial liquid to gas change, net radiation also adds essential mobility to individual water molecules. This kinetic energy breaks down the intermolecular bondage (described in section 4.1) and increases the separation and velocity of the water molecules. This increases their chance of passing through the restrictive membrane of the water surface, to facilitate their transfer into the adjacent atmosphere. As the faster molecules are the first to 'escape', so the average energy and temperature of those composing the remaining liquid will decrease and the amount of energy required for their continued release must become correspondingly greater. As a result, evaporation decreases the temperature of the remaining liquid

by an amount proportional to the latent heat of vaporisation, and is a well-tried domestic cooling technique.

It is now generally regarded that solar radiation is the most important single factor involved in evaporation and transpiration. In fact, all plant processes are accelerated by the duration and intensity of solar radiation, including photosynthesis and the circulation of water through the root–stem–leaf system. All vapour fluxes have close relationships with the diurnal and seasonal distribution of solar radiation, being greatest during the day (around solar noon) and summer (around the solstice month of June). Table 4.2 indicates the correlation between evaporation from three evaporimeters and the main meteorological variables involved. In all cases, solar radiation had a good correlation with the water loss (especially the high 0.77 correlation with the Black Bellani tank) and is generally the dominant atmospheric factor involved (apart from the Summerland tank, where vapour pressure deficit had the closest correlation).

Air temperature has been traditionally used as a significant factor involved in vapour flux processes, especially in theoretical estimations of PE. Since air temperatures are largely dependent upon solar radiation, it is expected that a reasonable correlation would occur between them and the rate of water loss. However, heat transfers (both

Table 4.2 Correlation coefficients between various evaporimeters and the daily variation of meteorological factors. (Adapted from Holmes and Robertson 1958)

Factor	Black Bellani tank	Summerland tank	4 ft tank
Solar radiation	0.771	0.603	0.601
Vapour pressure deficit	0.719	0.695	0.457
Air temperature	0.460	0.421	0.304
Wind speed	0.269	–0.012	0.232

Significance (N = 153 cases): 5% level r = 0.16; 1 : pc level r = 0.21.

lateral and horizontal), particularly in large, deep-water bodies, result in a considerable time-lag between maximum temperatures and evaporation rates in diurnal and seasonal terms. Indeed, as will be seen in the discussion on water characteristics (section 4.3), the storage of heat at great depths in deep-water bodies can be responsible for maximum evaporation rates occurring in the low-sun winter season, which is a complete reversal of the expected trend. Consequently, Table 4.2 indicates that the correlation between temperature and evaporation is poor, less than the critical 0.5 level in all cases.

It appears that the temperature differences between the water body and the surrounding air are much more pertinent to evaporative losses, compared with the air temperature alone, since they control the vital *vapour pressure deficit*. Vapour pressure is the independent partial pressure of the gas which varies from time to time, especially when fronts cross an area with a significant change of air mass (Ch. 12). The transfer of water molecules from the water body, soil or leaf into the air depends on vapour pressure variations which allow the molecules to flow from high to low vapour pressure areas. Thus, for evaporation and transpiration to take place, the vapour pressure at the water surface or in the leaf cells must exceed the atmospheric vapour pressure. When the gradient is reversed, and low vapour pressure exists at the evaporating surface (following water vapour advection in the ambient air), then the vapour flux is towards the surface which retards further vapour flux and actually causes condensation.

A further limiting condition is the absolute vapour saturation limit (called *saturation vapour pressure*, or SVP) since vapour flux cannot continue into a saturated atmosphere. A vapour pressure deficit represents the difference between saturated vapour pressure at the temperature of the water surface and the actual vapour pressure at the temperature of the surrounding air. The saturation vapour pressure changes dramatically with temperature (for example, from 1 mb at $-22\,°C$ to 100 mb at $46\,°C$) which controls the number of water molecules which are allowed to 'escape' into the atmosphere. Consequently, when the water body and air temperatures are identical, the vapour pressure deficit is minimal and evaporation proceeds slowly. For example, with a water and air temperature of $30\,°C$, the rate of evaporation will be $0.41\ mm\ hr^{-1}$ with a relative humidity of 60%. Conversely, when the water temperature is $6\,°C$ higher than the adjacent air

temperature, vapour pressure deficits are accentuated and the rate of evaporation is doubled to $0.82\ mm\ hr^{-1}$ at a relative humidity of 60% (Geiger 1965). Table 4.2 reveals the importance of vapour pressure deficit in evaporation, where correlation coefficients for two of the tanks were close to the 'good' association at the 0.7 level.

Relative humidity represents the proximity to saturation of the air and refers to the actual moisture level in the ambient air as a ratio of the amount held when saturated at that temperature (viz. 100%). Obviously, relative humidity will control the 'thirst' of the air since when it increases (following a temperature decrease), proportionally fewer water vapour molecules leaving the evaporating surface are held in the air so the rate of evaporation is gradually reduced. Conversely, a fall of relative humidity (associated with a rise in temperature) will increase the water vapour holding capacity of the air, which will increase the rate of vapour flux from the surface. Obviously, daytime warming is conducive to evaporation and transpiration whereas nocturnal chilling brings about the attainment of saturation and dew point, which favours condensation (section 5.1). With a water and air temperature of $30\,°C$, the rate of water loss will be $0.61\ mm\ hr^{-1}$ with a relative humidity of 40% compared with $0.20\ mm\ hr^{-1}$ at 80%, and zero evaporation at 100% (Geiger 1965).

The accumulation of water vapour molecules in the air overlying a water body or leaf will eventually lead to saturation of the lowest air layers and the consequent termination of evaporation. This situation would occur fairly soon in absolutely calm conditions when air stagnates over the evaporating or transpiring surface and vapour flux continues slowly between air molecules by *molecular diffusion*. However, in windy turbulent conditions, vigorous air movement mixes the lowest saturated layers with the drier overlying air and indeed this renewal of unsaturated air over the water body or leaf will accelerate the vapour flux. Clearly, the stronger the air flow, the more vigorous and effective will be this turbulent action or *eddy diffusion*, and the greater will be the rate of evaporation and transpiration. This represents the principle of fanning in hot, humid conditions where perspiration (and eventually evaporation, using up blood heat to cool the body) is accelerated when artificial turbulence/eddy diffusion renews the supply of dry air over the perspiring body.

However the turbulent movement of air is not

entirely due to the strength of the wind since surface roughness is important, especially over a land surface. Furthermore the relationship between wind speed and evaporation is only applicable up to a critical velocity, beyond which any further increase in wind speed leads to no further increase in evaporation. It is apparent that the turbulent removal of water vapour molecules from the air in contact with the water surface only enables evaporation to proceed at the maximum rate allowed by the other meteorological variables (especially solar radiation and vapour pressure). This leads to a poor relationship between wind speed and evaporation, as evident in Table 4.2 where the correlation coefficients for wind speed are well below the critical 0.5 level. Indeed, the Summerland tank coefficient was close to zero and the negative tendency indicated that evaporation actually decreased with higher wind speeds.

Wind speed does not appear to influence the actual initiation of the evaporation process but by removing vapour-laden air, permits a given rate of evaporation to be maintained. The relationship between wind speed and evapotranspiration appears to be even more tenuous since the crop surface tends to seal itself as wind speed increases, which results in a marked restriction of turbulent mixing with depth. It is apparent that the influence of wind speed on transpiration is not all that important, especially when compared with the influence of solar radiation. However, in given humidity conditions, the role of wind speed will tend to increase as the temperature falls.

4.3 Environmental factors and the rate of evaporation and evapotranspiration

The previous section emphasises the importance of atmospheric factors in all the vapour flux processes, which forces the environmental factors listed in Table 4.1 into a rather passive role. However, it is obvious that water, soil and plant characteristics will control the effectiveness of the actual water loss, although the actual contribution will depend on the process concerned. For example, PE is controlled almost entirely by meteorological factors with only two plant characteristics (viz. stomata size and plant type, which are independent of the permanent water surplus) able to compete with this atmospheric monopoly at certain times (e.g. stomata size dominates the

meteorological variables at night when the stomata are half-closed).

Water characteristics affect only evaporation, and include the quality, depth and size of the water body. The *salinity* of the water is important since it reduces vapour pressure, and evaporation decreases by about 1% for every 1% increase in salinity. Consequently, evaporation from sea water (with an average salinity of about 3.5%) is some 2–3% less than evaporation from fresh water. Water pollution appears to be an indirect limiting factor since changes in water colour and turbidity will change the surface albedo. Over time, this will modify the natural energy balance of the water body and change the heat storage, which could alter the normal evaporative losses. As was mentioned earlier, water depth has quite a considerable influence on evaporation rates, mainly associated with varying heat capacities. Shallow-water bodies have a restricted heat storage or *thermal capacity* so that the seasonal air and water temperatures are in harmony, which means that maximum water temperatures occur in midsummer with associated maximum evaporative losses at this time. For example, the Kempton Park Reservoir near London (9 m maximum depth) records about 125 mm evaporation in July compared with about 12 mm loss in December (Fig. 4.1(a)).

Conversely, deep-water bodies have a much higher heat storage which, coupled with the deep mechanical mixing activated by the winter water 'turnover', releases heat slowly during the low-sun season. As a result, the heat energy released at this time is made available for evaporation at a time when temperatures (and SVP) are much lower in the surrounding air. This develops a vapour pressure deficit which encourages a large amount of evaporation, which can exceed that of the high-sun summer season when air temperatures (and SVP) are higher. For example, Lake Superior in the USA (397 m maximum depth) records about 120 mm evaporation in December, compared with about 30 mm in June (Fig. 4.1(a)). The size of the water surface also influences the build-up of a protective vapour 'blanket' of saturated air, which decreases the rate of evaporation as it thickens. Large surface areas have the greatest 'blanket' development which will reduce the depth of water evaporated (Fig. 4.1(b)).

Soil factors control the effectiveness of evaporation and actual evapotranspiration since they determine the concentration of water molecules in films surrounding the soil grains and filling the spaces between them. These molecules escape into

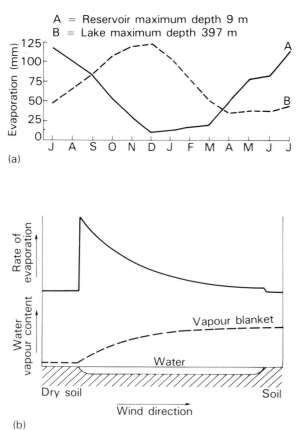

(a)

(b)

Fig. 4.1 (a) Evaporation and depth of water body. Graphs showing the annual evaporation regime for (A) Kempton Park Reservoir near London, England, and (B) Lake Superior, USA. (After Ward 1975); (b) evaporation and size of the water body, a simple illustration of the humidity increase and evaporation decrease associated with air movement over a large water surface. (After Ward 1975)

the atmosphere, when motivated by the meteorological factors described earlier, by direct vapour flux or indirectly through transpiration via the root–stem–leaf system of the plant. The main controlling factor must be the soil moisture content in the surface layer since this represents the potential for water availability or the so-called *evaporation opportunity* of the soil material. However, the relationship between soil moisture content and evaporation/transpiration is quite close, proceeding at a 100% 'evaporation opportunity' when the soil is saturated (at *field capacity*) and decreasing rapidly as the surface moisture content falls, until it is zero with a dry soil surface.

It is apparent that the moisture content of the top few centimetres of surface soil is the decisive factor since subsoil water may not affect the rate of surface vapour flux from shallow-rooted plants. In dry areas however, the role of subsurface water is influenced by the soil capillary features, which are responsible for the upward movement of soil water (against gravity) through the capillary tubes from wetted to dry grains (i.e. towards the dry surface soil). This movement is greatest (over a metre or so) in fine-grained soils, compared with only a few centimetres rise in coarse-textured material. Of course, the significance of subsurface water is increased when the *water table* (or upper level of subsurface water) is close to the surface, increasing the potential of groundwater availability for the 'evaporation opportunity'. Surface evaporation reaches maximum amounts when the water table is at the surface, averaging 5 mm day^{-1} at that time compared with 0.5 mm day^{-1} when the table is at a depth of 140 cm. It appears that evaporation decreases quite rapidly initially as the water table retreats downward. However, after a depth of about 1 m has been reached, any further decrease in water table height is accompanied by only a slight change in the evaporation rate (Ward 1975).

Additional soil factors include the colour of the material, which determines the surface albedo and heat storage in the soil, which represents the potential for the latent heat of vaporisation. For example, dark soils with an albedo below 10% should absorb maximum amounts of solar radiation and record highest surface temperatures and maximum amounts of evaporation. The degree of exposure of the soil surface is also important since, without a protective cover of vegetation, evaporation will proceed at an accelerated rate. For example, evaporation from bare soil is about twice as fast as that from soil under forest cover.

The last set of environmental controls involve plant factors, which naturally are only associated with transpiration rates. Transpiration is vital as a life function in that it causes a rise of plant nutrients from the soil and cools the leaves. The concentration of solvent molecules in cells of the plant roots exerts an *osmotic tension* of up to 15 atmospheres upon the water films between the adjacent soil grains. Furthermore, as these water films shrink, the tension within them increases and, if the tension exceeds the osmotic root tension, then the continuity of the plant's water supply is broken and wilting occurs (Barry and Chorley 1982). Four plant factors contribute to the transpiration losses from vegetation, viz. stomata size, plant type, plant colour and stage of growth. All four determine the effectiveness of actual evapotranspiration but PE only involves the first two. Stomata are small pores in the leaves, which are open during daylight. At this time, the actual pore size does not affect the rate of transpiration and, indeed, the vapour flux is governed by the prevailing meteorological factors. However, at night, when the stomata are half-closed, the pore size becomes the dominant factor which controls the nocturnal vapour flux.

The type of plant determines the availability of moisture and the degree of transpiration motivated by the atmospheric factors. Vascular and non-vascular plants are extreme cases of water potential, which is severely limited in the latter case. Non-vascular tundra plants, for example, are resistant to vapour flux and contribute largely to the polar desert environment. The colour of the plants controls the albedo of the vegetation and associated heat storage in the plant fabric, which is responsible for the latent heat available for transpiration. Light-coloured plants with a relatively high albedo (above 20%) reflect a considerable amount of solar radiation, and consequently have lower leaf temperatures, with minimum amounts of transpiration. The stage of growth of a plant will also determine the potential for moisture supply and the associated availability of water for transpiration. For example, maize is green in the early growing season, containing maximum amounts of plant water which encourages excessive amounts of transpiration under favourable meterorological conditions. Conversely, at the harvest stage, the crop is brown and desiccated with a greater resistance to vapour flux.

4.4 Retrospect

Evaporation and evapotranspiration are complex mass transfer processes, where considerable amounts of thermal and kinetic energy are re-quired by water molecules in order to 'escape' from the surface water body into the surrounding atmosphere. Apart from the energy expenditure involved, the vapour flux is controlled by a large number of atmospheric and environmental factors. However, it was shown that the relative effectiveness of these factors varied dramatically with each process, although meteorological elements (especially solar radiation and vapour pressure deficit) are vital for the initial separation and velocity of all the water molecules concerned. The environmental factors merely determine the effectiveness of the atmospheric controls by influencing the energy storage and water potential, which allow a given rate of vapour flux to be maintained.

Key topics

1. The mechanisms involved in evaporation and transpiration
2. Radiative energy and its effect on water vapour transfers.
3. The role of other atmospheric factors in vapour flux.
4. The modification of atmospheric controls by water bodies, soil and vegetation.

Further reading

Gray, D. M. (1973) *Handbook on the Principles of Hydrology*. Water Information Center; New York, pp. 3.1–3.66.

Penman, H. L. (1948) Natural evaporation from open water, bare soil and grass, *Proc. Roy. Soc.*, Ser. A, **193**, 120–45.

Shuttleworth, W. J. (1979) *Evaporation*. Report No. 56, Inst. Hydrol; Wallingford.

Tanner, C. B. (1957) Factors affecting evaporation from plants and soils, *J. Soil and Water Cons.*, **12**, 221–27.

Thornthwaite, C. W. and Holzman, B. (1939) The determination of evaporation from land and water surfaces, *Mon. Wea. Rev.*, **67**, 4–11.

Chapter 5 Condensation and precipitation

Condensation can be regarded as the second sub-system of the global cascade of water, since it is responsible for the return of water vapour to the earth's surface in a liquid or solid state. Technically, the term condensation applies only to the process in which water vapour is changed into liquid water. The change from vapour to solid (by-passing the liquid stage) is more correctly called sublimation, although this term is also used for the change from solid to vapour (Ch. 4). Together with evapotranspiration, condensation is responsible for colossal amounts of water exchange over time. For example, Chandler (1969) maintains that the evaporation–condensation exchange amounts to 19 million tons a second or 565 trillion tons per annum [*sic*]. Even so, he emphasises that if all this atmospheric vapour was suddenly condensed and released as precipitation, it would amount to only 25 mm over the entire earth's surface. In other words, the atmosphere contains the equivalent of only 25 mm of precipitation at any one time and, since the average global rainfall equals 915 mm, it is apparent that there are approximately 36 evaporation–condensation cycles each year. This means that a molecule of water vapour will remain in the atmosphere for an average of 10 days, drifting up to hundreds of kilometres before it is condensed and precipitated back to the earth's surface.

Chapters 2 and 3 indicate that condensation is a vital part of the energy cascade, since it releases considerable amounts of latent heat into the atmosphere. Furthermore, this heat release in rising air parcels initiates the saturated adiabatic rate (SAR) which leads to gross atmospheric instability and severe perturbations when this rate exceeds the environmental lapse rate (ELR). This chapter examines the condensation process in detail and describes the ways in which saturation is achieved in the atmosphere. This state is accomplished mainly by temperature decreases associated with surface contact cooling, air-mass mixing and adiabatic cooling. Considerable emphasis is placed on the forms of condensation resulting from these cooling mechanisms, ranging from the formation of dew, frost and fog (at or close to the surface) to clouds and precipitation within the troposphere, well away from the earth's surface.

5.1 The condensation process

As long as the vapour pressure of the air is less than the SVP over a liquid or frozen surface, then a vapour deficit will exist which favours a vapour flux (Ch. 4). However, when the current vapour pressure equals the SVP, then a state of equilibrium will occur. This prohibits evaporation and condensation since the rate of molecular transport across the water–air interface is the same in both

Table 5.1 Saturation vapour content (SVC) and air temperature

Temperature (°C)	SVC (gm m⁻³)
−1.1	4.4
4.4	6.5
21.1	18.3
26.6	25.0

directions. Finally, when the existing vapour pressure of the air exceeds the SVP, then the vapour flux is towards the surface which leads to condensation.

When unsaturated air rests on free water, soil or plant surfaces, the water steadily evaporates until the concentration of water vapour in the atmosphere (i.e. the humidity of the air) reaches saturation level. This level is maintained by the fact that there is a strict limit to the amount of water vapour that can be held in the air. After so much water has been evaporated into the confined space of the atmosphere, any further evaporation will produce an equal amount of condensation so that the amount of vapour present in the air will remain constant. This maximum value is referred to as saturation and the *saturation vapour content* (SVC or the water vapour capacity) of the air is controlled primarily by the temperature of the air (Table 5.1). Table 5.1 shows that cold air has a very limited capacity and, even at saturation point, it contains very little moisture. However, as the temperature rises, the capacity of the air for water vapour increases rapidly and furthermore, the rate of capacity increase itself grows greater as the temperature rises. In other words, a temperature rise of 5.5 °C is more than three times as effective at 21.1 °C, than at −1.1 °C, in increasing the amount of water than can be held in the air.

It is apparent that saturation is achieved by increasing the amount of water vapour in the air by continued evaporation/transpiration until the current vapour content equals the SVC at that temperature and the *saturation deficit* is eliminated. Alternatively, if the temperature falls, then the SVC will decrease accordingly until it is just equal to the actual amount of water vapour held by the air, thereby achieving saturation. This is the most important natural process which leads to the saturation of air. For example, at 26.6 °C and a current vapour content of 18.3 g m⁻³ (i.e. a 6.7 g saturation deficit), saturation will take place when continued vapour flux increases the vapour content to the SVC (25.0 g m⁻³) *or* the temperature can be lowered to 21.1 °C.

If the temperature continues to fall below the point at which the air becomes saturated, then there will be an excess amount of water vapour present compared with the SVC of the air at the new, lower temperature. Consequently, this excess vapour will be released by changing its state through condensation to a liquid or solid form and the temperature which forces this change is called the dew point (section 3.3). In Table 5.1 for example, when the temperature falls from 26.7 to −1.1 °C, then 20.6 g m⁻³ water vapour must be released through condensation to maintain the 4.4 g m⁻³ SVC at the lower temperature (cf. 25 g m⁻³ at 26.7 °C). Since the cooling of air is vital for the attainment of saturation, it is necessary to consider the reasons for this temperature decrease. The three major processes, which combine to cool air to and below the dew point, are as follows:

1. Contact cooling when air is chilled by conduction from a surface cooled by long-wave terrestrial radiation.
2. Air-mass mixing, especially when warm moist air is chilled by conduction during advection across a cold surface.
3. Adiabatic cooling as air parcels rise and expand in unstable situations and the temperature falls by the DAR and eventually the SAR.

The three cooling mechanisms were discussed in Chapters 2 and 3 and will be related to actual forms of condensation in the next two sections. However, before the cooling of saturated air can condense vapour into a variety of weather phenomena, non-aqueous *hygroscopic nuclei* (i.e. atmospheric particles which have an affinity for water) must exist in the air. These nuclei are essential for condensation since, in their absence, the air can become supersaturated and the relative humidity can exceed 100%. Laboratory experiments have indicated that pure, saturated air can be cooled to approximately −40 °C without condensation/sublimation occurring. Below this temperature, water vapour appears to sublime spontaneously to ice crystals in the absence of transient impurities (Cole 1970). Conversely, a large concentration of nuclei can promote condensation at 70 or 80% relative humidity, which partly accounts for the smog characteristics of urban and industrial areas.

Hygroscopic nuclei vary widely in composition although *salt* (sodium chloride) and combustion products (sulphur trioxide) have the best water-seeking properties. Salt occurs in the form of large nuclei with radii greater than 1.0 μm and is very widespread since it enters the atmosphere following the evaporation of ocean spray. However, it is present in small concentrations which are usually less than 10 per cm³. Combustion products are characterised by small nuclei (with radii less than 0.1 μm) which accumulate in the air following volcanic eruptions and burning of sulphur-containing fuel like coal and oil. Consequently, their main distribution is restricted to urban and industrial air masses where large volumes of sulphur products are released. However, despite their limited spatial distribution, they are present in large concentrations of up to 10^6 per cm³. The traces of sulphur dioxide (SO_2) are oxidised by sunlight into sulphur trioxide (SO_3) which combines with rain-water to become sulphuric acid (H_2SO_4), the notorious *acid rain* which has poisoned many lakes and rivers particularly in southern Scandinavia (section 30.4).

The enormous number of dust particles entering the atmosphere from arid lands is believed to be of secondary importance as condensation nuclei since the majority of the particles are non-hygroscopic, with a very weak water affinity. The size and concentration of all nuclei in the atmosphere determine the dimensions of water droplets forming on them which influence fog and precipitation characteristics, as described in the next two sections.

5.2 Condensation forms at or close to the earth's surface

Saturated air, with an abundance of hygroscopic nuclei, will condense at or close to the earth's surface when the air is chilled below its dew point by the contact cooling and air-mass mixing described earlier. The former cooling mechanism is responsible for the formation of *dew, hoar frost* and *radiation fog* and is associated with decreasing ground temperatures following the loss of heat by long-wave terrestrial radiation (sections 2.2 and 3.2). This infra-red heat loss to the atmosphere is accelerated on calm, clear nights and is confined to a very shallow layer of air in direct contact with the cold ground, since air is a very poor conductor and the cooling process is essentially by conduction. On cooling below the dew point, the excess water will condense in the air as radiation fog or is deposited on the surface of objects as dew, or as hoar frost when the object's temperature is

below freezing point with sublimation directly from the water vapour. The formation of dew and hoar frost is most uneven since dark objects (like vegetation, thermally insulated from soil heat storage by its foliage) always cool rapidly. They behave as the perfect black body described in section 2.2 which, at night, provide the coldest surfaces and encourage the greatest deposition of droplets or ice crystals.

The conditions which lead to the formation of dew and hoar frost are clear, calm nights with moist air close to the earth's surface (i.e. with a relative humidity preferably above 80% at sunset). Clear, cloud-free conditions are necessary to allow maximum long-wave radiation loss and ground cooling, since a 'blanket' of cloud will facilitate a greater 'greenhouse effect' and surface warming by counter-radiation (section 2.2). Windy nights prevent the warm moist air from stagnating and remaining in contact with the chilled earth or cold objects long enough to cool sufficiently, even with clear skies.

Radiation fog forms under the same conditions required by dew and hoar frost when, on clear nights, the earth's surface cools rapidly by long-wave radiation loss and chills the surrounding air by conduction. However, the air has to be cooled below its dew point over a longer period of time and a much greater depth than the 'skin' of air surrounding cold objects which is necessary for dew formation. Under these more advanced conditions, condensation will now form as dew on surface objects and on the hygroscopic nuclei of the adjacent air (normally up to 300 m). This condensation yields minute droplets suspended in the lowest 100–300 m of atmosphere and since the cooling of the air depends on ground chilling below by long-wave radiation, the resultant droplet accumulation is termed radiation fog. Whereas calm, stagnant air is necessary for dew formation, radiation fog requires a light wind (up to 3 m s⁻¹) gently to stir the cold air in contact with the ground and to give sufficient turbulence to spread the cooling upwards over a greater depth of air. Consequently, this turbulence thickens the fog from below when dew point chilling and droplets are diffused aloft. However, above 3 m s⁻¹, the vigorous air flow would prohibit the necessary air stagnation and contact cooling beyond the critical dew point level. Fog can exist at temperatures below 0 °C when the droplets are supercooled, which freeze on to cold objects to form fragile, low-density *rime ice*.

Radiation fog forms first and thickest in valleys

Plate 5.1 Radiation fog near Llangollen, North Wales, UK.

or hollows (like the Thames Valley in southern England) since cold, dense air drains downslope by gravity, and collects at the lowest elevations (Plate 5.1). This *katabatic air flow* is described in detail in Chapter 14 and is responsible for the characteristic valley fog which forms during stable anticyclonic weather, when the tops of telegraph poles and church steeples sometimes rise out of the cold, raw fog blanket. Radiation fogs are also thickened by temperature inversions (section 3.2) when cold dense air is trapped beneath a 'lid' of overlying warm (and fog-free) air. The fog's top now lies near the base of this inversion at a height of several hundred metres, with the crests of hills projecting upwards through the fog into clear, warm, dry air above. Consequently, radiation fog is normally extremely patchy, preferring sheltered valley locations, although it occasionally can become more widespread and regional.

After sunrise, the radiation fog is normally 'burnt off' quite rapidly by terrestrial heating and evaporation from the warming surface. The fog evaporates from below and gradually lifts during the early morning until the last upper remnant has the appearance of low, patchy stratus cloud extending outwards from the high ground. However, the evaporation and elimination of radiation fog after sunrise can be delayed in thick, high fogs forming under strong temperature inversions described above. Furthermore, when the concentration of smoke and aerosols (released by urban and industrial sources) destroys the transparency of these fogs, then the resultant opaque fogs can persist for many days as serious smog episodes. Perhaps the best example of this was the infamous smog over London, England, which developed in the stable weather of a winter anticyclone in early December 1952. The fog build-up was accentuated by the accompanying strong temperature inversion and fumigation-type pollution (section 3.2) and the resultant smog persisted for five days. The concentration of smoke and sulphur dioxide reached a peak on 7 December, at 1.6 mg m^{-3} and 0.75 ppm respectively, and the number of deaths reached a maximum (about 900) on the following day, totalling 4000 over the period.

Condensation associated with air-mass mixing produces *advection fog*, which develops when cold or warm airstreams move across warm or cold surfaces respectively. The first type of advection fog is related to the passage of cold polar air across a warmer sea surface which mixes with the warm moist air, providing a cooling–condensing mechanism. The warm vapours evaporating from the water body, concentrated in a shallow 'skin' or 'envelope' at the surface, are immediately chilled by conduction with the cold airstream and condense into a very shallow zone of fog. In fact, the sea surface appears to steam or smoke, giving rise to names like *steam fog* or *Arctic smoke*. They are very common in polar areas where very cold air drains off a snow-covered land mass over the warm water of exposed leads or pools in the sea ice.

The second type of advection fog is associated with the passage of warm, humid airstreams over a colder surface, especially a cold ocean current. It is the most common type of marine fog encountered, since about 80% of all such fogs owe their origin to this process (Donn 1965), and its formation depends on a significant contrast between air and water temperature, and a wind speed (and related turbulence) of about 4 m s^{-1}. On such occasions, the warm humid air is transported across the cold water and is chilled from below by conduction, leading to the development of a strong temperature inversion and stability (sections 3.2 and 3.4). Conductive cooling below the dew point causes condensation in the surface air and the necessary turbulence spreads the cooling upwards over a greater depth of air. However, if the air flow is too vigorous, exceeding 8 m s^{-1}, it will lift the fog to form a low stratus cloud with a base of 100 m or so.

These extensive marine fogs are particularly well developed in the Grand Banks sea area to the east of Newfoundland, where warm, humid southeasterly winds from over the Gulf Stream encounter the southward-flowing cold Labrador current.

Plate 5.2   Advection fog near Santa Barbara, California, USA.

The thermal contrasts and associated fog development is most pronounced in the summer months when it dominates two out of three days. These marine fogs also dominate California (Plate 5.2) in the San Francisco area, when warm, moist southwesterly winds flow onshore across the cold California current, which represents the upwelling of cold benthos water close inshore (section 34.3). Advection fogs can also be found inland when moist warm air flows over a cold land surface, especially when the surface is covered by thawing snow close to 0 °C. These fogs are more extensive than radiation fogs and will persist over large areas as long as the air flow is maintained. Over northern land masses, the more extreme winter cold (−30 °C) causes advection fogs to be composed entirely of ice crystals. Here, sublimation forms *ice fog* or '*diamond dust*' with accompanying rime ice deposition and surface optical phenomena (i.e. *mock suns* and *sun pillars*), due to sunlight refraction through the crystals.

lowing condensation. The distinctive base of the cloud is maintained at a uniform level because, if droplets fall below this dew point level where the air is not saturated, then these small droplets are readily evaporated. When condensation occurs at sufficiently high altitudes (normally above 6000 m) where temperatures are below 0 °C, then the vapour will sublimate directly into ice crystals. In the most basic classification, three principal cloud forms are recognised (viz. *cirrus*, cumulus and stratus) with a number of main sub-types (Fig. 5.1). Cirrus are high, thin ice crystal clouds in delicate feathery filaments or fibrous sheets (*cirrostratus*) although small ripples occur with turbulence and are called *cirrocumulus* or a '*mackerel*' *sky* (Plate 5.3). All cirrus/cirro clouds are associated with the mass advection of ice crystals above 6000 m, and refract the sun's rays to form *haloes*, mock suns, sun pillars, etc.

Cumulus are flat-based individual cloud masses with a pronounced vertical growth and rounded tops and, when they are associated with instability,

5.3 Condensation forms in the troposphere, away from the earth's surface

These forms are represented by clouds and precipitation and are associated with the adiabatic cooling of rising air parcels in unstable environments. This process was described in detail in section 3.3 and occurs when parcels of air are forced upwards by free or mechanical convection, and expand/cool independently of the surrounding atmospheric environment. Condensation takes place when the dew point is reached, assuming sufficient hygroscopic nuclei, and the DAR now gives way to the SAR, where the cooling by expansion is opposed by the latent heat released by condensation. The dew point elevation is also known as the lifting condensation level (LCL) and represents the base of the cloud. The parcels will continue to rise and accentuate cloud development as long as instability exists (ELR exceeds SAR). The top of the cloud will indicate an eventual change to stability when the parcel is now cooling at a faster rate than the surrounding air (see Fig. 3.4).

Clouds therefore result from the condensation of buoyant air parcels, when rising air currents keep the droplets and ice crystals in suspension, since they are very small and light. For example, the average size of a cloud droplet is 0.05 mm and this growth takes place in about 100 seconds fol-

Plate 5.3 Cirrocumulus cloud over eastern Baffin Island, NWT, Canada.

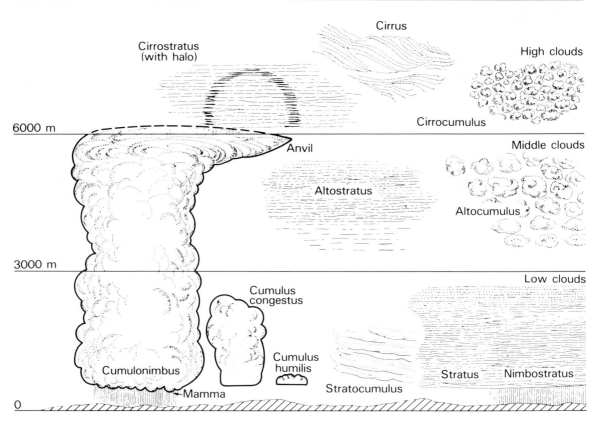

Fig. 5.1 Principal cloud forms and main sub-types. (Adapted from Cole–Pavey Academic Aids 1969)

they become 'heaped' congestus types (see Plate 3.2). Eventually, cumulonimbus clouds form when the highest part are composed of ice crystals, with the distinctive 'anvil' deformation by strong upper winds, and the cloud base deformed by strong perturbations into *mamma*-type undulations (Plate 5.4). Non-buoyant parcels have weak vertical growth and are called cumulus humilis (see Plate 3.1) or 'fair weather cumulus'. Compressed cumulus ripples between 3000 and 5000 m are called *altocumulus* (Plate 5.5), whereas at lower elevations (below 2000 m), the larger compressed globular masses are called stratocumulus. Stratus are extended sheets or shallow layers of droplets (or ice crystals when cirrostratus) covering most of the sky. They are associated with the mass advection of droplets under stable conditions below 2000 m (stratus) and between 3000 and 5000 m, where the cloud is called *altostratus*. Eventually, a thick mass of stratus/altostratus/cirrostratus

clouds (over 10 000 m extent) will produce precipitation to become *nimbostratus*.

Only the basic cloud types have been described in this section and for a complete, authoritative study of all species and varieties, reference should be made to the *International Cloud Atlas* (WMO 1956). However, the actual cloud form depends on the free or forced convective mechanism in operation (section 3.1) and whether stability or instability exists (section 3.4). For example, with free convection operating under stable conditions, vertical motion is prohibited either before the dew point is reached or soon after, leading to flattened cumulus humilis (see Plate 3.1). Conversely, with instability up to the tropopause or a high-level inversion at about 6000 m, free convection proceeds to develop clouds of great vertical extent such as cumulus congestus (see Plate 3.2) and eventually cumulonimbus. Forced convection is mainly associated with vertical motion related to orographic

Plate 5.4 Mamma cloud over Guelph, Ontario, Canada.

Plate 5.5 Altocumulus cloud over the Antarctic Peninsula, British Antarctic Territory.

and frontal displacement and the mass convergence of air in low pressure systems, where the actual cloud forms are mainly influenced by the strength of the uplift. For example, when air is forced to rise rapidly over a steep mountain slope, the strong vertical motion causes cumulus clouds to form which will proceed to congestus or cumulonimbus types if instability exists. Conversely, uplift over a gently sloping mountain flank produces a more gradual ascent which favours the formulation of layered stratus clouds, and altostratus above 2000 m elevation.

The development of clouds along a front depends mainly on the buoyancy of the warm air that is being displaced. For example, at a warm front with non-buoyant air, a considerable thickness of layer clouds dominates including cirrostratus, altostratus, stratus and (eventually) nimbostratus. This development is analogous to the gradual uplift over gently sloping mountain flanks described above. On the other hand, if the displaced warm air becomes increasingly buoyant, then the ascent could lead to conditional or convective instability which encourages the development of cumuliform clouds, possibly cumulonimbus. Similarly, cold frontal cloud development is controlled by the equilibrium tendency of the rising warm air. With instability, cumulus clouds form rapidly and develop into towering congestus or cumulonimbus (which compare favourably with the cloud formation over steep mountain slopes). Conversely, if the displaced warm air is associated with absolute stability, then cloud development is restricted to less than 3000 m and stratocumulus clouds are more common.

Cloud development in the troposphere will ultimately lead to precipitation since the essential difference between cloud and rain droplets is one of size and weight, in terms of the actual strength of the vertical motion. Whether or not droplets, ice crystals and *hailstones* will be precipitated as *hydrometeors* or will remain suspended in the air as cloud, depends entirely upon their weight in relation to the velocity of the upward currents of air which support them. For every size of droplet or ice crystal, there is a velocity of air which is just capable of holding it at its own level. This is known as its *terminal velocity*, which varies considerably as the radius/size of the hydrometeor changes (Table 5.2).

Since clouds are formed in upward-moving air currents, sometimes of considerable velocity, it follows that there must be processes within clouds which lead to the growth of droplets and ice crys-

Table 5.2 Terminal velocity of hydrometeors. (Adapted from Pedgley 1962)

Hydrometeor	Radius (mm)	Terminal velocity (m s^{-1})
Cloud droplet	0.05	0.25
Drizzle	0.25	2.0
Raindrops	0.50	3.9
	1.50	8.1
	2.50	9.1
Snowflakes	5.00	1.7
Graupel	5.00	2.5

tals to a size large enough to overcome the lifting forces and fall to the earth's surface by gravity. Table 5.2 indicates that the size of the largest raindrop (2.50 mm) is about 50 times the size of the average cloud droplet and it is apparent that special processes must operate in a cloud from which hydrometeors are released. These four growth mechanisms are as follows.

Firstly, *coalescence* of droplets within clouds when large droplets (0.09 mm), usually forming around large hygroscopic nuclei, fall with greater velocity than small droplets around the smallest nuclei (0.001 mm), since their terminal velocity is considerably greater (i.e. 70 cm s^{-1} compared to 0.01 cm s^{-1}). The large droplet descending through a cloud of smaller ones 'sweeps up' and coalesces with a large number of them lying in its path. The fusion takes place at the rear of the falling large droplets, where the vortex is greatest, and the final size to which a drop will grow in this way depends on a number of basic conditions which determine the actual rate of growth. One of the main factors concerns the fusion efficiency of droplets, which is influenced by their size (i.e. the minimum radius required is 0.02 mm, which is more common in maritime clouds). Also, the liquid water content of the cloud should be large (at least 1 g m^{-3}) and the updraught of air must be moderate (1–5 m s^{-1}), not exceeding the terminal velocity of the droplet. Finally, the cumulus cloud concerned should have an adequate thickness of several thousand metres and should persist for at least 30 minutes (Pedgley 1962).

Secondly, *aggregation* of ice crystals (with branching plates) forms snowflakes when they interlock on collision, because of their complex stellar and dendritic branch structures. It appears that this process is aided by the freezing and cohesion of supercooled water-like films on the surfaces of the crystals which join together when the surfaces come into contact. This cohesion diminishes as both temperature and the degree of supersaturation decrease so that aggregation occurs most readily at temperatures between 0 and −4 °C. Furthermore, the process is very common in clouds composed largely of ice crystals, such as cirrostratus and the uppermost parts of cumulonimbus.

Thirdly, the *Bergeron process* was proposed in 1933 by the Norwegian meteorologist T. Bergeron when he emphasised the significance of the coexistence of both supercooled droplets and ice crystals for hydrometeor growth, in clouds with temperatures below 0 °C. In simple terms, the role of ice particles is associated with their greater attraction to wandering water vapour molecules (due to the resultant supersaturation and deposition on to the particles), compared with that exerted by supercooled water droplets. However, the actual ice crystal growth is accomplished by simultaneous evaporation from supercooled droplets and sublimation–glaciation upon the ice particles as a result of related vapour pressure differences. Figure 5.2 illustrates this growth mechanism, where the vapour flux from droplet to ice particle is stimulated by the gradient associated with the low vapour pressure vortex over ice and high pressure divergence over the droplet. For example, at −12 °C the corresponding vapour pressures are 2.128 mb over ice and 2.398 mb over

the droplet and this represents a 'peak' flux gradient.

This theory is supported by the fact that showers become most severe when cumulus congestus clouds grow into cumulonimbus with their characteristic ice crystal tops and the juxtaposition of droplets and crystals in the uppermost parts of the cloud. It also accounts for the success of cloud seeding with dry ice (solid carbon dioxide crystals) and silver iodide (freezing nuclei) in portions of cloud with an abundance of supercooled droplets. However, ice crystals can only grow large enough to overcome the terminal velocity when the ice crystal concentration is small (for example, one crystal per thousand or million supercooled droplets) and when the cloud's liquid content is small, with cloud temperatures in the range −10 to −30 °C. Under these conditions, this large size (several millimetres across) will be rapidly attained in about 10–30 minutes (Pedgley 1962). The process cannot be applied to hydrometeor growth in the tropics where towering cumulus clouds frequently yield torrential rainfall without reaching the freezing level. Here, coalescence of coexisting warm and cold droplets is more important, possibly associated with the development of an adequate vapour pressure gradient between the warm droplet (high vapour pressure) and cold droplet (low vapour pressure).

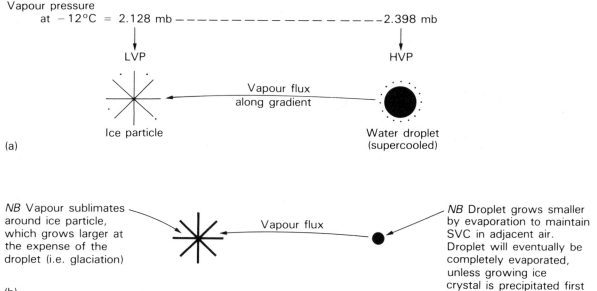

Fig. 5.2 Hydrometeor growth by the Bergeron process (LVP – low vapour pressure; HVP – high vapour pressure): (a) initial condition; (b) after 10–30 minutes of coexistence.

Finally, *accretion* can also take place in clouds containing both supercooled water droplets and ice crystals, and occurs when ice crystals are swept aloft by updraughts into a zone of supercooled droplets with a concentration of less than 1.0 g m^{-3}. On striking the flake, the droplet freezes almost immediately on its surface to produce a *rimed snowflake*. If accretion is more pronounced in clouds with a liquid content well in excess of 1.0 g m^{-3}, then the frozen droplets can accumulate as rime ice to form soft, fragile *graupel*. Also at $-10\,°C$ the droplets can freeze directly into hard, durable clear ice as *ice pellets* or hailstones, an integral feature of thunderstorms described at the end of this section and in section 14.4.

There are definite associations between the type of hydrometeor and the free or forced convective mechanism in operation and whether or not stability or instability exists (in the same way as cloud forms were controlled, as discussed earlier). For example, in stable conditions, hydrometeors are characterised by their small size, steady fall and a tendency to persist over a long period of time as drizzle, gentle rain or ice needles. These types of precipitation are particularly common at stable warm fronts and with an updraught of air over gentle mountain slopes. The best indicator of stability is the deposition of *freezing rain* or *glaze ice* since its formation depends on a fairly strong temperature inversion normally associated with a winter warm front. Here, temperatures are above $0\,°C$ in a shallow zone of advected warm air (say between 1000 and 2000 m) away from freezing conditions at the earth's surface.

Unstable conditions lead to the formation of hydrometeors with a large size, intense fall and intermittent behaviour. They develop as large raindrops, hailstones and large snowflakes, and the resultant weather in buoyant free convection systems is dominated by severe, showery *squalls*. These types of precipitation are also experienced at unstable warm and cold fronts and with the violent updraught of air along steeply sloping mountain flanks. The most outstanding effect of instability is the continued growth of cumulonimbus clouds, resulting in thunderstorms with the release of torrential rain and hailstones.

This storm development is illustrated as a single cell in Fig. 5.3, where the vigorous updraughts and downdraughts (reaching 1800 m min^{-1}) cause intense friction between the air and hydrometeors, which leads to a very high voltage build-up. The positive charges are concentrated in the upper parts of the cloud, since they have an affinity for the ice crystals, whereas the negative charges build up in the lower parts. The latter results in a positive charge at the earth's surface which reverses the fair weather gradient. When the negative–positive voltage potential difference becomes high enough then an electrical discharge (*lightning* flash) takes place within the cloud and from cloud to ground. The rapid expansion and contraction of intensely heated/cooled air along the discharge path produces the explosive sound called thunder which reverberates or rumbles because of the time differences required for sound to reach the observer from various parts of the flash. The light from the flash is received almost instantaneously whereas the slower sound waves can only travel 1.6 km in five seconds, producing the time-lag between flash and accompanying thunder.

The violent updraughts and downdraughts in a thundercloud prevent ice pellets from reaching the ground. They are carried aloft and grow by the accretion of supercooled droplets as rime or opaque ice in the coldest, higher parts where the liquid water content is small, and glaze (clear ice) when abundant large droplets freeze rapidly. A pellet which passes through several parts of the thundercloud with varying liquid water content acquires alternate layers of clear ice (high content) and opaque ice (low content) forming a typical multi-layered hailstone growing as large as a golf ball. Eventually, its terminal velocity becomes so large that the updraughts cannot keep it aloft in suspension, and it falls rapidly to earth, normally in the early stages of a cumulonimbus development. The actual circulation and life-cycle of a thunderstorm are discussed later in section 14.4.

5.4 Retrospect

Condensation occurs when the concentration of water vapour in the atmosphere reaches saturation level, when any further evaporation will produce an equal amount of condensation (assuming an adequate supply of hygroscopic nuclei) and latent heat release. The SVC is controlled primarily by air temperature since the capacity of the air for vapour decreases rapidly as the temperature falls. Consequently, a temperature decrease is the most important natural process which leads to saturation and, with a continued fall below the dew point, then considerable amounts of excess water

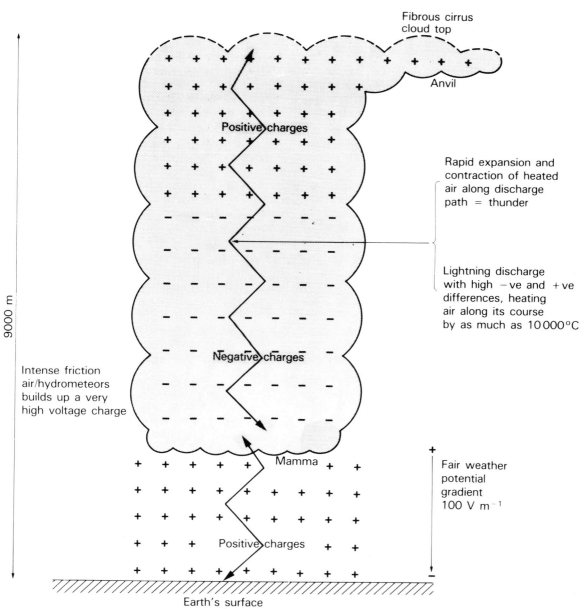

Fig. 5.3 Cumulonimbus cloud and thunderstorm development.

vapour present must be released as a condensation phenomenon to maintain the saturation vapour content of the air. The necessary cooling mechanisms are contact chilling associated with terrestrial long-wave radiation (which forms dew, frost and fog), air-mass mixing when an airstream advects across a warmer or colder surface (producing marine fogs) and the adiabatic cooling of buoyant air parcels in unstable air, which produces clouds and precipitation.

Key topics

1. The factors controlling the saturation of air.
2. The meteorological phenomena resulting from radiative/advective cooling and associated condensation at or close to the earth's surface.
3. The dynamics of cloud formation.
4. The conditions under which precipitation occurs in the troposphere.

Further reading

Barry, R. G. and Chorley, R. J. (1982) *Atmosphere, Weather and Climate*. Methuen; London and New York, pp. 74–102.

Durbin, W. G. (1961) An introduction to cloud physics, *Weather*, **16**, 71–82, 113–25.

Gray, D. M. (1973) *Handbook on the Principles of Hydrology*. Water Information Centre; New York, pp. 2.1–2.111.

Mason, B. J. (1962) *Clouds, Rain and Rainmaking*. Cambridge University Press; Cambridge.

Ward, R. C. (1975) *Principles of Hydrology*. McGraw-Hill; London and New York, pp. 16–53.

Chapter 6 Subsurface water transfer in the hydrologic cycle

Section 2.1 indicates that although the bulk of continental fresh water (74.61%) is locked up in relatively inaccessible ice sheets and glaciers, most of the readily available fresh water is stored in porous rocks as groundwater (25% of continental water). Soil moisture storage is quite minimal (0.06%) and merely represents a temporary phase in the percolation of surface precipitation towards the water table. Collectively, both storage processes involve subsurface water transfers and represent a vital subsystem in the global mass cascade. However, since the quantity of water concerned remains unchanged over the long term, this mode of transportation is not conspicuous in the average annual hydrologic cycle (see Fig. 2.1). Conversely, over the short term, subsurface water represents an important component of the water balance (see Table 2.3) and, furthermore, artificial abstraction of groundwater by human activities has resulted in serious depletion problems. For example, in the eastern part of the London Basin in southern England, the water table was lowered by some 75 m in the confined chalk aquifer up to 1965, which has resulted in increased well depths and pumping costs (Ward 1975).

This chapter examines the distribution of subsurface water and describes the factors which control the infiltration of precipitation. The different types and occurrence of groundwater are also discussed and emphasis is placed on the groundwater budget, where atmospheric and terrain factors interact in order to balance the discharge and recharge.

6.1 Subsurface water and infiltration

Subsurface water is found in four main moisture zones (Fig. 6.1) below the earth's surface. Precipitation first enters the *soil zone*, where the water is held temporarily in the roots of plants and pores of the thin layer of soil at the ground surface. Water is removed from this zone either through evapotranspiration into the air or by downward percolation through the *intermediate zone* towards the water table. Both these upper zones make up the *zone of aeration* in the ground which, as unsaturated soil and subsoil layers, represents actual soil moisture. In the zone of aeration, where the pore spaces are filled with both water and air, the *capillary suction forces* (associated with soil water held by cohesion within the pores) predominate and the pressure of water is less than atmospheric.

The lower two zones comprise the zone of saturation where the pore spaces are almost completely filled with water and where the water pressure is equal to or greater than atmospheric pressure. The *capillary fringe* occurs immediately above the water table and represents the zone of

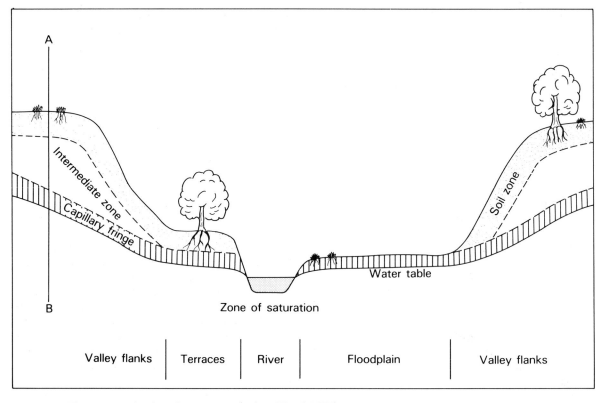

Fig. 6.1 The zones of subsurface water. (After Ward 1975).

movement where a substantial number of pores are filled with water which has moved upwards by suction forces against the force of gravity. Below the water table is the main zone of saturation where nearly all the rock *interstices* (i.e. the pores and voids) are full of water. This represents actual groundwater and technically becomes an *aquifer* when the porous rocks yield significant quantities of water to become a great storage reservoir, through which water slowly moves.

The vertical water movement in the intermediate zone is controlled by the *infiltration capacity* of the soil and subsoil layers (i.e. the maximum rate at which water enters the soil under specified conditions). Initially, water enters the soil surface due to the combined influence of gravity and capillary forces at a rate determined both by total porosity and pore size. Both forces act in a vertical direction to cause downward percolation, although capillary forces also divert water laterally from the large pores (which act as so-called feeder canals) to capillary pore spaces, smaller in dimension but more numerous. As the process continues, the capillary pore spaces become filled and

the percolation of gravitational water to greater depths encounters increased flow resistance. This is associated with a reduction in the extent or size of the flow channels, and the occurrence of impermeable barriers. At the same time, the soil surface becomes more resistant to the inflow of water since a surface-sealing takes place, due to raindrops breaking down the soil aggregates and the subsequent infill of pores with this finer soil material. This causes a rapid reduction in the actual rate of infiltration in the first few hours of a storm, after which the rate remains nearly constant for the remainder of the period of storm rainfall excess (section 19.2).

The actual infiltration capacity of a given soil depends on the type, intensity and duration of precipitation (section 19.2 and Fig. 19.3), together with a variety of environmental factors. These include soil surface conditions, vegetation cover, the temperature and chemical composition of the water, the height of the water table and the physical properties of the soil. These latter properties are associated with porosity, grain/pore size, moisture content and the *hydraulic conductivity* (*K*), which represents a measure of the ease of water flow through a *permeable* material, when subjected to a pressure difference due to a head of water. It appears that maximum rates of infiltration will occur during long-duration and low-intensity rainfall, especially when the soil surface is sun-cracked or unconsolidated. However, the rate will decrease when the surface is sealed by pore infilling associated with rainwash (described above) or frost action, and when the depth of surface water decreases. In terms of soil grain size, measurements reveal that infiltration in initially wet soils without a vegetation cover was $0-4$ mm hr^{-1} in clays, $2-8$ mm in silts and $3-12$ mm in sands (Kirkby 1969).

Vegetation cover also increases the rate of infiltration by restricting water movement and reducing the impact of raindrops on soil aggregates, which minimises the associated sealing process. Experimental evidence indicates that infiltration is higher beneath a forest than beneath grass, although the presence of ground litter has a more pronounced control than does the actual plant type. The actual role of plant cover on infiltration is evident from measured rates of 57 mm hr^{-1} under old permanent pasture compared with 6 mm hr^{-1} in bare, crusted soil (Kirkby 1969).

6.2 The occurrence and types of groundwater

The previous section indicates that the zone of saturation or groundwater represents the most significant accumulation of subsurface water, where nearly all the openings or interstices in the rock are filled with water. As will be discussed later, this saturation causes the retention and movement of groundwater to be governed by the laws of saturated flow. The accumulation of groundwater is mostly associated with precipitation, which percolates through the intermediate soil zone towards the water table. The infiltration rate of this *meteoric water* is controlled by the atmospheric and terrain factors described in the previous section but it is apparent that the surface catchment area rarely corresponds with an underlying groundwater catchment. For example, the River Itchen in Hampshire, England has a groundwater catchment area of 430 km^2 which is approximately 20% greater than that of the surface catchment (Ward 1975). Groundwater is also supplied by *influent seepage* from surface water bodies (as indirect

meteoric water, particularly in arid lands) and as saline intrusions from the oceans. Minor supplies include the condensation of water vapour from air held in pores and interstices in the zone of aeration, *connate water* trapped in sedimentary rocks at the time of formation and *juvenile water* from below the aquifer.

The accumulation of water in the zone of saturation is controlled by the hydraulic conductivity or permeability of the rocks concerned which, in turn, is influenced by the absorption and transmission of water by the rock interstices. These openings can be original, when they are created between individual particles at the time the sedimentary rock was formed, such as intergranular spaces in well-sorted valley gravels. Alternatively, the interstices are secondary features created by later geologic and climatic activity, which form more massive interconnected openings like fault fractures, crevices or joints. In Carboniferous limestone, these are enlarged by solution weathering (Ch. 16) into shafts and caverns which can transmit considerable amounts of water to the aquifer. Secondary openings also result from shrinkage and gas bubbles in solidifying igneous material like lava.

Aquifers represent the largest reservoirs of groundwater where the bedrock or superficial deposits are permeable with interconnecting interstices, which are able to absorb and transmit precipitation to subsurface layers. This transmission regulates the hydrologic cycle by maintaining *baseflow* to the river in dry weather when no surface runoff occurs (section 7.1). *Aquicludes* are semi-permeable formations which collect groundwater slowly in porous rocks which can slowly absorb water. However, without interconnecting interstices, they are unable to transmit this water in sufficient quantities to the river bed, springs and wells. Finally, *aquifuges* are *impermeable* rocks which are unable to absorb and transmit subsurface water.

The occurrence of groundwater depends mainly on geological factors, particularly those associated with the alternation of permeable and impermeable strata in a synclinal structure (Fig. 6.2). In simple terms, three types of groundwater occur although, in complex geological structures, it is sometimes impossible to differentiate between them. Firstly, *perched groundwater* occurs when isolated impermeable or semi-permeable beds (i.e. aquifuges or aquicludes respectively) are located well above the main water table and groundwater zone, forming shallow and very localised water

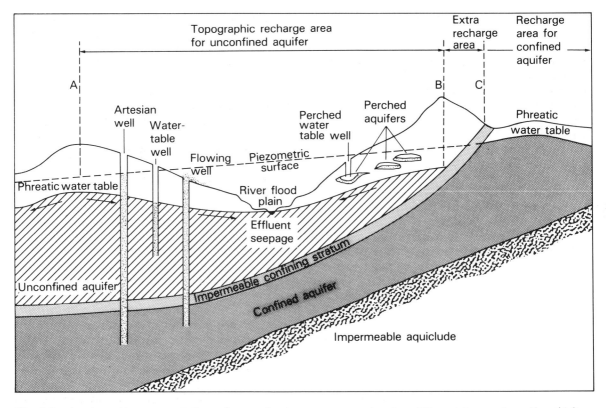

Fig. 6.2 A schematic representation of groundwater occurrence showing different types of aquifer. (After Smith 1972)

bodies. Secondly, *unconfined groundwater* or *water table aquifers* occur when permeable rocks outcrop at the surface. Here, the upper level of the aquifer is represented by the water table, which rises and falls in response to the groundwater balance changes outlined in the next section. In fairly uniform strata (like an uncomplicated syncline structure), the water table assumes a distribution which reflects the surface contours in a subdued form (Smith 1972). The base of the unconfined aquifer is represented by an impermeable confining stratum or aquifuge.

The third type of groundwater storage is represented by a *confined or artesian aquifer*, where impermeable strata occur above and below permeable bedrock. In this case, the main aquifer recharge is not as rainfall percolation from above (as in the unconfined aquifer) but may take place many kilometres away when permeable rocks are exposed on the surface (Fig. 6.2). However, the upper confining impermeable bed rarely forms an absolute barrier to groundwater flow (especially if it is an aquiclude) so that there is usually some

water transfer from the unconfined aquifer above (Ward 1975). Figure 6.2 also indicates that the water table in the recharge area is situated at a higher elevation than the main storage area of the confined aquifer. Consequently, the confined groundwater is under a pressure that is equal to the difference in equilibrium or *hydrostatic level* between the two elevations (which exceeds atmospheric pressure) and the water table represents the hydrostatic pressure level to which, theoretically, the confined water would rise if released through a deep borehole. The water would then rise to the level of the water table (which represents the actual hydrostatic head) projected into the recharge area as the *piezometric surface*. When this surface occurs above ground level, a borehole into the confined aquifer will flow continuously as a *flowing artesian well* at a considerably higher elevation than the *in situ* water table (Fig. 6.2). Conversely, a well sunk into the unconfined aquifer will be independent of the piezometric surface and the water will only accumulate up to the water table level as a *water table well*.

6.3 Groundwater balance

The storage of water in a groundwater system will remain constant only if the inflow of water exactly balances the outflow. Conversely, a change of storage ($\triangle S$) will occur when the inflow or recharge (Q_r) does not balance the outflow or discharge (Q_d), so that the groundwater balance equation is simply:

$$\triangle S = Q_r - Q_d \qquad \textbf{[6.1]}$$

The recharge of the aquifer takes place mainly by the infiltration of precipitation intercepted at the earth's surface which is, in turn, controlled by the infiltration rate constraints discussed in section 6.1. For example, it is generally accepted that rainwater infiltration is greater with fine, prolonged drizzle than with short, violent downpours, although observations on the Canadian prairies do emphasise the importance of very intense rainfall (Freeze and Banner 1970). Infiltration recharge is also greater when the water table is close to the surface in a shallow zone of aeration characterised by a high hydraulic conductivity. Snow ablation also appears to be a significant recharge factor in areas with an appreciable winter snow pack, especially with the springtime melt in topographic basins which act as receiving sites.

Other recharge mechanisms include *influent seepage*, *leakage* and *artificial water injection*. The former inflow process occurs when surface water bodies, such as lakes and rivers, are in direct contact with groundwater, although it does alternate over a few hours with *effluent seepage* which (as discussed shortly) acts as a groundwater discharge mechanism. Influent seepage is important in desert areas especially from ephemeral streams where, for example, seepage recharge can be up to 10 times more effective than rain infiltration into the aquifers. However, the rate and amount of recharge from this process depend partly upon channel characteristics (such as shape, wetted perimeter length and permeability) and partly upon water characteristics (for example, temperature and depth). Normally, ephemeral stream seepage in arid areas is relatively shallow and builds up *groundwater mounds* beneath the channel. Furthermore, these mounds have a much reduced salt content compared with aquifer water at a greater distance from the seepage zone. Here, the infiltration of rainwater is more effective, with greater salt concentration in the percolate which results from the excessive surface evaporation.

Leakage acts as a recharge process when water is transferred slowly from adjacent aquifers, normally through an intervening aquiclude. However, this slow transfer of water does not make a significant contribution to the total recharge. Similarly, artificial recharge makes a negligible contribution despite some leakage from canals and irrigated surface water, where the infiltration rates of a few centimetres or metres per year are low (Nace 1969). A preferable alternative to the large-scale inundation of tens of thousands of hectares of valuable land is to inject water artificially through boreholes extending to the aquifer, especially in the rainy season when excess surface water is available. However, problems here are associated with the suspended sediment which clogs the aquifer and the chemical reaction between oxygen-rich percolate and 'natural' groundwater which introduces laxative-type characteristics.

Groundwater discharge is the antithesis of recharge, and the processes operating are directly opposite to those discussed above, namely evaporation–transpiration, effluent seepage, leakage into adjacent aquifers and *artificial water abstraction*. The former meteorological process is most effective where the water table and/or capillary fringe are close to the surface. Then, especially in low-lying floodplains, the groundwater loss by evapotranspiration around solar noon (with excessive amounts of net radiation necessary for the vapour flux) exceed the infiltration rate, which is responsible for a fall in the water table (Fig. 6.3). Conversely, at night with zero or negative net radiation, groundwater inflow greatly exceeds the negligible evapotranspiration and the water table rises. Over the long term, the effects of evapotranspiration will be paramount only if the nocturnal recovery rate does not exceed the daytime drawdown. However, eventually the falling water table and capillary fringe will move out of the zone of direct evapotranspiration, so that the process becomes increasingly ineffective over time.

Effluent seepage or outflow from the aquifer transfers groundwater directly into a surface water body although, as described earlier, this process tends to alternate over a few hours with influent seepage. However, it does appear to be the main discharge mechanism and is particularly effective as natural spring flow when the water table intersects with the surface, especially along an escarpment where perennial *springline* emission has favoured the development of early settlements. Leakage into adjacent aquifers via an intervening aquiclude must naturally follow on from the earlier discussion on leakage as a recharging process,

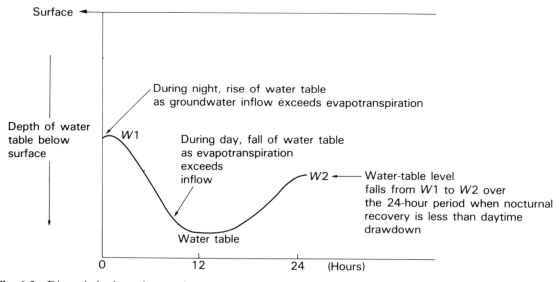

Surface

Depth of water
table below
surface

During night, rise of water table
as groundwater inflow exceeds evapotranspiration

W1

During day, fall of water table
as evapotranspiration
exceeds
inflow

W2 ← Water-table level
falls from W1 to W2 over
the 24-hour period when nocturnal
recovery is less than daytime
drawdown

Water table

0   12   24   (Hours)

Fig. 6.3 Diurnal rhythm of groundwater recharge and discharge.

although both transfers are slow and generally insignificant. Finally, artificial abstraction from wells and boreholes has been responsible for dramatic falls of the water table and piezometric surface, especially in aquifers like the London Basin which have been used as a major source of domestic and industrial water for many centuries (see the introduction to this chapter).

The actual movement of the groundwater in the recharge–discharge balance is largely in response to the hydraulic conductivity of water-bearing materials and the prevailing *hydraulic gradient* or *hydraulic grade line* (i.e. the slope of the water table). For example, the movement is slow (less than 1 mm day^{-1}) in fine-grained porous rocks increasing to a rapid 5500 m day^{-1} in well-fissured chalk. Furthermore, the hydraulic conductivity of groundwater is controlled mainly by aquifer characteristics like the geometry of the pore spaces (size and shape of grains), rock particles (surface roughness influences velocity of flow) and secondary geological structures (faulting, fissuring and anticlinal folding) which all accelerate groundwater flow. Less significant factors are related to the fluid characteristics, especially the density and *viscosity* (i.e. the inherent property of fluids to resist flow past obstacles) of the water. These physical properties of groundwater are influenced by its temperature and salinity. For example, temperature inversely affects viscosity, and influences the speed of groundwater flow which approximately doubles with an increase of temperature from 4.5 to 32.5 °C.

The velocity of groundwater movement follows *Darcy's Law* (1856) for saturated condition, which is expressed as follows:

$$V = -K \nabla \phi \qquad [6.2]$$

were V is the macroscopic water velocity (m s^{-1}), $-$ is movement in the direction of decreasing head or potential, K is hydraulic conductivity (ranging from 10^{-11} m s^{-1} in undisturbed clays to 10^{-4} m s^{-1} in silty clays) and $\nabla \phi$ is the gradient of total potential or head (suction less gravitational flow). Apart from minor deviations (especially at extreme high and low flow rates), Darcy's Law can be successfully applied to all normal cases of groundwater flow in both confined and unconfined aquifers (Ward 1975).

6.4 Retrospect

Subsurface water behaviour is a complex process since its accumulation and movement are controlled by complicated physical laws. However, this chapter has avoided these complications and has emphasised the atmospheric and terrain factors which dominate the rate of infiltration and the occurrence and characteristics of groundwater. Furthermore, as a geographical process, the most significant part of the study of groundwater concerns the factors controlling the actual groundwater balance and water transmission, where the the recharge and discharge are dominated by precipitation/infiltration and effluent seepage at the springline respectively.

Key topics

1. The controls on surface water infiltration and subsurface water accumulation.
2. The factors determining the occurrence and movement of groundwater.
3. The influence of geology on groundwater distribution.
4. The role of hydraulic conductivity in effecting groundwater movement.

Further reading

Bouwer, H. (1978) *Groundwater Hydrology*. Mc-Graw-Hill; London and New York.

Davis, S. N. and De Wiest, R. J. M. (1966) *Hydrogeology*. John Wiley and Sons; New York and London.

Domenico, P. A. (1972) *Concepts and Models in Groundwater Hydrology*. McGraw-Hill; London and New York.

Kirkby, M. A. (1979) *Hillslope Hydrology*. John Wiley and Sons; New York and London.

Maxey, G. B. (1969) Subsurface water-groundwater, *The Progress of Hydrology*. University of Illinois; Urbana, Vol. 2, pp. 787–815.

Todd, D. K. (1964) Groundwater, in Chow, V. T. (ed.) *Handbook of Applied Hydrology*. McGraw-Hill; London and New York, section 13.

Chapter 7 Surface water transfer in the hydrologic cycle

Surface water transfer represents the gravity flow of water across the earth's surface (known as runoff) when precipitation exceeds both evapotranspiration and infiltration/groundwater storage. Runoff takes place in sheets (as overland quasi-laminar flow) or, more normally, in a confined channel varying in size from small rivulets to large rivers like the Amazon. Section 2.1 reveals that only a minute percentage of the continental fresh water (0.03%) is transferred by river and stream channels. Even so, surface runoff is a vital subsystem in the global mass cascade and, indeed, can be regarded as the final or 'closing' stage of the hydrological system. At this time, all surplus surface water (i.e. 13 cm yr^{-1} in Fig. 2.1) is transported back to the oceans to be available for the initiation of the next cycle by evaporation.

This chapter examines the runoff process and the way it is organised into different types of river flow. However, the material concentrates on the physical processes which control runoff distribution, namely meteorological elements and catchment characteristics, which include topography, geology, soil, vegetation, fauna and watercourses.

Human factors are also briefly discussed since they control runoff through deliberate or inadvertent cultural activities. The associated transfer of sediments across a slope surface or in channels will be discussed in chapters 19 and 20.

7.1 The runoff process

Runoff is also referred to as *streamflow* and *catchment yield* where the flow of water across the surface is derived mainly from excess rainfall. However, this flow is most intermittent since its occurrence depends entirely upon the rainfall rate exceeding the loss rate by evapotranspiration and infiltration, which varies dramatically from day to day. Snow ablation also provides an important source of runoff in cold continental climates. In Canada, for example, streamflow from snowmelt constitutes about 30–40% of the total runoff from the drainage basins of the Fraser and Saskatchewan Rivers and its effect may persist for several months in spring/early summer (Bruce and Clark 1969). Groundwater also contributes to runoff and, since this storage is sustained by the infiltration of rainwater and snowmelt (section 6.1), the contribution of this subsystem does represent a form of indirect precipitation. Furthermore, this basal flow continues as long as the water table reaches the river channel and is termed permanent or *'fair-weather' runoff* since it persists during drought conditions (section 7.2).

The runoff process is therefore controlled by the three forms of precipitation discussed above. However, the role of direct rainfall is determined by the amount of rainwater lost in *surface reten-*

tion, through *interception* by the vegetation canopy and *depression storage* in surface hollows. When the rainfall first reaches the earth's surface in humid climates, it meets a dense layer of vegetation which intercepts a fairly constant proportion of the rainfall. For example, Norway spruce trees intercept 58% of the rainfall compared with 31% for Lodgepole pine, 33% for wheat, 27% for tall grass, 14% for corn/thinned pine forest and 12% for clover (Newson and Hanwell 1982). As the rainfall continues, the *interception capacity* of the vegetal cover will become satisfied and the water will be released to the ground either by dripping or *throughfall* from leaves and branches or by *stemflow* down the main trunks of trees and shrubs. Interception losses depend on the prevailing weather conditions, since high winds accompanying rainfall appear to decrease the amount of water stored this way. However, if the rain persists over many hours, then the increased rate of evaporation related to the higher wind velocity (section 4.2) would actually increase the total volume of rain intercepted over time. Interception of snow by trees can be quite significant and, for example, a dense stand of Ponderosa pine in Idaho was found to intercept and dissipate some 25% of the seasonal snowfall (Bruce and Clark 1969).

The rain that reaches the surface directly or via throughfall and stemflow, immediately infiltrates into the soil material and percolates into the bedrock to replenish the groundwater storage. However, section 6.1 indicates that eventually, when the soil becomes saturated, the infiltration rate is exceeded by the rainfall rate and *overland flow* occurs (section 19.2). Some of this surface water is retained by storage in surface depressions and the remainder flows across the surface as a sheet or follows micro-channels which coalesce with others of increasing size. This flow of surface water represents the saturated overland flow component of the runoff process which combines with *channel flow*, *interflow* (or *through flow*) and *groundwater flow*, to complete the input to total runoff from a drainage basin. Figures 7.1 and 7.2 represent the total runoff process where the *drainage basin* precipitation (excluding interception, storage and other losses) over a heterogeneous catchment area is divided into these four flow components. Channel flow is associated with the direct precipitation on to the water surface which makes an immediate, direct contribution to runoff through *quickflow*. However, since the water surface area for most catchments is less than 5% of the total area,

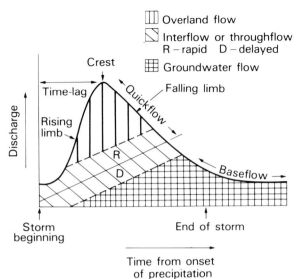

Fig. 7.1 The theoretical runoff components of a typical storm hydrograph.

the contribution by channel precipitation is normally small.

Overland flow has been described above and represents surface runoff, in sheets or micro-channels, which fails to infiltrate the soil due to saturated or frozen surface conditions and rainfall intensity exceeding infiltration capacity (i.e. Hortonian flow in section 19.2). It is a relatively rare occurrence which is most conspicuous in Arctic areas when snowmelt takes place over impermeable permafrost, particularly in hollows and at the base of slopes (section 19.2; Knapp 1978). At this time, the rate of flow can exceed 270 m hr^{-1} and provides most of the direct surface runoff, which combines with channel flow to generate quickflow from a catchment. Interflow or throughflow is responsible for most of the total runoff and experimental evidence indicates an 85% contribution (Ward 1975). It takes place when rainwater and snowmelt infiltrate the soil and, when reaching a less permeable soil layer at shallow depth, are forced to move laterally through the upper layers. Interflow occurs mainly as saturated flow in the perched groundwater zones described in section 6.2, which proceeds slowly through the soil at some 5 m hr^{-1}. However, some of this soil water will eventually reach the river channel as *rapid interflow* where it contributes directly to surface runoff and quickflow. Conversely, *delayed interflow* reaches far greater depths and forms the main part of the *subsurface runoff* (Fig. 7.2) which

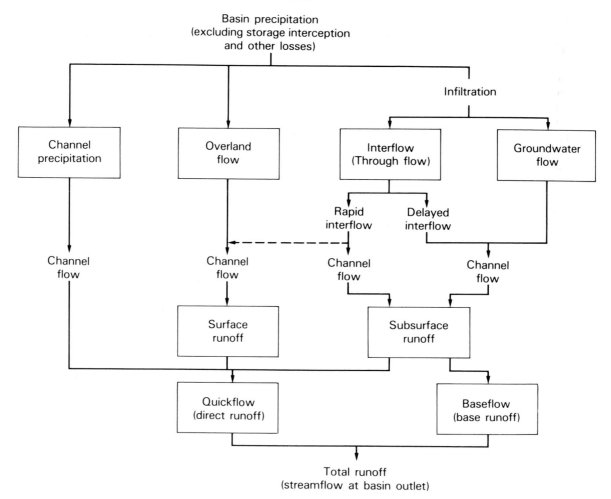

Fig. 7.2 Diagrammatic representation of the runoff process. (After Ward 1975)

continues to percolate downwards to recharge the groundwater.

Finally, when the water table intersects the stream channel, this groundwater becomes a component of runoff through baseflow or effluent seepage (section 6.3). This 'fair-weather flow' is characterised by a considerable time-lag between the precipitation event and the actual outflow and will continue through drought conditions to represent the main long-term component of total runoff. It is apparent that quickflow is virtually instantaneous during storms and makes a major contribution to flooding. However, over the long term, it only contributes about 20% of the total runoff in the forested eastern USA, compared with the 80% contribution of baseflow (Ward 1975).

7.2 Types of river flow

It is apparent that the long-term relationship between the component parts of the runoff process, particularly quickflow and baseflow, will determine the flow characteristics of a particular stream or river. For example, rivers influenced only by instantaneous storm quickflow will have very occasional 'flashy' runoff compared with the steady, continuous flow of rivers 'fed' by permanent baseflow. This relationship in fact produces three types of river regime, i.e. *ephemeral flow*, *intermittent flow* and *perennial flow*. The first type is characterised by the 'flashy' temporary quickflow following exceptionally heavy rain or rapid snow ablation. As a result, the stream's bed remains dry for most of the year since the water table is well below the surface and baseflow does not occur. The best example is the *wadi* of the arid lands where flash-flooding occurs on a most irregular basis.

Intermittent flow is also termed semi-permanent or seasonal flow. It occurs during wet periods when runoff is related to both quickflow from the rain event and/or baseflow from a temporary rise of the water table to the level of the river channel. For example, the *bourne* or *lavant* of the English chalklands represents a river flow where the upper reaches of a dry valley flood during the wet winter season. Conversely, by late summer, the river bed is normally dry following the reduced rainfall and a falling water table. Finally, perennial flow is a permanent, continuous flow since the water table is always at the level of the river channel.. The river has a constant baseflow at all times, but is occasionally supplemented by quickflow after heavy rainstorms, with flood events. A good example is the River Thames at Reading, England, when, even during the 1976 drought, runoff continued albeit at a reduced rate. The chalk bourne described above will have perennial flow along the lower reaches, below the springline, compared with the intermittent flow in the upper reaches. In fact, most rivers will represent a combination of two of these three types of flow along some part of their course.

7.3 Factors affecting runoff distribution

The total runoff distribution over time in a catchment or drainage basin (see Fig. 20.1) is controlled by the prevailing meteorological conditions and basic environmental characteristics. The former conditions are mainly associated with the actual precipitation and storm behaviour, since section 5.3 indicates that the type, duration and distribution of hydrometeors depend mainly on the type of free or forced convection involved and the degree of instability concerned. In terms of precipitation type, the storage of a snow pack over the winter season will obviously produce a considerable time-lag for runoff, which will only occur with the spring-time ablation. However, even at this time, the initial meltwater is stored at depth in the snow pack as part of its *densification* process, absorbing a considerable volume of water for future rapid quickflow (especially if the soil surface layers are frozen and impermeable).

Rainfall makes a more direct contribution to runoff but its actual role depends on a number of other behavioural characteristics. For example, the type of storm is important since violent, localised convective systems and thunderstorms produce torrential rain and peak flows in small drainage basins whereas more extensive hurricanes and wave depressions (Ch. 13), with distinctive fronts, are generally more effective in flooding large basins. Furthermore, the direction of storm movement is important since storms moving down an elongated catchment will result in all tributary runoff leaving the area at the same time, which produces a single flood peak over a short period of time (i.e. *hydrograph* B in Fig. 7.3). Conversely, when a storm moves up-catchment, it allows the tributary discharge to be transferred from various parts of the catchment at different times. Therefore, hydrograph A in Fig. 7.3 is characterised by a reduced magnitude of runoff with four distinctive discharge phases spread over a much longer duration.

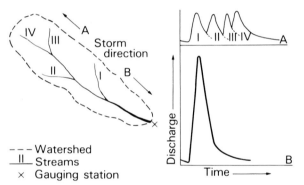

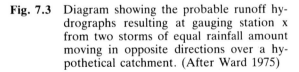

Fig. 7.3 Diagram showing the probable runoff hydrographs resulting at gauging station x from two storms of equal rainfall amount moving in opposite directions over a hypothetical catchment. (After Ward 1975)

Rainfall intensity and duration also govern the amount and type of runoff and the actual peak flow rate. The 'rule-of-thumb' is that increased rainfall intensity will increase the peak discharge and volume of runoff provided the infiltration rate of the soil is exceeded (section 6.1). Rainfall duration is particularly important in flat, low-lying areas when, if sufficiently prolonged, infiltration will raise the surface of saturation to the ground surface itself. This reduces the infiltration rate to zero and causes a rapid increase in runoff in the form of quickflow. Conversely, on steep slopes, maximum potential runoff occurs with rainfall which is characterised by a much shorter duration.

The assessment of environmental factors is much more complex since it is difficult to establish the net control of the individual variables concerned, which influence the various runoff components in markedly different ways. For example, a flat surface favours infiltration and baseflow at the expense of quickflow although the actual infiltration rate is soon terminated by an impermeable rock type, which now leads to rapid quickflow. However, it does appear that the topography, geology, soils and vegetation of a catchment control the runoff although the actual relationships are rather difficult to measure. When the topography is characterised by gentle slopes or flat surfaces, infiltration and baseflow will dominate until, of course, the eventual rise of the water table to the surface reduces the infiltration rate to zero. Conversely, steep slopes cause excessive quickflow and rapid interflow at the expense of baseflow. Similarly, the actual shape of the catchment is important since a radial drainage pattern (as found in desert depressions or *bolsons*) will experience simultaneous discharge from all tributaries and a high runoff peak over a very short period, via the ephemeral desert wadis discussed earlier.

The catchment geology is also an important influence on runoff in terms of rock type and structure. Impermeable rocks (like clay) encourage a well-developed vein-like surface drainage pattern

with a rapid disposal of overland flow and interflow. This is evident in Fig. 7.4 where the hydrograph for the River Enborne is characterised by a 'flashy' quickflow with high discharge rates. Conversely, the relatively few, broad tributaries on the permeable chalk of the River Winterbourne have a very restricted discharge which is related entirely to baseflow. Figure 7.4 reveals that on 1/2 December 1975, this baseflow discharge equalled 4.6 m^3 s^{-1} compared with the 360 cusec quickflow recorded on the River Enborne. The structure of the catchment influences runoff in that a synclinal structure has a reduced time-lag between rainfall and baseflow compared with more horizontal strata, even though the actual rock types are similar in both structures.

Soil types affect the infiltration rate and the disposition of rainfall as either overland flow, interflow or baseflow. For example, coarse-grained open-textured sandy soils have higher infiltration rates than fine-grained closely compacted silty soils (section 6.1; Kirkby 1969) and have smaller volumes of quickflow. Also, the location of impermeable layers within the soil profile markedly reduces the vertical hydraulic conductivity (section 6.3) which results in rapid interflow and eventual overland flow. Conversely, deep uniformly permeable soils have extensive infiltration and baseflow dominance.

Surface vegetation controls runoff by retaining rainfall through interception (section 7.1) and evapotranspiration (section 4.1). Consequently, forested catchments have drier soils, greater infiltration capacity and exaggerated baseflow (at the expense of quickflow) compared with non-vegetated basins. Research by the Institute of Hydrology at Plynlimon, West Wales, indicates that the annual streamflow from the Wye catchment (1.2% forest cover) averaged 1964 mm between 1970 and 1978, compared with 1398 mm from the adjacent Severn catchment (67.5% forested) and the general conclusion is that stream flow from the forested catchment is some 15% lower than that from grassland (Institute of Hydrology, 1984). The influence of fauna is also important when soil is compacted along animal trails which accelerates quickflow and soil erosion. Research in the Luanga Park, eastern Zambia between 1950 and 1973 indicates that with a fourfold increase in elephant trails, the total length of overland flow gullies increased from 77 to 7077 m and the associated soil loss increased from 7 to 21 cm yr^{-1}.

Human activities control runoff deliberately with the construction of reservoirs which act like

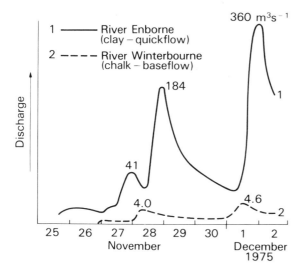

Fig. 7.4 Runoff hydrographs over different rock types in southern England. (After Thames Water Authority, Reading)

natural water bodies and absorb peak flows, which delay runoff and eliminate flooding. Also, channels are straightened to accelerate discharge from low-lying flat areas and this technique has been used successfully on the Rivers Welland, Nene and Great Ouse in the English Fenlands. Agricultural practices can deliberately modify runoff through contour ploughing and levee construction which store water and increase both evaporation and infiltration at the expense of surface runoff. Land-use changes bring about more inadvertent changes through deforestation and afforestation which modify the interception and evapotranspiration of rainwater (section 7.1). However, urbanisation represents the most dramatic example of inadvertent human control of runoff with the creation of the so-called urban-*karstic* environment. In urban areas, surface-sealing by concrete and bitumen and drain/sewer construction are responsible for the acceleration of quickflow and an increase in the magnitude of peak flow with a greater flood frequency in adjacent rural areas. For example, in Connecticut, USA, the average annual runoff from an urban watershed is twice the State average and flood peaks from Hartford City are about three and a half times the mean rate for the State (Bruce and Clark 1969).

7.4 Retrospect

Runoff is a complicated process and its distribution represents the integrated result of all hydrological and meteorological factors operating in a drainage basin. However, the actual integration is modified by catchment controls and human activities which are difficult to quantify. Runoff is the most important water resource in the United Kingdom where the ratio of abstraction from surface sources compared with groundwater supplies is about 9 : 1. The north and west of the country have water balances where a surplus runoff dominates and the only aquifers of note are in the relatively dry southeast of England.

Key topics

1. The sources of runoff and surface retention of precipitation.
2. The ways in which water moves over and through the earth's surface.
3. The role of atmospheric factors on runoff distribution.
4. The modification of atmospheric controls on runoff by topography, geology, soils, biota and humans.

Further reading

Bauer, W. J. (1969) Urban hydrology, *The Progress of Hydrology*. University of Illinois; Urbana, Vol. 2, pp. 605–37.
Kirkby, M. A. (1979) *Hillslope Hydrology*. John Wiley and Sons; New York and London (especially pp. 227–93).
Leopold, L. B. (1974) *Water – A Primer*. W. H. Freeman and Co.; San Francisco, pp. 34–62.
Selby, M. J. (1982) *Hillslope Materials and Processes*. Oxford University Press; London and New York; pp. 83–116.
Smith, K (1972) *Water in Britain*. Macmillan; London and New York, pp. 90–9.

Part II
Atmospheric circulation systems
Chapters 8–14

Variations in the mass and energy exchange budgets of different latitudes give rise to atmospheric circulations, which are discussed in the following seven chapters. Atmospheric circulations are systems involving energy inputs and transformations which are subject to feedback regulation. For convenience, their horizontal components (surface and upper winds) are treated first and the inevitable linkage with vertical components is considered in the chapter on pressure systems. Furthermore, these circulations operate on various scales, from the global (or primary) through the synoptic or secondary (represented by weather maps) to the mesoscale (or tertiary), to each of which a chapter is devoted. The characteristic circulation of middle latitudes, involving air-mass and frontal behaviour, receives particular attention.

Satellite view of the global atmospheric circulation, Meteosat 16 May 1979.

Chapter 8 Horizontal air flow near the surface

The movement of air is a fundamental process in atmospheric systems and horizontally moving air (wind) is one of the most obvious aspects of weather to the man in the street. It is less obvious that the ground on which he is standing is also in motion and if the air is moving in exactly the same way as the ground beneath (i.e. in the same direction and with the same speed), then he will experience completely calm conditions. Wind is, in fact, the movement of air relative to the underlying earth's surface. Like any other body, air will move only in response to an external force. In fact, at least two forces are involved in horizontal air movement (and often as many as four) and the actual motion represents the resultant of these separate forces. These will now be examined in relation to air flow near the earth's surface.

8.1 Atmospheric pressure and the pressure gradient

The pressure of the air, measured by some form of barometer, strictly represents the force exerted on a unit area of surface and is expressed in millibars (mb): 1 mb equals a force of 100 N on 1 m². It can effectively be considered as the weight of the total air column bearing down on the surface. 'Average pressure' is 1013.25 mb (supporting a column of mercury 760 mm or 29.92 in high; hence the original measurement of pressure in terms of length), assuming a reading is taken at sea-level at latitude 45° and a temperature of 0 °C. All mercury barometer readings have to be corrected to a standard temperature (to allow for the expansion of mercury) and latitude, because the value of gravity and consequently the weight of the mercury vary with latitude. The adjustment to sea-level is essential for comparative purposes, since pressure reduces rapidly with elevation. The high compressibility of air dictates that the mass of the atmosphere is concentrated in the lower layers. In the latitude of Britain, half the weight of the atmosphere is found on average below an altitude of 5.5 km and, at the height at which a long-distance jet aircraft flies, the pressure is about one-quarter of that at sea-level.

Inequality of pressure constitutes a first and most obvious reason for air movement. When pressure varies with distance there is said to be a *pressure gradient* and this would be expected to exert a force urging the air from high to low pressure until the inequality is removed. A pressure gradient would appear to be a situation high in potential energy, analogous to that of a U-tube in which the level of liquid in one limb is artificially kept higher than that in the other. When the obstacle is removed, liquid flows horizontally and the levels become equal. On this reasoning alone, a weather map should show surface winds blowing at right angles to the *isobars* (lines of equal pressure) representing the pressure gradient. Furthermore, it would be expected that the steeper the pressure gradient, the greater would be the impelling force and the stronger the surface winds.

Figure 8.1 shows part of an actual weather map in which only the mean sea-level isobars and the surface wind arrows at the various stations are represented. It will be seen that where the pressure gradient is steepest (i.e. in the northwest of the British Isles), the winds are indeed strongest. Furthermore, they fall away to light or calm towards the southeast where the gradient becomes slack (the isobars are few and far between). However, the wind directions, far from being down the pressure gradient (across the isobars), are much more nearly (but not quite) parallel with the lie of the isobars. Clearly, while the pressure gradient is a primary cause of air movement, it cannot be the only one.

The notion of an equilibrium between two forces may be readily understood in relation to the vertical pressure gradient. In the lowest kilometre of the atmosphere, pressure falls with height at a rate of about 10 mb in 100 m, which represents a pressure gradient many times steeper than that in the horizontal direction. For example, over Britain, in the particular example of Fig. 8.1, a change of 10 mb occurs in something like 350 km. That the air near the ground is not impelled upwards at high speed is due to the countering force of gravity. This state of balance or *hydrostatic equilibrium* can be disturbed temporarily within the narrow limits of the troposphere by density changes which give rise to vertical currents (section 3.4). The analogous force that is capable of balancing the horizontal pressure gradient is a less immediately understood consequence of earth rotation.

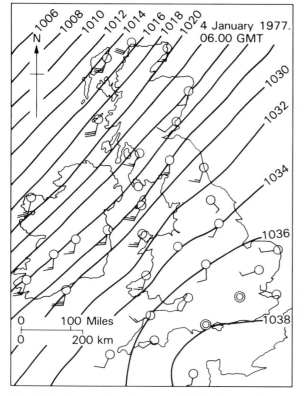

Fig. 8.1 The relation between MSL isobars and surface winds. Isobars are drawn at 2 mb intervals. Arrows blow with the wind: a short feather represents 5 knots, a long feather 10 knots: a double circle indicates calm. (1 knot = 0.5 m s⁻¹ approximately)

8.2 Earth rotation and the Coriolis force

The *Coriolis force* (named after the French physicist Coriolis, 1792–1843) arises from the fact that an object on or above the earth's surface, which is set in motion along a straight path as seen by a fixed observer in space, will appear to an observer on the rotating earth to be following a

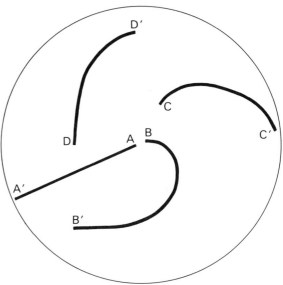

Fig. 8.2 Attempts to reproduce a straight line on a rotating disc.

curved path. The validity of this notion may be confirmed by experiments with a record player turntable. Figure 8.2 is drawn from a photograph of a paper disc fitted on to the turntable, a hole having been pierced for the spindle. A straight-edge was placed across the cabinet containing the turntable so that it could be held in a fixed position, initially immediately above the spindle. While the turntable was stationary, there was of course no difficulty in moving a felt-tip pen along the straight-edge so as to produce a straight line, such as AA′. When, however, the turntable was rotated manually in a counter-clockwise direction, a line drawn along the straight-edge from B took the curved path BB′. CC′ and DD′ represent attempts to draw straight lines on the rotating disc starting from points other than the centre. In each case, the path traced on the disc represents a deflection to the right of the original direction as represented by the straight-edge. It is clear that nothing has happened to cause this deflection except the rotation of the disc and the fact that points nearer the centre are moving more slowly than points nearer the outer edge.

A rotating flat disc is only a partial representation of the rotating earth, yet the analogy holds in many respects. A point on the equator travels the circumference of the earth in 24 hours, i.e. at a speed of about 470 m s^{-1}. However, a point at latitude 60° travels its shorter path at about half that speed while the North and South Poles, like the turntable spindle, rotate on the spot. A parcel of calm air at the equator shares the speed and direction of the underlying surface. If this parcel is urged northwards over slower-moving latitudes, it would appear to an observer at the starting-point to be accelerating away to the right of the meridian along which he has fixed his sights. A parcel at the North Pole, which is impelled equatorwards over faster-moving latitudes, will similarly exhibit a deflection to the right from the standpoint of an earth-bound observer. It will be clear that the turntable experiment described above, specifying an anticlockwise rotation, represents a northern hemisphere situation. However, if the turntable is rotated clockwise, then the deflection is to the left, as indeed it is in the southern hemisphere.

The deflection effect due to earth rotation is thus an apparent force, yet one to be reckoned with. It can be given a magnitude which (per unit mass) depends on the angular velocity of spin about the vertical (ω) and the horizontal speed (V) of the air. It has a maximum value at the North or South Poles, where the vertical is coincident with the earth's axis of rotation. At increasingly lower latitudes, however, ω necessarily reduces as the horizontal surface, above which the air movement takes place, approaches nearer and nearer a plane parallel to the earth's axis: ω thus depends on the sine of the latitude (which varies from sin 0° = 0 to sin 90° = 1). The magnitude of the Coriolis force (f) at any point is given by:

$$f = 2\omega V \sin \phi \qquad [8.1]$$

where ϕ is the latitude, and thus reaches a maximum at the North and South Poles and has zero value at the equator.

8.3 The geostrophic balance

Consideration of the Coriolis force alone would suggest that air flow must invariably be along curved paths, as the air is urged more and more to the right (or left) of its initial direction of movement. Figure 8.1, however, shows comparatively uniform wind directions over western and northern districts of Britain and this was only part of a broad stream of air blowing steadily on that occasion from mid-Atlantic to northern Scandinavia. When air follows a straight and steady path, this implies a state of equilibrium between the two forces already described, namely the pressure gradient, urging the air from high to low pressure, and the Coriolis force acting in the opposite direction. Figure 8.3 represents the situation in the northern hemisphere. As a result of this balance, the actual wind must blow parallel to the isobars, as shown by the solid arrow and the relationship is expressed in the so-called *Law of Buys Ballot*

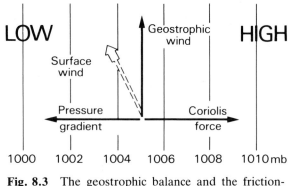

Fig. 8.3 The geostrophic balance and the frictionally weakened surface wind (northern hemisphere).

(where, if an observer stands with his back to the wind, the low pressure is to the left) formulated in 1857. In the southern hemisphere, the low pressure is to the right but, for convenience in any further discussion in this chapter, the northern hemisphere will be assumed. The wind that blows in this balanced situation is known as the *geostrophic wind* (geostrophic meaning 'earth turning') and the balance of forces implied is represented by the *geostrophic wind equation*, which can be expressed as:

$$\frac{\mathrm{d}p}{\mathrm{d}x} = 2\omega V \sin \phi \qquad [8.2]$$

where dp/dx represents the change of pressure p with distance x at right angles to the isobars, i.e. the pressure gradient.

It can be seen that at any given latitude and assuming a uniform isobar interval (i.e. dp, ω and ϕ being constant), the wind speed V is inversely proportional to the distance between isobars on the weather map. One of the tools of the meteorologist's trade is the *geostrophic wind scale*. This is a Perspex scale on which is etched a set of isobar spacings corresponding to various wind speeds, appropriate to a particular latitude band, map scale and isobar interval. If the scale is laid across the isobars at right angles to them, the geostrophic wind corresponding to that isobar spacing may be read off. The scale may also be used to estimate the speed of movement of fronts (section 12.2).

It should not be thought that the geostrophic condition is always achieved. Figure 8.1 shows a set of fairly straight isobars over northern Britain and it may be correctly supposed that there is a level at which the actual wind corresponds quite closely to the geostrophic value. However, this certainly does not apply to the surface winds, which in fact blow at an angle to the isobars (though this angle is remarkably constant over much of the country). Consequently, the Law of Buys Ballot is only approximately valid for an earth-bound observer. This is an example of *ageostrophic flow* which, for near-surface conditions, is due to the intervention of yet another force associated with the frictional drag of the earth's surface itself on the movement of the air.

8.4 Surface friction and the Ekman spiral

The effect of surface *friction* is to slow down the movement of the lower-air layers. Since the wind velocity is an element of the Coriolis force, this side of the geostrophic balance (equation [8.2]) is necessarily weakened and the stronger pressure gradient is able to turn the wind direction somewhat towards the low pressure side. Thus, in Fig. 8.3, the broken-line arrow represents the surface wind direction and this oblique relationship with the isobars is well illustrated in the weather map of Fig. 8.1. The frictional effect, which is a function of the roughness of the surface and the condition of the lower-air layers, is felt in a variable zone up to a height of 500–1000 m. Above this level, given straight isobars and no pressure changes, the geostrophic balance occurs and the real wind blows along the isobars at a speed closely approximating that obtained from the geostrophic wind scale.

Below that level, within what is sometimes referred to as the friction layer (section 3.1), the wind speed reduces and the angle with the isobars increases as the surface is approached. This results in a wind spiral (Fig. 8.4) known as the *Ekman spiral* after V. W. Ekman (1905) who examined the analogous behaviour of ocean currents with

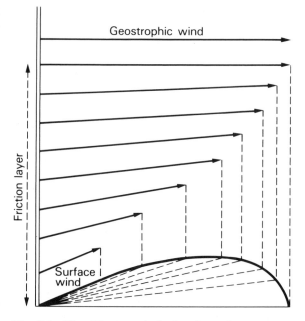

Fig. 8.4 The Ekman spiral. A schematic representation of the change of wind speed and direction within the friction layer. If the ends of the vectors representing the winds at different levels are joined, they form a spiral.

depth. The exact shape of the spiral varies with surface roughness. For example, over the smooth sea, the surface wind may have two-thirds of the geostrophic speed and may be backed from the isobars (i.e. blow from a more anticlockwise direction) by only 10°. At the same time over the rougher land, the surface wind may have only half the geostrophic wind speed and the angle may be 20°–30°. Furthermore, at night over land, the surface wind may drop to a very low velocity and the angle may increase to 40°–50°. Over southern England in Fig. 8.1 (before dawn in January), the geostrophic wind has a value of 7–8 m s^{-1}, but surface wind speeds do not exceed about 2.5 m s^{-1}, and some stations report calm. In general, when the pressure gradient is slack, surface winds are light and variable in direction, being vulnerable to various mesoscale influences (Ch. 14).

8.5 Curved flow and the gradient wind

The discussion so far has been confined to the case of straight isobars and steady air flow in the same direction (i.e. geostrophic flow) above the friction layer and backed winds nearer the surface. When air pursues a curved path, since this implies an acceleration, conditions depart from the geostrophic, as yet another factor comes into the reckoning. With curvature of flow, whether in low or high pressure systems, there is an inward acceleration or *centripetal* force (*c*) maintaining the air in its curved path. It has the magnitude:

$$c = \frac{mV^2}{r} \qquad [8.3]$$

where m is the mass of moving air, V its velocity and r the radius of curvature of the flow: it is sometimes referred to as the *cyclostrophic* term.

In a low pressure (cyclonic) system, the play of forces is as shown in Fig. 8.5(a). The resultant air flow in the lower layers is along a curved path in a counter-clockwise direction (as may be confirmed by application of Buys Ballot's Law). The Coriolis force, directed outwards, is weakened by the centripetal (inward) acceleration. This means that the wind velocity is less than in the geostrophic case (equation [8.2]) and wind speeds in cyclonic flow are therefore described as *sub-geostrophic*. In a high pressure (anticyclonic) system, in which the curved flow is in a clockwise sense, the forces are aligned differently (Fig. 8.5(b)). This time, it is the pressure gradient which is

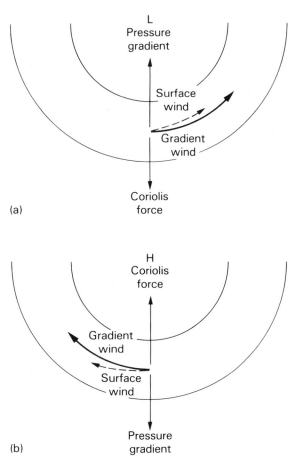

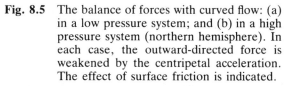

Fig. 8.5   The balance of forces with curved flow: (a) in a low pressure system; and (b) in a high pressure system (northern hemisphere). In each case, the outward-directed force is weakened by the centripetal acceleration. The effect of surface friction is indicated.

weakened and the wind velocity is *super-geostrophic*. The importance of this effect is reduced in the case of the anticyclone because the pressure gradients are generally slack and the radius of curvature (*r*) is large. In an intense cyclonic circulation, in which *r* is small, or in curved flow at upper tropospheric levels where velocities are large (see Ch. 9), the effect may be considerable. In all cases of curved flow, the wind represented by the isobar pattern is maintained by a balance of three forces and is referred to as the *gradient wind*, rather than the geostrophic, when only two forces are involved.

The above considerations refer, of course, to frictionless flow at heights above 500–1000 m. Near the surface, frictional retardation comes into play and the surface winds are as shown by the broken-line arrows in Fig. 8.5. They are directed inwards across the isobars at the base of a low pressure system and outwards at the base of a high. These surface winds in fact blow in response to the pull of no less than four influences.

8.6 The restless wind

This chapter is concerned with predominantly horizontal air flow, but truly laminar flow (i.e. in straight parallel lines) is rare in the troposphere. Above a very thin skin (only a few millimetres thick) of still air covering all surfaces, the air flow is more or less turbulent, with the horizontal motion overlain by a series of eddies or small-scale circulations (section 3.1). This is partly due to the frictional drag exerted on the air movement by the underlying surface. However, the intensity of this mechanically induced turbulence and the size of the eddies generated depend on the wind speed, the stability conditions of the air and the roughness of the surface.

The nature of this kind of turbulence can be appreciated by observing the form of a smoke plume issuing from a chimney stack, on a cloudy day with moderate wind. The diffusion of the plume in the form of an elongated cone downwind of the chimney, with its apex at the orifice, gives some idea of the size and behaviour of these eddies. The effect of an obstacle such as a large tree or building in the path of the air flow is to induce vertical eddies comparable in size to the obstacles themselves. Thermally induced turbulence (free convection) due to heating from below results in larger-scale circulations, often reaching to middle and upper tropospheric levels. This depends on stability as well as surface conditions. A chimney plume in the unstable conditions of a warm summer afternoon, shows large loops of smoke curling downwind which graphically trace the form of the large eddies generated.

What is embodied in the everyday term 'wind' is in fact a complex set of processes. The superimposition of eddying motion makes the wind an extremely fidgety element, blowing in gusts and lulls as the eddies alternately strengthen and weaken the horizontal flow. The environmental significance of turbulent flow is enormous, for it is the most effective mixing mechanism in the vertical direction. It is by this means that heat energy and water vapour are transferred upwards from warm or evaporating surfaces (Chs 2, 3 and 4). It is also the main process responsible for the removal of air pollution for, contrary to what might be supposed, pollution is not simply blown horizontally through and away from a town. In fact, pollutants are diffused upwards by turbulence and are removed by the stronger horizontal flow at higher levels. The more turbulent regions of the earth are fortunate from this point of view while those where persistently stable conditions damp down turbulence may have exacerbated pollution problems (section 3.4). Since wind strength falls off near the ground and is affected by obstructions, it is necessary (for comparative purposes) to accept a standard height for wind measurements. This is internationally agreed as 10 m above open, level ground and a 'surface' wind refers to at least that height. However, where there are adjacent trees and buildings, it is necessary to go even higher.

8.7 Retrospect

Horizontal air flow is a component of circulation systems of various scales and this chapter has concentrated on conditions near the surface. It has been shown that straight, frictionless flow results from a balance between the pressure gradient and the Coriolis deflection and this so-called geostrophic wind may prevail at a height of 500–1000 m above the surface. Departures from the geostrophic conditions occur below that level, due to frictional drag, so that surface winds are both weaker than and backed from the geostrophic wind. This also occurs when curvature of flow introduces a centripetal component of movement. The wind, now adjusted to the curved isobars, is known as the gradient wind. The surface wind in the case of curved flow is subject to all four influences.

The relationship between pressure and wind is one of mutual adjustment. In a situation of initial geostrophic balance, if the pressure gradient tightens, the wind speed increases and so does the Coriolis parameter. Pressure developments result from ageostrophic flow, but the tendency is always to return to equilibrium. A pressure gradient is a necessary condition for the creation of wind but equally it can be argued that, in effect, pressure gradients are maintained by the air movement. At the equator, the Coriolis parameter has zero value and geostrophic conditions cannot obtain. It is for this reason that tropical circulatory systems such as hurricanes do not form within about 5° of the equator (section 10.3).

transfers and the removal of pollutants in the troposphere.

Key topics

1. The relationship between pressure gradient and surface wind strength/direction.
2. The role of Coriolis force in air movement.
3. The influence of friction on winds.
4. The significance of turbulence for energy/water

Further reading

Barry, R. G. and Chorley, R. J. (1982) *Atmosphere, Weather and Climate*. Methuen; London and New York (especially Ch. 3).
Panofsky, H. A. (1981) Atmospheric hydrodynamics, in Atkinson, B. W. (ed.) *Dynamical Meteorology*. Methuen; London and New York, pp. 8–20.

Chapter 9 Upper-air flow

In Chapter 8, discussion was focused on the lowest kilometre or so of the troposphere, where the air-flow conditions are depicted by the mean sea-level (MSL) isobars and the wind arrows of the surface synoptic chart (such as Fig. 8.1). However, surface air flow forms part only of broader circulation systems, which also involve horizontal flow at higher tropospheric levels, represented by upper-air charts which are just as important to the meteorologist as the better-known surface maps. In fact, one of the major advances in the development of meteorology dates from the period (broadly of the Second World War) when new technologies such as radio- and radar-sounding made possible a three-dimensional view of tropospheric structure and air-flow patterns. Interest then shifted from the surface to middle and upper tropospheric levels, mainly because of the forecasting needs of high-flying aircraft. However, the change was also associated with the recognition that the broader flow at such levels could be more significant for the understanding of atmospheric behaviour than the detailed patterns near the disturbing influence of the surface.

Careful observation of the sky when several cloud types are present will often show that the high cirrus clouds (at perhaps 10 km elevation) or even medium clouds like altocumulus (at perhaps 5 km) are moving in directions different from that of the cumulus or stratocumulus with their bases not far from the 1 km level. They are no doubt moving at different (usually greater) speeds too, but it is difficult to judge relative speeds from ground level. This implies that there are pressure gradients at these upper levels which are markedly different from those observed at the surface. It can be shown that these are related to horizontal temperature gradients in the troposphere and the fact that pressure reduces more slowly with height in warm air than in cold.

9.1 Thermal winds and the upper westerlies

Figure 9.1 represents schematically a portion of the troposphere in the form of a cross-section along the line AB. In (a), it is assumed that the pressure is exactly the same (say 1010 mb) at every point along the surface, i.e. there is no surface pressure gradient and therefore no surface wind. It is further assumed that the surface temperature is everywhere the same and that the fall of temperature with height (the ELR of section 3.2) is identical throughout the cross-section. Successively lower pressure levels will therefore be reached at uniform heights, as shown by the isobars (where the pressure surfaces are cut by the

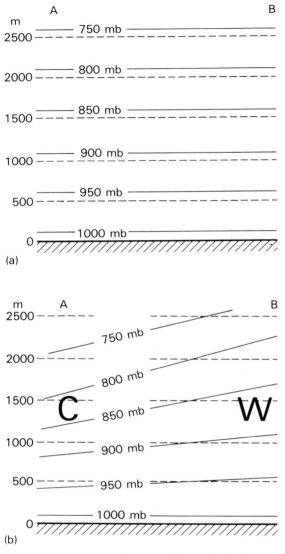

(a)

(b)

Fig. 9.1 Illustrating the principle of the thermal wind. In these schematic atmospheric cross-sections (a) assumes no horizontal temperature gradient, and (b) assumes a temperature increase from A to B.

cross-section), which are necessarily parallel with the height lines. In this hypothetical situation, there is no reason for air movement at any level.

In Fig. 9.1(b), it is assumed that the air at B has been warmed, as indicated by a large W, while that at A has been cooled (large C). The air layers are now expanded in the warmed section and contracted in the cooled section and the isobaric surfaces tilted from B down towards A. As a result,

there is now a pressure gradient at all levels except the surface (for example at 1500 m the pressure at B is about 870 mb and at A about 800 mb) and this effect increases with elevation until such levels at which the horizontal temperature gradients change (not shown in the diagram). Because of this pressure gradient, a flow of air would be expected from B to A in Fig. 9.1(b), with a speed increasing with elevation. This horizontal motion is wholly dependent on temperature differences and the slower rate of fall of pressure with height in warm air than in cold and is appropriately referred to as a *thermal wind*.

This theoretical situation, in which a wind blows at height but not at the surface, cannot last long since the redistribution of air horizontally will cause surface pressures to rise at A and fall at B, inducing a return surface wind. This is, in fact, the basis of thermal circulation cells which exist at various scales (see Chs 10, 11 and 14), but other factors must be taken into account. A persistent thermal wind, like any other, responds also to the Coriolis acceleration, so that the upper wind in Fig. 9.1(b) will be deflected to blow into the paper (assuming the northern hemisphere). The direction of a thermal wind can be ascertained by a variant of Buys Ballot's Law: if the observer stands with his back to a thermal wind, the cold air lies to his left (again, in the northern hemisphere). With a persistent horizontal temperature gradient, thermal winds persist.

The situation depicted in Fig. 9.1(b) broadly parallels that found on average in the northern hemisphere with B representing the equatorial side and A the polar, with the result that westerly winds are frequently found in the middle and upper troposphere. The same applies in the southern hemisphere, since here the directions of both the thermal gradient and the Coriolis deflection are reversed. For simplicity, in Fig. 9.1, the initial assumption was a zero surface pressure gradient, but the argument is not destroyed if there are surface pressure gradients and winds to begin with. As long as the temperature distribution is maintained, the thermal-wind effect is present and, in vector terms, the actual wind at any upper level is compounded of the surface wind plus the thermal wind. So it is that the dominant upper-air flow in most latitudes is westerly and the term *upper westerlies* is in regular use. This flow clearly derived its energy from the horizontal temperature gradient and represents an example of the conversion of potential to kinetic energy.

9.2 Upper-air charts and their meanings

Charts depicting air flow at upper levels are of two kinds; constant-level charts, which show isobars for specified heights and are thus strictly comparable with the MSL isobaric charts, and constant-pressure charts, which show contours of specified pressure levels. For practical reasons, the latter *contour charts* are more commonly used. It will be remembered that the aircraft altimeter or the aneroid element in a radiosonde apparatus measures pressure, and height has to be calculated from a knowledge of sea-level pressure and the mean temperature of the air column concerned. With a constant-level chart, air density becomes a factor in the measurement of the geostrophic wind at the specified level and a different geostrophic wind scale is necessary for this purpose at different levels. With a constant-pressure chart, density is not involved. It can be shown that wind speed depends only on the slope of the isobaric surface (i.e. the closeness of the contours), and one scale can be used for different pressure levels. Contour charts are usually drawn for the 700, 500, 300 and 200 mb surfaces.

Contours of a pressure surface are completely analogous to those of the ground surface. The significance of the topography of a pressure surface may be seen from Fig. 9.1(b). Where contour values are high (i.e. the pressure surface is elevated), this means that the air layer beneath it is warmed and expanded. Conversely, low contour values of a given pressure surface imply a cold, contracted air layer. Contour patterns thus mirror the distribution of mean temperature throughout the horizontal layer of atmosphere below the level specified and, as has been shown, horizontal temperature gradients largely control air flow at such levels. Contours are also isobars, by definition, since the 3 km contour line on the 700 mb pressure surface is exactly equivalent to the 700 mb isobar on the 3 km height surface. Contour charts are thus flow charts and, given equilibrium conditions, air flow is along the contours in a direction given by Buys Ballot's Law ('low' now meaning low contour values) and at a speed depending on the contour spacing. Hence the use of an appropriate geostrophic wind scale and the value of the contour chart as a forecasting tool for aviation purposes.

The relationship between surface isobaric and upper-air contour patterns is illustrated in Fig. 9.2. In the case of a northern hemisphere surface

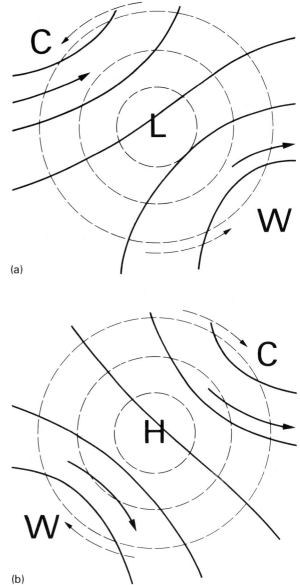

(a)

(b)

Fig. 9.2 The relationships between sea-level isobar and upper-level contour patterns in the case of: (a) a surface low; and (b) a surface high (northern hemisphere). Broken lines represent MSL isobars, solid lines 500 mb contours, broken-line arrows surface wind directions, solid-line arrows directions at 6–7 km.

low, the coldest air blows into the north rear portion and this is where the 'low' in the contour pattern is to be found (C in Fig. 9.2(a)). With a surface high, the warmest air invades the south rear portion and the upper 'high' (W in

Fig. 9.2(b)) is situated here. Thus, there is a displacement between surface and upper patterns, with the axis of a low tilting upwards to the cold side and the axis of a high to the warm side. Changes of wind direction in the vertical are also illustrated in the diagram. In general, the flow structure at middle and upper tropospheric levels is simpler than that near the surface, so that the intricate pattern of closed circulations commonly found there broadens out into open troughs and ridges in the upper flow.

Applying these considerations on a global scale, it can be seen that the cold polar caps are represented by low contour values and the warmth of low latitudes by high contour values. Figure 9.3 shows mean contours of the 500 mb pressure surface for both northern and southern hemispheres. Although the northern hemisphere chart contains some interesting details (which will be referred to later), the main impression is that of a concentric pattern obeying the dictates of the temperature distribution. The mean air flow adjusted to this contour pattern, and prevailing at 5–6 km elevation in most latitudes, is the broad circulation of the upper westerlies: this is also referred to as the *circumpolar vortex*.

Yet another method of depicting the thermal pattern is the so-called *thickness chart*. It follows from Fig. 9.1 that the vertical thickness of an air layer between two pressure levels (for example, 1000 and 500 mb) is a measure of the mean temperature of that layer and that high thickness value represent an expanded, or warmed, layer. Thickness isopleths (lines of equal thickness) thus directly represent the distribution of warm and cold air within the layer. They are equivalent to mean isotherms, while at the same time depicting the thermal winds. Thickness charts have played a significant role in forecasting, especially since frontal lows tend to move at the speed and in the direction of the thermal wind above their centres, a principle known as *thermal steering* (Sutcliffe 1947). Clearly, the 1000–500 mb thickness chart will generally closely resemble the corresponding 500 mb contour chart.

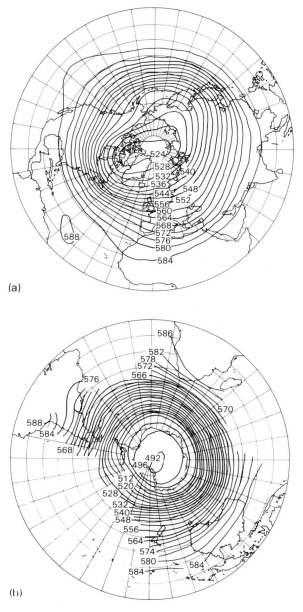

(a)

(b)

Fig. 9.3 Annual mean 500 mb contours: (a) northern hemisphere; and (b) southern hemisphere. Contour values are in decametres. (After Lamb 1964)

9.3 Rossby waves

The mean contour charts of Fig. 9.3 give an impression of a near-circular flow of air around the polar caps, but the averaging process, to a large extent, masks a rather different reality. Figures 11.5 (a) and 11.6(a) reproduce actual contour charts from which it can be seen that the pattern, far from being concentric, is rather one of ridge (extension of higher contour values) and trough (extension of lower values), and the actual air flow has a sinuous or meandering character in response to this pattern. These alternating ridges and troughs may number between three and six, according to situation, and follow one another in a west–east progression around the North and South Poles. They constitute the *long waves* in the upper westerlies but, alternatively, they are known as *Rossby waves*, after the Swedish-American C. G. Rossby, who pioneered their investigation.

Experience has shown that the weather of middle and high latitudes is largely dictated by the number and behaviour of the Rossby waves. Not only does the number of waves vary from time to time, but also correspondingly their amplitude and wavelength. Their movement can be erratic although generally it is in a west–east direction and much slower than the winds that blow through them. For example, some waves become stationary and some even move east–west against the general westerly current. Furthermore, the wave pattern is not necessarily symmetrical and individual troughs or ridges may distort to such an extent that the end portions become detached. The significance of these waves can be appreciated if it is remembered that the troughs represent tongues of colder air extending equatorwards from polar regions, while the ridges are extensions of warmer air from lower latitudes.

The surface expression of the movement and behaviour of the Rossby waves is therefore the changing mosaic of air masses and fronts as experienced on the ground and depicted on surface synoptic charts (see sections 12.1 and 12.2). It will also be shown (section 10.3) that cyclonic development is associated with the forward limb of a trough and anticyclonic development with the rear limb. The consequence of wave distortion may be an isolated body of cold air in warmer latitudes – a so-called *cut-off low* (see Fig. 11.5(a)) – or a detached body of warm air in colder latitudes. Both situations have significance for the weather of the regions concerned.

To understand the nature of these waves, Rossby invoked the principle of the conservation of absolute vorticity, as will be explained in section 10.2. This interpretation requires a mechanism to impose a northward or southward component of motion on part of the general westerly flow: once initiated, the meandering pattern tends to persist. It seems difficult to escape the conviction that this mode of mid- and upper-tropospheric behaviour is dictated by the topography of the underlying earth and the mean contour maps of Fig. 9.3 provide a hint that this may be so. The northern hemisphere chart shows a distinct trough in the contour pattern over eastern North America and another, less marked, over easternmost Asia. These two regions lie in the lee of the great mountain systems of the Rockies and the Tibetan block respectively, whose peaks reach and exceed the height of the 500 mb pressure surface. Such formidable obstacles seem bound to disturb the westerly flow by a lee-troughing effect (section 10.3), which may initiate a train of waves downstream. It would be difficult to ignore also the effect of the large-scale seasonal pressure systems resulting from thermal contrasts between continent and ocean. There would appear to be sufficient justification for regarding the behaviour of the upper westerlies as a reaction to earth surface features.

On the other hand, an alternative view is encouraged by the evidence of the well-known 'dishpan' experiments, first carried out at the University of Chicago in 1951 and repeated elsewhere since. A shallow circular dish containing water is placed on a turntable so that it can rotate about its centre. The vessel is heated at the rim to represent the equator and cooled at the centre to represent the North or South Pole, thus simulating atmospheric conditions in one hemisphere. A sprinkling of powder or filings on the water renders its motion visible and a cine-camera mounted above the pivot (and rotating with the dish) records the evidence. Different patterns of motion are observed with different rates of rotation but, at a certain range of speeds, wave-like movements are produced which closely resemble Rossby waves.

There are obvious differences between a flat, uniform dish and a hemispherical earth's surface with topographical and thermal irregularities, but the dishpan experiments have proved capable of simulating other features of the global circulation and will be referred to again in Chapter 11. In the present context, they suggest that Rossby waves might occur even above a uniform earth, given the temperature gradient between low and high latitudes. Yet the standing troughs shown in Fig. 9.3(a) cannot be ignored, any more than their virtual absence from the corresponding chart of Fig. 9.3(b). In the southern hemisphere, where ocean surfaces dominate and high land is of small area, Rossby waves are less well developed than their northern counterparts and this too argues for the significant role of earth topography.

9.4 Jet streams

Long before the introduction of upper-air charts as a routine tool of meteorological analysis, there had been hints of wind velocities at great heights which far exceeded those at the surface, except in the most vigorous systems such as tropical hurricanes or tornadoes. In 1923, a balloon released at a fête in Calshot, Hampshire, England, came to earth at Leipzig in Germany four hours later, having covered the distance of 917 km at an average speed of 64 m s^{-1}. Even earlier, First World War Zeppelins had been blown off course by unpredicted strong winds and there were more frequent reports from Second World War aircraft of unexpectedly strong headwinds or tailwinds at heights above 5–6 km which were associated with rather inaccurate bombing raids. By the end of the Second World War, it was well established that there existed at such levels ribbons of fast-moving air, of no great width (often only 2–300 km), within which velocities reached 50–100 m s^{-1}. In 1947, these high-velocity flows were christened *jet streams* by Rossby and his associates at the University of Chicago.

The existence of jet streams is revealed clearly enough in contour charts of the upper troposphere. Even the mean northern hemisphere 500 mb chart (Fig. 9.3(a)) shows regions of particularly tight gradient over the western Atlantic and western Pacific Oceans. Figures 11.5(a) and 11.6(a) show that jet streams, represented again by close-packed contours, are in fact embedded in the Rossby waves. Analogous features may be reproduced in the flow patterns developed in the dishpan experiments described earlier. Close examination of upper-air charts at various levels establishes that there are several types of jet stream phenomena, with different origins, locations and roles in the general circulation. Figure 9.4 (adapted from Sawyer 1957) shows three contour charts of the northern hemisphere for the same situation, using a polar projection and omitting coastlines for the sake of clarity. In addition to the contour lines, there are shaded areas where the wind speeds at the relevant level exceeded approximately 25 and 50 m s^{-1}, as ascertained either from direct observation or from use of the geostrophic wind scale. Figure 9.4(a) gives the 500 mb contours and the flow at 5–6 km, and a pattern of Rossby waves can be seen with a large trough extending roughly along longitude 60 °W (east coast of America) and another along 120 °E (over eastern Asia), which represent typical trough locations. A belt of strong winds (more than 25 ms^{-1} but exceeding 50 m s^{-1} in places) meanders from subtropical latitudes in the Caribbean and the Pacific to very near the North Pole over northern Russia. There are also detached regions of strong winds over northwest Africa and southwest Asia, around latitude 30°. Figure 9.4(b) shows the 300 mb contour pattern, at elevations now around 9 km. Many features are similar to those at 500 mb but the belt of fast-moving air transgressing the lines of latitude is strengthened and the belt hugging subtropical latitudes is both stronger and more continuous. At yet a higher level, 200 mb or around 12 km (Fig. 9.4(c)), the belt twisting into high latitudes has weakened but that encircling the hemisphere at about 30 °N is not only stronger but also quite continuous.

Comparison of many such charts shows that there are at least two major jet stream belts. One, which fairly consistently hugs subtropical latitudes, is relatively broad and persistent enough to show up on mean monthly contour charts of the appropriate pressure surface. It has its maximum development around the 200 mb level and is known as the *subtropical jet stream* (STJ). Another is so variable in behaviour that it is not apparent on mean contour charts. It is apt to meander across latitude lines, sometimes joining forces with the STJ, sometimes swinging into very high latitudes. Furthermore, while there is evidence of it at the 500 and 200 mb levels, the core is usually found at around 300 mb. From its location generally in higher latitudes, this has been called the *circumpolar jet stream*, but an examination of its origins justifies the alternative title of the *polar front jet stream* (PFJ). Both these major jet stream belts have their counterparts in the southern hemisphere, although the meanderings are less pronounced.

Air flow in the middle and upper troposphere is dominated by horizontal temperature gradients and it is to be expected that strong winds should be related to steep gradients. In middle and high latitudes, the steepest gradients are found in the transition zones between warm and cold air masses, i.e. at *fronts*. Full consideration of air masses and fronts is left to Chapter 12, but, for the moment, it is sufficient to remind the reader that air masses of contrasting origin and temperature are separated by sloping transition zones, where the warmer air overrides the colder. Figure 9.5 represents a cross-section of the atmosphere along the line AA' in Fig. 9.4(a) (over North America), which has been constructed from the

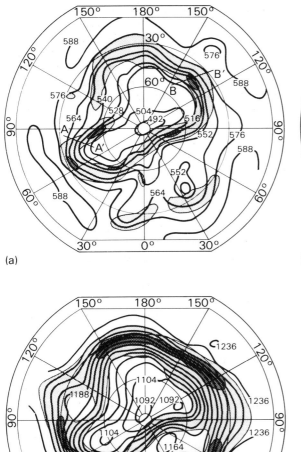

(a)

(b)

(c)

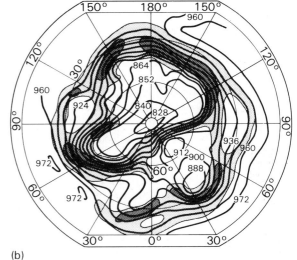

Fig. 9.4 Contour charts of the northern hemisphere for: (a) 500 mb; (b) 300 mb; and (c) 200 mb levels, 19 December 1953, 0300 GMT. Contour values in decametres. Lightly shaded areas – wind speed > 25 m s⁻¹; heavily shaded areas – wind speed >50 m s⁻¹. (After Sawyer 1957)

available upper-air data for the period of the contour chart. The isotherm pattern reveals the presence of warm air in the southwest and cold air in the northeast and the sloping frontal zone is clearly located by the sharp downward dip of the isotherms. The isokinetics (lines of equal wind speed) show a jet maximum of more than 60 m s⁻¹ at about 350 mb. This is the PFJ, blowing out of the paper (in accordance with Buys Ballot's Law), with its core lying about 1000 km ahead of the surface front on the cold air side. The jet stream occurs just below a marked change in tropopause level.

The circumpolar jet stream (or PFJ) is thus a zone of highly concentrated kinetic energy which obviously derives from the enhanced potential energy inherent in a frontal structure. As such, it accompanies the polar front (the major middle-latitude frontal belt), moving with it, though always displaced on the cold air side, and becoming involved in the distortions that occur with the development of frontal disturbance (section 13.2). Local jet streams may be found in association with detached or isolated frontal belts in middle or high latitudes.

Figure 9.6 is a similar atmospheric cross-section along the line BB′ in Fig. 9.4(a) in the region of Japan, again for the same period. This clearly shows two jet streams: one situated above latitude 45 °N, and with its core at about 300 mb, is related

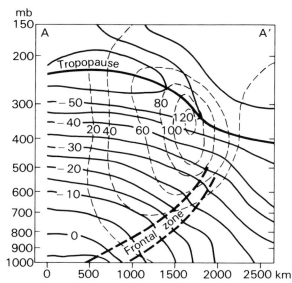

Fig. 9.5 A SW–NE cross-section of the PFJ over North America, 19 December 1953, 03.00 GMT. Continuous lines are isotherms (°C), broken lines are isokinetics (expressed in knots; halve values to give approximate equivalents in m s^{-1}). (After Sawyer 1957)

to a frontal zone and is readily identified as the PFJ. The other, above latitude 33 °N and with its core at 200 mb, is clearly the STJ. It is not associated with a frontal zone and indeed, as other analyses have confirmed, seems little related to surface synoptic developments. However, it does coincide with a tropopause break and with a region of marked horizontal temperature gradient,

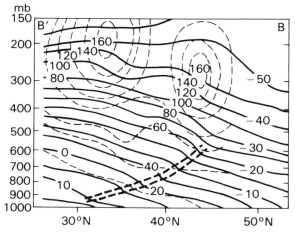

Fig. 9.6 A NNE–SSW cross-section of jet streams near Japan, 19 December 1953, 03.00 GMT. Key as for Fig. 9.5. (After Sawyer 1957)

and the origins of the STJ must be sought in the context of the global circulation (Ch. 11). Indeed, both the PFJ and STJ are part of the general westerly flow of the middle and upper troposphere. An *easterly tropical jet stream* (ETJ) has also been recognised lying over peninsular India and extending to West Africa during the northern hemisphere summer. Further reference to these various jet stream types will be made in Chapters 11 and 12.

9.5 Retrospect

The flow pattern in the middle and upper troposphere, as depicted by contour charts for the appropriate pressure levels, is dominantly one of a circumpolar swirl in a west-to-east direction. In detail, this is compounded of a progression of Rossby waves moving in an irregular and often distorted pattern in the general stream, within which individual air parcels may move northwards or southwards over tens of degrees of latitude. Embedded in the general flow are the belts of fast-flowing air known as the jet streams, of which there are several types. The STJ, being apparently related to the Hadley cell (sections 12.2–12.4), shows little seasonal migration, except that it disappears when the cell breaks down over southern Asia in summer. On the other hand, the PFJ, while generally a middle- and high-latitude feature, is so variable in location and vigour as to disappear from sight in mean monthly charts. Other jet streams are of a more seasonal or ephemeral nature.

Whatever the vagaries of the Rossby waves or jet streams, the pressure fields and associated air flow in what is loosely called 'the upper air' are much simpler than those at the base of the troposphere, yet are related to them, as later chapters will show. What happens at 300 or 200 mb may seem remote to the ground-based observer and the affairs of human beings, but in fact there are important linkages between upper-air events and surface weather. The broad upper patterns would appear to dictate the behaviour of surface systems, not only by a steering effect which controls the movement of surface lows and highs, but also often by encouraging their development in certain circumstances and locations. In turn, some surface features may become so powerful as to be capable of distorting the upper-air pattern, so that cause and effect may ultimately be difficult to unravel.

Key topics

1. The significance of middle and upper tropospheric flow in the global circulation.
2. The contrasting pressure changes with height in warm and cold air.
3. The genesis of Rossby waves and their influence on the behaviour of air masses and fronts.
4. The causes and effects of surface and upper-air flows, including jet streams.

Further reading

Barry, R. G. and Chorley, R. J. (1982) *Atmosphere, Weather and Climate*. Methuen; London and New York (mainly Ch. 3).

Chapter 10 Development of pressure systems

The total climatic system may be seen as embracing the atmosphere and its underlying land and ocean surfaces in a series of cascading systems linked by flows of mass or energy (Ch. 2). These flows, acting together with the distinctive properties of different parts of the system, produce the global or primary circulation of the atmosphere. However, within this overall circulation system are embedded numerous subsystems which share certain common attributes, for example of vertical and horizontal components intimately linked. These features appear on the weather map, or may be inferred from appropriate satellite imagery, as a complex and repeated pattern of pressure systems. It will be convenient to examine the characteristics of high and low pressure systems before consideration of the necessarily more generalised concept of the global circulation.

A suitable starting-point is a reminder that pressure differences seen on a surface synoptic chart reflect an uneven distribution of the mass (weight) of air overlying that part of the earth's surface. If atmospheric pressure at a station increases from say 980 to 1020 mb, as may happen in a day or two in the changeable weather regime of the British Isles, then this implies the addition of some 4% in the mass of overlying air: a comparable pressure decrease means a withdrawal of that amount of air. It will be seen that in the development of highs and lows, such apparently small pressure differences are in fact net changes resulting from very considerable horizontal transports of air at different levels in different directions. Furthermore, these horizontal flows are linked by vertical motions which (since these are essentially tropospheric systems) are shallow by comparison.

10.1 Basic themes

In Fig. 10.1* the basic mechanisms of high and low pressure systems are illustrated schematically in both plan and cross-sectional views. Low pressure (deficit of air) may be induced by various processes, as will be shown shortly, but, however initiated, must be accompanied by an inward or convergent flow of air at low levels in obedience to the pressure gradient (Fig. 10.1 (a)). The equation of continuity (which restates the principle of conservation of mass) demands that a body of air compressed horizontally must expand vertically and in this case the air has no alternative but to rise. This is the meaning of *convergence* in the meteorological, rather than the everyday sense: it implies both horizontal and vertical motion. At upper tropospheric levels, where further ascent is

*In the following discussion, the northern hemisphere is assumed.

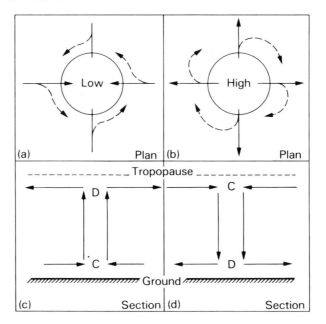

Fig. 10.1 Basic air-flow patterns in high and low pressure systems. In (a) and (b), the continuous-line arrows point towards lower pressure; the broken-line arrows indicate the effect of the Coriolis deflection on the low-level flow. In (c) and (d) the lengths of the horizontal arrows are schematically proportional to the rates of inflow and outflow.

limited, the air column is effectively compressed vertically which means that it spreads out horizontally: this is *divergence*.

All low pressure systems are, in fact, characterised by convergence below and divergence aloft, with rising air at mid-tropospheric levels between (Fig. 10.1 (c)). The subsequent fate of the system depends on the relative magnitude of inflow and outflow: with a persistent or deepening low, the latter over-compensates the former. If the reverse applies, the low 'fills'. Pursuing the argument in reverse, it is easy to see that a high pressure system (addition of air) is characterised by divergence below and convergence aloft and that the vertical motion is one of descent or subsidence (Fig. 10.1 (b) and (d)). High pressure is maintained by upper-level inflow more than compensating low-level outflow: otherwise the high will collapse.

In reality, of course, the horizontal flows depicted in Fig. 10.1 must be deflected by the Coriolis force (section 8.2) so as to create the characteristic directions of low-level circulation, anticlockwise (cyclonic) in the case of the low, clockwise (anticyclonic) in the case of the high (in the northern hemisphere). What is not indicated in the diagram is that, in both systems, the sense of circulation necessarily reverses with height. It will be remembered that active development requires ageostrophic conditions. These are always to be found in the lowest kilometre or so of the troposphere (the friction layer) where the frictional inflow at the base of a low, or the frictional outflow at the base of a high, may help to maintain these systems even in the absence of cross-isobar movements elsewhere.

10.2 Vorticity

Another insight into the circulation of these systems is provided by the concept of *vorticity*, which is an expression of the rotational property or spin of fluid particles. In the present context, vorticity (strictly vertical vorticity) is concerned with the horizontal motion, in either direction, of an air column about a vertical axis: it has the magnitude of twice the angular velocity (2ω). The principle of conservation of angular momentum has it that the product of angular velocity and moment of inertia (I) is constant. The moment of inertia is:

$$I = \Sigma \, \mathrm{m} r^2 \qquad \qquad [10.1]$$

where m is the mass of each element of the body and r its distance from the axis. It describes how the mass is distributed in a rotating body, which is large if the mass is distributed far from the axis and small if it is concentrated near it. To take a simple example, if a weight is attached to a fixed point by a length of string and made to rotate around it, the rate of rotation is increased if the length of the string is reduced (small I). Similarly, the spinning ice-skater begins his rotation slowly with arms outstretched, body bent and a leg extended (large I) and speeds up by drawing himself in as close as possible to his axis of rotation (small I).

Applying this notion to the air columns involved in Fig. 10.1, it follows that if a column is stretched vertically, as the result of convergence, its rate of spin must increase and if it shrinks vertically and spreads out horizontally (divergence), its spin rate decreases. However, the air in question is not devoid of vorticity even before these developments take place. The earth itself obviously possesses vorticity, spinning as it does from west to east about its axis, and the magnitude of this varies from zero at the equator to a maximum at the North or South Pole. Viewed from above the North Pole, the earth spins in an anticlockwise sense, which is to say cyclonically. The argument does not differ if the view is above the South Pole, but here a clockwise rotation is cyclonic.

The atmosphere as a whole rotates with the earth and must share this cyclonic vorticity, which is therefore in a sense fundamental and is given a positive sign in mathematical treatments. If the development of a low begins with an area of zero pressure gradient and completely calm air, this is equivalent to saying that the air possesses exactly the same vorticity as the underlying earth's surface. Convergence in the lower layers (column stretching) increases the spin relative to the earth, which manifests itself as a cyclonic circulation. On the other hand, divergence at upper levels (column shrinking) implies a reduction of cyclonic spin below that of the earth's surface beneath, which becomes apparent as an 'anticyclonic' rotation. The argument can readily be applied in reverse to the case of a developing high.

It follows from the preceding discussion that a body of air may have zero spin relative to the underlying earth's surface, while possessing cyclonic spin relative to a point in space outside the earth–atmosphere system. Alternatively, such a body may exhibit both *absolute vorticity* (with respect to absolute space) and *relative vorticity* (with respect to earth). Vorticity is usually designated by the symbol ζ and it is useful to distinguish absolute and relative vorticity as ζ_a and ζ_r. The difference between these two is clearly the spin of the earth itself about the local vertical, which is represented by the Coriolis parameter (section 8.2). If this is labelled f, then

$$\zeta_a = \zeta_r + f \qquad \qquad [10.2]$$

This relationship, together with the principle of conservation of absolute vorticity, helps to explain the Rossby waves described in section 9.3. If a mass of air travels across the lines of latitude and its absolute vorticity remains constant, its relative vorticity must change since the rotation of the earth's surface itself changes with latitude. In Fig. 10.2, it is assumed that there is one latitude B where the rotation of the air and that of the underlying earth's surface are equal ($\zeta_r = 0$). If some of this air is displaced polewards, it will travel over surfaces with increasing cyclonic spin: the air will then develop negative relative vorticity (i.e. anticyclonic curvature) and will eventually, at some latitude A, turn back equatorwards. If its momentum carries it beyond B towards C, its cyclonic spin will exceed that of the earth's surface and cyclonic curvature will develop. These oscillations, which characterise the broad westerly flow of middle latitudes (section 11.4), play an important role in the genesis of many of the travelling depressions of those regions (section 13.2).

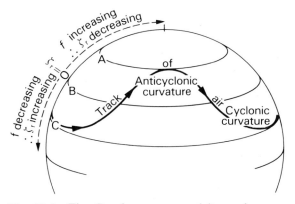

Fig. 10.2 The Rossby wave: vorticity and curvature. (A) latitude of maximum anticyclonic relative vorticity; (B) latitude of zero relative vorticity; (C) latitude of maximum cyclonic vorticity. (Adapted from Scorer 1957; Harwood 1981)

10.3 Variations on basic themes – low pressure systems

Attention hitherto has focused on the necessary attributes of these circulating systems. The actual processes that generate them in the atmosphere must be capable of setting in train the basic mechanisms of Fig. 10.1 and it seems that the required energy inputs can operate at any point in the system. A particularly simple example is the thermal or *heat low*, in which the energy input into the system is thermal and is provided by the underlying surface. This will happen readily over islands and peninsulas in summer, when a considerable temperature contrast develops between heated land and cooler surrounding sea. The process is sufficiently illustrated by Fig. 10.1(c), assuming an island to exist under the ascending air column and, for simplicity, initially still conditions. Air above the heated surface becomes buoyant and expands vertically. Below a certain level, there will be accumulation of air and hence a pressure gradient at that elevation resulting in outflowing air. This is the divergence depicted by D in the diagram. The outward flow (with anticyclonic curvature) reduces the total mass of air above the island, thus creating the low and the inevitable convergence (C) with its accompanying cyclonic vorticity.

The thermal low is clearly a system in which radiant energy is converted into thermal energy at ground level, which in turn transforms into the potential energy embodied in the rising air currents, which in their turn give rise to the kinetic energy of the associated horizontal flows (Ch. 2). Apart from the steady attrition by friction and heat loss by long-wave radiation, the system ceases to function when the solar energy input is cut off. Of course, if cloud formation and precipitation occur, then latent heat transfers and mass flow are also involved. The release of latent heat when condensation occurs may be a significant energy boost in any type of low, but is likely to be important in the thundery lows (for example, the storms that develop over Spain and drift northwards over France and Britain in summer) because of the warmth and high water vapour content of the air.

In yet another variant of the heat low, considerable release of latent heat may provide the major energy input. When polar or Arctic air masses move from the *source region* (section 12.1) over warmer ocean waters, the air becomes unstable and frequent showers develop. Sometimes, shallow troughs or even closed low pressure centres

appear in areas of particularly intense showery activity and move embedded in the airstream. In these so-called *polar air lows*, the origins of which are not completely understood, it may well be that much of the energy input enters the system literally in mid-air.

To some extent, this applies also to the intense low pressure systems (*cyclones, hurricanes, typhoons*) of tropical latitudes (section 13.5) but these are not solely thermal in character, having more complex and composite origins. It is true that they form only over very warm ocean waters (with surface temperatures exceeding 27 °C) and the air converging into their bases is not only warm but richly laden with water vapour, which yields a prodigious release of latent heat when condensation occurs. However, their formation is also associated with pre-existing areas of low-level convergence, usually with the troughs in the *trade wind flow* which are known as easterly waves (section 13.5). They do not form within 5° of the equator because of insufficient Coriolis force there to create a circulation (section 8.2). Their paths are usually confined to the seas, since overland the latent heat supply is cut off, but there have been some notable examples of highly destructive passage inland. Tropical lows are more fully described in Chapter 13.

Unlike the types so far considered, the typical travelling depression of middle latitudes develops not in homogeneous air but astride a front separating two air masses (Chs 12 and 13), in which the initial solar energy input has become converted into differing air temperatures and hence densities. The proximity of warm, less dense air and colder, denser air represents a situation high in potential energy. In fact, the development of the system through a typical life-cycle is largely concerned with the replacement of warm air by cold at the surface and the conversion of this potential energy into the kinetic energy of a stronger circulation. The frontal origin of the system already presupposes the low-level convergence needed to satisfy one of the basic requirements.

However, both theory and experience point to the need for this low-level convergence to be coincident with, and over-compensated by, high-level divergence. This (and indeed the high-level convergence required for a developing high) can readily be supplied by the nature of the upper tropospheric air flow in middle latitudes. The Rossby waves already described provide such opportunities. It was shown in Chapter 8 that in the axis of a trough (where cyclonic curvature is maximum) the wind is sub-geostrophic, while in the downwind limb of the trough (where the flow has straightened out) the speed increases to geostrophic. This acceleration, with more air leaving a given area than entering it, is effectively a divergence. Similarly, it can be seen that a slowing-down of speed from the axis of a ridge (maximum anticyclonic curvature and super-geostrophic winds) to its downwind limb creates a piling-up of air or convergence. Thus, the forward or downwind limb of a trough in the Rossby wave pattern is a favoured area for cyclonic development at low levels and the rear or upwind limb (which is the forward limb of the following ridge) favours anticyclonic development below (Fig. 10.3 (a)).

Figure 10.3(b) illustrates another pattern of upper tropospheric flow which seems capable of creating convergences and divergences at those levels. In middle and high latitudes, jet streams are associated with frontal systems (Chs 9 and 12) and therefore with areas of bad weather in general. However, the so-called *jet entrance* and *exit* are regions especially favourable to low-level cyclonic development because of lack of adjustment between wind strength and pressure gradient. At the entrance, where the pressure gradient tightens rapidly, the wind is sub-geostrophic and there is

a consequent transport of air across the contours from high pressure (warm air) to low pressure (cold air) side. On the right-hand side of the entrance (looking downwind), a deficiency of air results (upper-level divergence) which tends to generate convergence and cyclonic development at low levels while, to the left of the entrance, low-level anticyclonic development is favoured. At the jet exit, where the pressure gradient slackens and the wind is super-geostrophic, the pattern is reversed.

Finally, the orographic low or *lee depression* (or trough, since a closed circulation may not occur) owes its origin to the shape of the ground. When a broad stable stream of air moves over a high mountain wall, a ridge of high pressure forms on the windward side and a low or trough on the leeward side. While this can be regarded simply as a piling-up of air on one side and a deficit on the other, it also follows from vorticity considerations that air ascending the windward slopes suffers vertical shrinking. The air now develops anticyclonic (negative) spin (it being assumed that stability suppresses further ascent), while air descending the lee slopes redevelops its initial vorticity. A lee depression is often found south of the Alps in northern Italy (the so-called Genoa Low), while air of polar origin may be accelerated into it, through the major relief gap of the Rhône Valley, as the *Mistral*.

10.4 Variations on basic themes – high pressure systems

It will be obvious that the various types of low-pressure system have their broad counterparts among the high-pressure systems and indeed it is often difficult to avoid mentioning both in the same context. *Orographic highs* are a case in point and require no further mention. Ageostrophic conditions in the upper troposphere of middle latitudes generate closed highs or ridges as well as lows and these are equally part of the changing weather pattern of those regions. There are *thermal highs* which form as a result of the cooling of a land surface, as over an island during the night or, on a much bigger scale, over a continental interior in winter. The highest MSL pressures (often over 1070 mb) are recorded in the Siberian or Eurasian winter high, which is also the most extensive. In response to radiational cooling, the lower layers shrink vertically and therefore spread

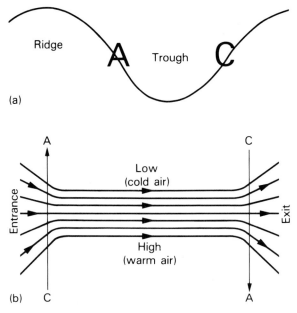

Fig. 10.3 Schematic contour patterns of: (a) a Rossby wave, and (b) a westerly jet stream, indicating areas of favoured cyclonic (C) and anticyclonic (A) development at low levels.

horizontally to give the divergence that sets the rest of the mechanism in train. With the excess weight due to a concentration of cold air near the surface, this type is classed as a *cold anticyclone*. These are essentially shallow features and the overlying cyclonic circulation (often no more than a trough) may be discernible at a height of only 2 km. Transitory cold anticyclones, formed in polar air, often separate the travelling depressions of middle latitudes.

Warm anticyclones are of quite a different origin. The air is relatively warm at lower and middle tropospheric levels and, as a result (unlike cold highs), these are deep systems being still strongly evident on 500 mb contour charts. The excess weight is attributed to a high tropopause level and relatively cold air in the upper troposphere. Warm highs, of which the *subtropical highs* (like the Azores High) are the major examples, cannot be fully understood except in relation to the global circulation (see Ch. 11).

All anticyclones, of whatever origin, embody subsiding air. It will be recalled (section 3.3) that subsidence results in adiabatic warming, which manifests itself as a temperature inversion on a temperature–height graph or a tephigram. This inversion, which first appears at some height but lowers progressively to within 1 or 2 km of the ground (sometimes difficult to separate from a nocturnal surface inversion), explains many features of anticyclonic weather. As a warm dry layer, it effectively limits vertical motions in the underlying air and on summer days, this often means clear skies or shallow cumulus or stratocumulus cloud at most whereas in winter, dull or foggy weather may persist. Subsidence sometimes blurs the distinction between cold and warm anticyclones since it is capable, through its conversion of potential energy into thermal energy, of transforming a more persistent cold high into a warm type.

10.5 Retrospect

Low pressure systems are of various origins but are all characterised by convergence and cyclonic circulation at low levels, divergence and anticyclonic flow at upper levels, with ascending air between. The latter feature explains the cloudy and rainy weather associated with all lows. Their role as rain-bringers is less appreciated in those regions like western Europe and western North America, where they are regular and commonplace visitors, than in regions like northeastern peninsular India, where the arrival of the monsoon depressions that bring much of the summer rain becomes a matter of life or death.

High pressure systems involve low-level divergence and anticyclonic circulation and upper-level convergence and cyclonic flow, with subsiding air between. This latter characteristic explains the dry (not necessarily fine) weather typical of all types of high. Anticyclones provide welcome spells of fine weather in the unsettled summers of middle-latitude regions (e.g. the UK and northwest Europe in 1983) but become drought-bringers if too persistent, as in the notorious English summer of 1976. The stagnant air and inversion lid (section 3.4) also encourage high concentrations of urban and industrial pollution. In London (until recent anti-pollution measures became effective) smog (section 5.2) was historically an accompaniment of winter anticyclones. On the other hand, the equally notorious (but chemically different) Los Angeles smogs require also the high sunshine hours associated with the summer North Pacific High to encourage the photochemical reactions that convert vehicle exhaust emissions into dangerous gaseous pollutants.

Both highs and lows are found on various scales, ranging from mesoscale features waxing and waning diurnally to continental features varying seasonally and to global features of a semi-permanent nature. They are broadly but not always precisely counterparts. All highs and some lows develop in uniform air but the frontal lows of middle latitudes involve two air masses in their genesis and growth. Lows are frequently vigorous systems, fast-moving and with strong circulations, while highs, though generally more extensive, tend to be sluggish in movement and with a large calm centre: their lower portions are in effect left with weak cyclonic spin.

The two systems, however, are clearly complementary. In Fig. 10.1, their structures have been depicted separately but even in such a schematic representation, it is obvious that the divergence in the upper regions of a low feeds the convergence required by an adjacent high and that at lower levels, the outflow from the latter contributes to the inflow to the former. Highs and lows are inexorably knit together in the overall scheme of global circulation, which is explored in the next chapter.

Key topics

1. The significance of convergence and divergence in atmospheric pressure systems.
2. The role of absolute and relative vorticity in the generation of Rossby waves.
3. The contrasting dynamics of low and high pressure systems.
4. The influence of lows and highs on surface weather conditions.

Further reading

Barry, R. G. and Chorley, R. J. (1982) *Atmosphere, Weather and Climate*. Methuen; London and New York (mainly Ch. 3).
Crowe, P. R. (1971) *Concepts in Climatology*. Longman; London and New York (Ch. 5).

Chapter 11 The global circulation

Any scheme or conceptual model of the global (primary or general) circulation of the atmosphere must satisfy two basic requirements. It must accord reasonably with the observed facts of the circulation, as built up over several centuries of observation from the early reports of seafarers to modern techniques of measurement and mapping. It must also recognise that the atmosphere has a fundamental problem that it must solve and obviously does solve, whether or not meteorologists yet have the wit to appreciate exactly how the trick is done. This problem was in part stated in Chapter 2. The atmospheric circulation is clearly a so-called closed system, in which the input of energy is from outside, i.e. from the sun. But the distribution of solar radiation input over the globe (see Fig. 2.2) is manifestly uneven, for astronomical reasons.

Furthermore, when account is taken of the output of terrestrial radiation, it becomes evident that the earth's surface in each hemisphere is a heat source only equatorwards of about latitude 35° and a heat sink to polewards (section 2.2 and Table 2.4). If no other processes are involved,

then there is nothing to prevent low latitudes from becoming steadily hotter and middle and high latitudes steadily colder. That this does not occur implies that some negative feedback mechanisms are in operation. which successfully transport excess thermal energy from low to high latitudes. This is one aspect of the atmospheric problem.

11.1 The observed circulation

The mapping of pressure and winds (Leighly 1949) over the surface of the globe dates mainly from the middle of the nineteenth century, although as early as 1686, Edmund Halley (more widely known for the comet whose orbit he calculated) had produced a chart of winds of the tropical oceans. The first maps of mean monthly and annual pressures for the whole globe were drawn by Alexander Buchan in 1868–69. This notable contribution (more important than the so-called Buchan Spells with which he is popularly associated) formed part of the famous *Atlas of Meteorology* prepared by Bartholomew and Herbertson in 1899, which became a standard work for many decades. Two Americans (Matthew F. Maury and James Coffin) mapped global winds between 1848 and 1860. World maps of pressure and winds also appeared in the earliest meteorological atlas, that published by the German Julius Hann in 1887. Maury deserves attention too, with his charts of ocean currents published in the 1850s for, as will be shown, atmosphere and ocean cannot be separated in any consideration of the general circulation.

The main features of the mean surface pressure

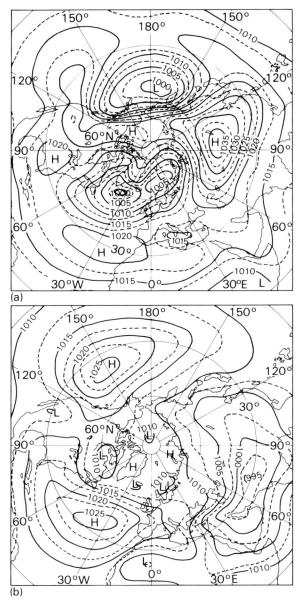

(a)

(b)

Fig. 11.1 Average MSL pressure over the northern hemisphere in: (a) January; and (b) July, 1951–66. Isobars at 5 mb intervals. (After Lamb et al. 1973)

North America. In July, this feature is confined to the oceans in the form of extensive high pressure cells, separated by low pressure over the land masses. A similar high pressure belt is located around 30°S and is much more continuous in the ocean-dominated southern hemisphere. Known by various regional names (i.e. the Azores High, the North Pacific High, etc.) this feature, generalised as the subtropical high, is relatively permanent, though varying in position and intensity, and must be considered an essential component of the global circulation. It has long been recognised that its intensification and extension over the land masses in winter, and its interruption by continental low pressures in summer, are due to the development of thermal highs and lows related to the contrasted thermal properties of land and water bodies.

The subtropical highs of both hemispheres flank an *equatorial trough* of relatively low pressure. This rather vaguely defined zone tends to move with the migrating overhead sun but, in the northern summer, becomes merged with the huge thermal low that develops over southern Asia and the lesser one over the southwest of the USA. On the poleward side of the subtropical highs are located the so-called *subpolar lows*. In winter these comprise the Icelandic Low of the North Atlantic and the Aleutian Low of the North Pacific, but in summer the features are ill-defined and tend to be 'swallowed up' by the continental lows. In the southern hemisphere, however, the corresponding subpolar low is virtually continuous about latitude 40°. The subpolar lows are statistical features, since they have no identity as such on individual synoptic charts, but represent the high frequency of travelling low pressure systems in these latitudes. In July, a shallow high can be seen over the polar cap. Sometimes called the *North Polar High*, this weak feature is absent from the January chart, where the highest pressures are found over northernmost Canada and Siberia.

Comparison of Fig. 11.1 and Fig. 9.3 will show how the complexities of the surface patterns give way to the relative simplicity of the middle and upper troposphere, where the cold shallow highs, whether of the winter continental or the weak polar variety, are no longer represented. Only the subtropical highs persist into these upper levels, where their axes are found somewhat nearer the equator, i.e. displaced towards the warmer side (section 9.2).

It is customary, in discussion of the surface circulation, to eliminate seasonal and other differ-

distribution of the northern hemisphere in midwinter and midsummer respectively are shown for a recent period in Fig. 11.1. Most striking in the January map is a broad belt of high pressure at around latitude 30°, interrupted only over the Mediterranean and to some extent in the central Pacific, but intensified and extending into higher latitudes over the continental masses of Asia and

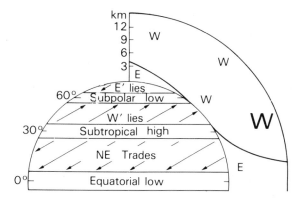

Fig. 11.2 The idealised distribution of surface pressure and winds in the northern hemisphere and schematic mean zonal wind directions in cross-section (E – easterly winds, W – westerly winds; large W – STJ).

ences by idealising the pressure and wind patterns as shown in Fig. 11.2. This hypothetical planetary circulation scheme has the merit of simplicity and draws attention to the relationship between pressure distribution and wind direction. It presumably is something like what would prevail on a globe of uniform surface. It has the demerits of giving the impression of a static pattern of pressure and wind belts, some way from the realities of constant variation and of a cellular pressure pattern in equilibrium with a complex pattern of air flow. However, it is convenient to visualise the *trade winds* or low-latitude easterlies (frictionally turned into northeasterlies in the northern hemisphere) as blowing between the subtropical high and the equatorial trough.

The subtropical high also feeds the *Westerlies*, actually directed somewhat polewards as southwesterlies towards the subpolar low, which also receives the *polar easterlies* blowing outwards from the polar high. The opportunity has been taken to add to Fig. 11.2 a schematic cross-section which serves as a reminder of the broad relationship between surface and upper flow and of the dominance of westerly winds at height. The large W clearly represents the STJ maximum.

11.2 Classical circulation models

The evolution of ideas concerning the global circulation provides some fascinating examples of perceptive reasoning on the basis of the data available at the time and of changing insights, even revolutions in thought, as advances in observational techniques brought fresh facts to light. The process is not yet over (see Crowe 1971 or Barry 1969).

Much of the early thinking focused on the trade winds, which were interpreted by some sixteenth- and seventeenth-century philosophers as air left behind (in view of its lightness) by the west–east movement of the earth's surface and so constituting an easterly wind. Halley, who had mapped these winds, was well aware of the notion of a thermal circulation, as his famous (1686) paper to the Royal Society of London showed. Indeed, he was the first to interpret the Asiatic summer monsoon as a large-scale thermal low and sea-breeze effect, a concept with a germ of truth, although now recognised as over-simplified. The logic of Halley's thinking, applied to the global circulation, was that the low-latitude heat source expanded the overlying air layers, creating a high-level pressure gradient which moved air polewards as a southerly flow. In these high latitudes, the air descended over cold surfaces and returned equatorwards as a northerly flow at low levels (Fig. 11.3 (a)). Unaware of the significance of earth rotation, Halley explained the westward flow of the trades as yet another thermal effect, with the air being impelled towards a low pressure zone under the westward-moving sun.

It was almost half a century later that George Hadley (1735), in another classical paper, incorporated the effect of earth rotation. Consequently, the poleward flow of the Halley scheme became one of upper westerlies (in good accord with later observations) and the return surface flow was now easterly Fig. 11.3 (b)). This left large areas of surface westerlies in higher latitudes to be explained, but the notion of what has become known as a *Hadley cell* persists to this day, although now restricted to lower latitudes. During the second half of the nineteenth century, the concept of circulation cells took a strong hold and Ferrel (first in 1856 but more fully argued in 1889) was the earliest advocate of a *three-cell model*.

Ferrel explained the subtropical high in terms of conservation of angular momentum, as an accumulation of poleward-moving air (convergence) at high levels, having acquired maximum westerly momentum. Descent and low-level divergence feed not only the trades returning equatorwards but also the surface westerlies of middle latitudes. These, in turn, form the low-level limb of a second cell with rising air in subpolar latitudes. Yet a

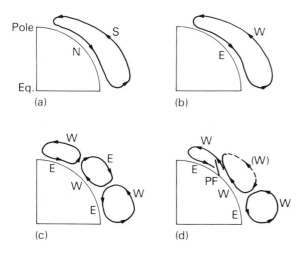

Fig. 11.3 Some early models of the general circulation due to: (a) Halley; (b) Hadley; (c) Ferrel; and (d) Rossby. PF – Polar Front, W and E as Fig. 11.2. (Adapted from Barry 1969; Crowe 1971; Rossby 1941)

third cell involves descending air in the polar high and surface easterlies (Fig. 11.3 (c)). The low-latitude and high-latitude cells are both thermal or direct, driven by supposed temperature differences between equator and subtropics and between polar caps and subpolar regions respectively. Furthermore, the direction of air flow, given the Coriolis deflection, is consistent with Fig. 11.2. The middle-latitude cell was less clearly explained but Ferrel was aware that surface winds could not have the same direction all over the globe because that would affect the rate of rotation.

The final elaboration of the three-cell model (Fig. 11.3 (d)) was due to Rossby (1941). He retained the notion of thermally driven (direct) cells in low and high latitudes and regarded the middle-latitude cell as indirect or frictionally driven. Logic demanded that its upper limb should have an easterly direction of flow, while observations showed this to be a region of strong westerly flow. He explained this by suggesting that the easterly flow was masked by the imposition of westerly momentum from the flanking cells on either side through the agency of the Rossby waves. More crudely put, the air was frictionally dragged in a westerly direction by the adjacent westerly streams

Owing, as it did, much to Ferrel and others, Rossby's model explained many observed features of the general circulation. It incorporated three cells which he called the trade wind (Hadley) cell, the middle cell and the polar front cell. By

this time, the air-mass and frontal concepts of the Norwegian School (fully considered in Ch. 12) were well established. The subtropical highs could readily be seen as major source regions feeding subtropical air polewards (the surface westerlies) as well as equatorwards (the trades). In middle latitudes, the warm air masses encountered cold air originating over polar regions at the so-called polar front, along which convergence in the frontal disturbances created one of the world's stormy and rainy belts. In the equatorial trough, the trade wind streams from both hemispheres converged at what was initially termed the intertropical front, which represented another major rain belt. The high pressure zones of subsiding air were the arid regions of the earth.

11.3 Collapse of the classical models

Despite the considerable vogue enjoyed by the three-cell model, which was still featured in some textbooks of the 1960s, Rossby had jettisoned his own scheme in a paper of 1949. This was only partly because of the difficulties attending the interpretation of the suggested middle cell. He recognised that there was often insufficient thermal gradient to drive the trade wind cell (indeed, in the northern summer in some parts of the globe, the arid subtropics were much hotter than the cloudy equatorial zone). Also, he noted that a system of discrete circulation wheels did not necessarily effect the transfer of thermal energy from low to high latitudes. It was also partly a reaction to the notion that the complexities of the global circulation could be adequately represented by any single steady-state model, meant to apply everywhere and at all times. In addition, studies initiated as far back as the 1920s, by Sir Harold Jeffreys, among others, had suggested an alternative approach.

The classical approach, entrenched over two centuries, focused on mean conditions, eliminating the everyday details of the synoptic chart. It also sought explanations in meridional cells in which air moved polewards at one level and equatorwards at another. The new approach looked for answers in the synoptic details themselves, refusing to ignore them as the mere 'noise' in the system and, at any rate from the 1950s onwards, was able to claim support for its conclusions with some quantitative calculations. Jeffreys (1926) had shown that the classical cells alone were not capable of transporting the surplus heat polewards. This essential task, he considered, was carried out by the large-scale horizontal eddies (i.e. the travelling lows and highs of middle latitudes) which shifted deep streams of warm air polewards and cold air equatorwards. These migrating systems were thus seen as undertaking a vital mixing function and attention shifted to middle latitudes for the key to a more complete understanding of the global circulation.

The problem was not seen purely in terms of a thermal energy transfer. As Ferrel was aware, the easterly winds of low latitudes exercise a frictional braking effect on the earth's surface, while the westerly winds of middle latitudes have an accelerating effect and, as earth rotation proceeds at a constant rate, some kind of balance between these two torques must be achieved. The law of conservation of angular momentum states that the total sum of angular momentum in the entire earth–atmosphere system must be considered constant. Therefore, it follows that the angular momentum imparted in low latitudes by the faster-moving earth to the slower-moving (i.e. easterly) air above, must be transported polewards to be returned from faster-moving (i.e. westerly) air to slower-moving earth. In addition, there is an obvious requirement from the point of view of the moisture balance, in that water vapour must be transferred from regions of high evaporation (mainly the oceanic subtropical highs) to regions of high precipitation (i.e. the equatorial trough and the disturbed middle latitudes). The general circulation must satisfy all these requirements.

11.4 The fluctuating circulation

Statements concerning the general circulation, whether verbal or diagrammatic, are now more cautious than those offered before the 1950s. It is recognised that the circulation does not have the same characteristics in different longitudes and may well vary with time, on either a seasonal or a synoptic scale, at any one longitude. Modern schematic models like that of Palmén (1951) retain the Hadley cell nevertheless, as the necessary mechanism for transferring heat energy and angular momentum polewards in lower latitudes, whether the cell is considered to be thermally or frictionally driven or both. Some difficulties are avoided by specifying that the model applies to the northern hemisphere in winter. It is assumed that the upward movement of heat energy in the equatorial zone takes place essentially in the large cumulus and cumulonimbus towers associated with the low-level convergence of that zone. Here, as elsewhere, the upward heat flux is in the form of latent rather than sensible heat.

There are difficulties, not yet resolved, concerning the interpretation of the downward limb of the Hadley cell and the associated subtropical high. Reasonable calculations of the rate of heat loss by radiation in the upper poleward flow ($1-2 \,°C \, day^{-1}$) would suggest sinking at around 30° latitude. The conservation of angular momentum in this southerly flow, following Ferrel and others, would have the same effect. Furthermore, the related notion that the development of maximum westerly momentum (before sinking occurs) is represented by the SJJ is also attractive. It has also been pointed out that the developing depressions of middle latitudes tend to be steered polewards and the neighbouring anticyclones to be steered equatorwards. This conspicuous steerage had been related to their frequent origin under the forward and rear limbs respectively of a trough in the Rossby wave pattern (Fig. 10.3 (a)). Furthermore, there is ample other synoptic evidence to show that the subtropical highs are periodically reinforced by transitory cold highs which merge with them.

In middle latitudes, the emphasis is on horizontal mixing as the main agent of transport of heat and momentum across the lines of latitude. This is the region of the polar front, the transition zone between air of subtropical and of polar origin (section 12.2). If this front could be visualised as both continuous and undisturbed, it would constitute a distinct barrier to the cross-latitude air movements necessary for heat and momentum exchange. But the polar front is neither since it is frequently disturbed by the development astride it of frontal depressions (section 13.2). In these systems, warm air is thrown polewards and upwards and cold air, while subsiding, thrusts equatorwards. The whole comprises what has been described as a 'slantwise' convective system. The upper tropospheric waves, with which these disturbances are linked, are another aspect of the same mixing mechanism. The vital function of the middle latitudes in the general circulation is difficult to convey in a single simple diagram. However, Fig. 11.4, mainly after Palmén and Newton (1969), attempts to summarise the main features of the winter circulation of the northern hemisphere.

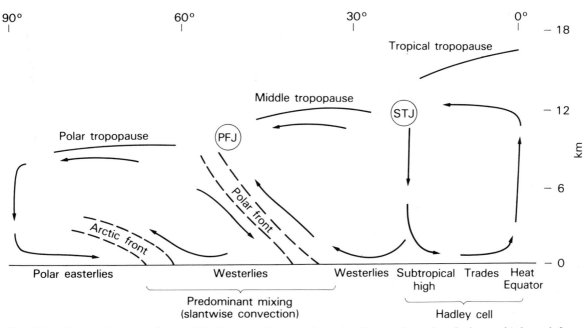

Fig. 11.4 Some features of the global circulation in winter in the northern hemisphere. (Adapted from Palmén and Newton 1969)

Both physical and numerical models support this view of the general circulation. The famous dishpan simulations seem capable of reproducing a variety of circulation patterns depending on conditions. If the heat differential between rim and centre is applied without rotation, a simple Halley-type cell develops (as in Fig. 11.3(a)), with the surface water flowing inwards and the bottom water outwards. Applying rotation in the correct (cyclonic) sense results initially in the surface flow acquiring a 'westerly' momentum, while the bottom flow has an easterly component, i.e. a Hadley-type cell. With faster rotation, a pattern of long waves with embedded jet streams and vortices develops, simulating middle-latitude conditions to a remarkable degree. With even faster rotation, the motions break down into a chaos of small convection cells and the model then ceases to represent reality. Computer models adopt in effect a similar approach. As the basic parameters are successively fed in (e.g. the correct thermal gradient between high and low latitudes, rotation in the correct sense and, at appropriate rates, even frictional and moisture considerations) so the results approximate the realities of the circulation, with a low-latitude Hadley cell and the disturbed wave-oscillation pattern of the middle latitudes (Fig. 10.2).

If the essential role of the general circulation is one of transporting energy polewards by means of cross-latitude exchanges of air, synoptic experience shows that this function is discharged more vigorously at some times than at others. For example, when the flow of air in middle latitudes, whether considered at surface or at upper-air levels, is predominantly *zonal* (i.e. along lines of latitude) there is obviously little energy exchange. Conversely, when the flow is more dominantly *meridional* (along lines of longitude), there is the maximum opportunity for these necessary exchanges to take place. Synoptic analysis by American workers around 1950 established that atmospheric behaviour tends to alternate between periods of strong zonal and marked meridional flow and this tendency has been systematised in the concept of the *zonal index*.

This index measures the average pressure gradient (from either the surface or an upper-air-contour chart) between latitudes 35 and 55 °N. A high zonal index obtains when the gradient exceeds 8 mb and represents a strong circumpolar vortex, with disturbances following one another in a generally west–east direction and unsettled weather for regions in their path. It also implies poorly developed Rossby waves and minimal fulfilment of the circulation's basic role (Fig. 11.5). A low zonal index has a mean pressure gradient less than 3 mb and at such times the vortex is weak and the

97

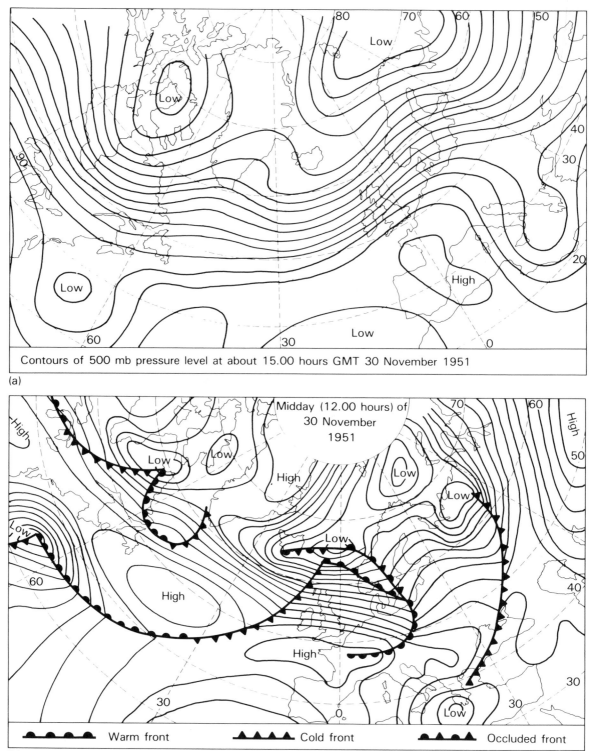

(a) Contours of 500 mb pressure level at about 15.00 hours GMT 30 November 1951

(b)

Warm front ● Cold front ▲ Occluded front

Fig. 11.5 (a) Typical 500 mb contour pattern; and (b) corresponding surface chart, illustrating a high zonal index type. (After HMSO 1951)

Rossby waves attain maximum amplitude. The surface synoptic charts show huge *blocking anticyclones* (i.e. blocking westerly flow) on whose flanks warm air penetrates into unusually high latitudes and cold air sweeps into the subtropics. Depending on location with respect to these meridional flows, specific regions may enjoy unseasonably warm or suffer unseasonably cold weather, sometimes for weeks on end. Low index periods are clearly the times when maximum poleward transport of energy is effected (Fig. 11.6).

11.5 Global circulation and world climates

The overall earth–atmosphere system is too complex to be categorised as any one single type. However, from a particular point of view, it can be regarded as a process-response system, in which the various processes involved in atmospheric behaviour invoke a response in the pattern of climatic regions over the surface of the earth. A full description of climatic regions is beyond the scope of this book, but Table 11.1 is designed to give a broad indication of the way that prevailing circulation regimes are translated into regional climatology.

There are several climatic classifications of a *genetic* character, i.e. based on atmospheric circulation. For example, the scheme adopted here in Table 11.1 broadly follows that of Flohn (see Barry and Chorley 1982) which recognises four constant regimes (i.e. having the same circulation type all year) and three regimes characterised by a seasonal change. Unlike classifications proposed by Köppen and Thornthwaite, Flohn's scheme does not set out to delimit vegetation zones. However, there is a broad relationship, which is indicated in Table 11.1 (see also Ch. 28 and Fig. 28.4). A few explanatory notes now follow, to supplement Table 11.1.

For reasons discussed in section 12.7, the original notion of an Intertropical Front (ITF) between the converging trade wind streams from both hemispheres has been discarded and reference is made in Table 11.1 simply to the equatorial trough. The trade wind regime, which dominates the tropical marginal zones in winter, is charac-

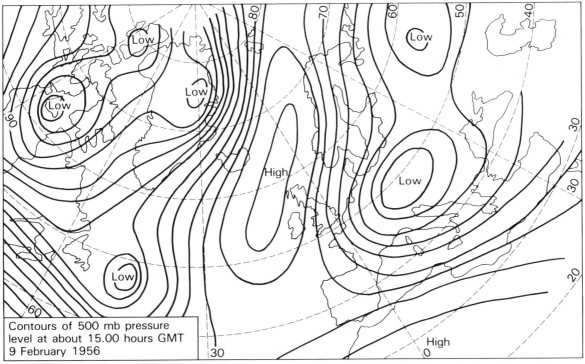

(a)

Fig. 11.6 (See p. 100 for caption and Fig. 11.6(b)

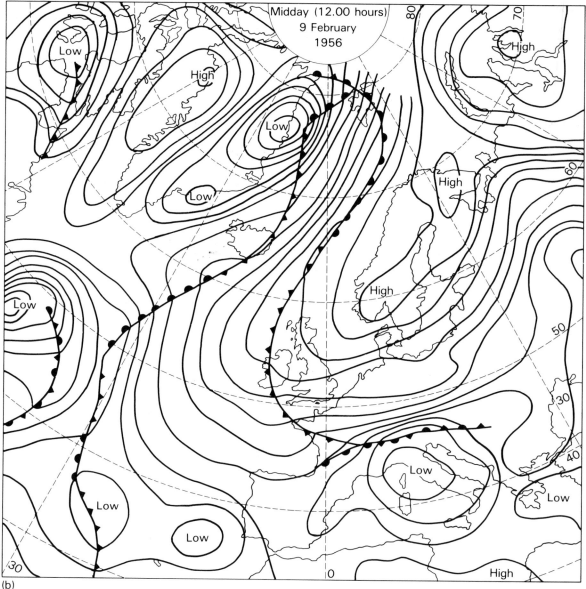

Fig. 11.6 (a) Typical 500 mb contour pattern; and (b) corresponding surface chart, illustrating a low zonal index type. The blocking high during this spell sent warm Tm air well north of Spitsbergen and cold Arctic air into the Mediterranean. (After HMSO 1956)

terised by subsidence and a related inversion which limits convection. A cumulus sky typifies the winter period, with the chance of scattered showers increasing equatorwards, where the inversion weakens. The easterly waves referred to in Table 11.1 are discussed in section 13.5: they are significant because some develop into tropical storms.

An important variant of the tropical marginal type occurs in the *monsoon* lands of Southeast Asia, where summer rainfall amounts may be large, especially where enhanced by orographic effects. In winter, much of China is influenced by cold continental air spreading from the thermal (Siberian) high of interior Asia. However, India, the classical monsoon region, is protected from this air by the Tibetan mountain barrier and its 'northeast monsoon' involves air of subtropical

Table 11.1 World climatic zones (Mainly after Flohn)

Zone	Circulation type	Typical weather	Typical vegetation
Equatorial (C)	Equatorial trough (convergence): light variable winds or 'Doldrums': westerlies in some regions	Rain all year, often heavy: convectional, orographic, local disturbances	Luxuriant tropical rainforest (selva)
Tropical margins (S)	Equatorial trough in summer: trade winds in winter (inversion)	Mainly summer rain, often from easterly waves: orographic influence important	Tropical grassland (savanna) – rainforest in wet uplands
Subtropical arid (C)	Subtropical highs (inversion) or trade winds	Generally arid	Desert, steppe
Subtropical winter rain 'Mediterranean' (S)	Subtropical highs (summer) alternating with middle-latitude westerlies (winter)	Winter rain (disturbances on polar or Mediterranean front)	Evergreen forest – degenerated to scrub (chaparral)
Moist temperate (C)	Middle-latitude westerlies	Rain all year (disturbances on polar front) on western margins; more settled spells from blocking highs: summer convectional rain in interiors	Deciduous and mixed forest (western margins): grassland (interiors)
Subpolar (S)	Subpolar low, polar easterlies	Little rain but all year (disturbances on polar and Arctic fronts): interiors have winter snow and summer convectional rain	Coniferous forest degenerating to tundra
Polar (C)	Polar highs, polar easterlies	Meagre snowfall from occasional invasion by polar or Arctic/Antarctic front disturbances	Ice desert

C = constant regime.
S = seasonally changing regime.

origin. In summer, a thermal low develops over the Asian interior, but the lowest pressure is found over northwest India. This feature replaces the equatorial trough and the Hadley cell is destroyed at this season in these longitudes. Eventually, warm and very moist equatorial air (which may be of southern hemisphere origin) is drawn into the region, depositing the rainfall which is vital for its rural economy. The actual mechanism of the summer monsoon is complex and involves the replacement of the STJ above these latitudes by an ETJ at 200 mb. This develops on the southern flank of an upper tropospheric anticyclone which is thought to be thermally induced by the relatively high surface temperatures achieved by the elevated Tibetan plateau (compare this development with that depicted by Fig. 9.1).

The subsidence inversion typical of the subtropical highs is responsible for some of the world's great deserts and where cold water currents reinforce the stability, certain western coasts (e.g. Peru) are rainless for years on end (section 3.2). The so-called 'Mediterranean' climate is the only one to have the bulk of its rain in winter. This is basically a western marginal type and the region from which it takes its name is in many ways the least typical (see section 12.4 for a fuller discussion). Both the moist temperate (also known as the extra-tropical westerly zone) and the sub-polar zones are, strictly, confined to western margins. Certainly the description 'temperate' cannot be applied to the continental interiors, where winter thermal highs generating cold, stable air alternate with summer thermal lows, under whose influence hot spells and convectional rain are common.

11.6 Retrospect

In the complex system known as the general circulation, the unequal receipt of the solar energy input creates a heat source in low latitudes and a heat sink in middle and high latitudes. In addition, over most of the globe, the atmosphere acts as a heat sink to the heat source of the earth's surface. These heat sources and sinks do not become steadily warmer and colder respectively and this implies a circulation that embodies negative feedback functions. Thermal energy must be transported upwards and polewards. Earth rotation creates a momentum source in the fast-rotating low latitudes and a momentum sink in middle and high latitudes and this fact too requires a circulation of similar character. Warm air in low latitudes and colder air to polewards constitute a situation of high potential energy, which is translated into the kinetic energy of horizontal air motion. Reduction of thermal energy by long-wave radiation and destruction of kinetic energy by friction are losses which are made good by continuous input of more solar energy. In general, a Hadley cell seems to be the main agent of heat and momentum transfer in low latitudes (though not always operative) and the large-scale eddies perform that function in middle latitudes.

Some 70% of the base of the atmosphere is underlain by ocean and it is obvious that ocean water, with its mobility and capacity for storing thermal energy, must play an important part in atmospheric circulation (Ch. 3). Warm ocean currents carry thermal energy polewards, while evaporation transfers it as latent heat to the atmosphere and, in this way, the hydrological cycle (section 2.1) 'locks' into the circulation system. The wind belts play their part in creating the circulation of ocean currents (Ch. 34) and in fact, there is every justification for thinking in terms of an overall ocean-atmosphere system (Perry and Walker 1977). Estimates of the relative contribution of atmosphere and ocean to the poleward transport of heat vary. It has been put as two-thirds to one-third but obviously there are some longitudes (e.g. 60–80° W, where the Gulf Stream is found) where the role of the ocean is particularly important.

The global circulation system finds regional expression in the pattern of world climates and, through the operation of another set of linkages (the ecosystems of Ch. 28), in the world vegetation zones. Where the circulation regime is repeated, analogous regional climates are found, though possibly with variations due to differences in the distribution of land and sea or the lie of high ground. Within each climatic region, the daily succession of weather events is related to smaller-scale atmospheric systems which develop as part of the master-scheme. These secondary and tertiary systems, which comprise the weather experience of the inhabitants of those regions, are the subject-matter of the next three chapters.

Key topics

1. Global air circulation as a negative feedback control.
2. The mean surface pressure distribution and its simplification into the idealised circulation scheme.
3. Middle-latitude disturbances as the key to understanding the global circulation.
4. The significance of the zonal index in the fluctuations of the global circulation.

Further reading

Barry R. G. (1969) Models in meteorology and climatology, in Chorley, R. J. and Haggett, P. (eds) *Physical and Information Models in Geography*. Methuen; London, University Paperback. Useful treatment of much material in Chapters 11, 12 and 13 from the standpoint of the models.

Crowe, P. R. (1971) *Concepts in Climatology*. Longman; London and New York (especially Chs. 6 and 7).

Perry, A. H. and Walker, J. M. (1977) *The Ocean–Atmosphere System*. Longman; London and New York.

Chapter 12 Air masses and fronts

The now familiar concepts of air masses and fronts are usually attributed to the Norwegian meteorologists of the First World War and their intensive synoptic studies during a time of much reduced exchange of weather information. However, these perceptive ideas have earlier origins. Luke Howard, writing in 1833, was aware of the clash of two currents – a southerly and a northerly, which 'may raise each other, the colder running in laterally under the warmer current' – and its implication for rainfall. The notion was revived by the German Dove (1862), and Ferrel (1889) knew of the tendency of warm and cold air to 'keep apart', separated by a line of temperature, pressure and wind change. What the Norwegians contributed was systematisation of these concepts (Bjerknes and Solberg 1922, among others), fitting them into existing schemes of the general circulation and providing the framework for an enhanced understanding of the weather of middle latitudes.

They conceived the air mass as a large body of air, extending horizontally for some thousands of kilometres and vertically to the tropopause, having approximately uniform properties of temperature, humidity and stability in the horizontal direction. A more modern definition would refer to a large area of 'slack meteorological gradients', but the implication is the same: within each air mass, weather is substantially similar throughout, allowing for some variation due to latitudinal or local influences. Neighbouring air masses are separated by gently sloping transition zones which the Norwegians termed 'fronts' by obvious analogy with the military dispositions of the war. The fronts were seen as comparatively narrow zones (perhaps 1000 km in width) of discontinuity in air-mass properties. They became recognised as belts of sharp meteorological gradients and also as wet-weather zones because of the tendency of warmer lighter air to ride up, or be forced up, above colder, heavier air. The Norwegians were also able to link the middle-latitude disturbances with developments on the major fronts (Ch. 13). Their original concepts, based largely on surface observations and middle-latitude experience, have not survived unmodified over the 60 years since their formulation, but nevertheless retain much of their value for descriptive if not for analytical purposes.

12.1 Air-mass genesis and classification

Air masses are described in terms of their basic properties, i.e. temperature and humidity and the important derived property of stability (section 3.4) which depends on the distribution of temperature and humidity in the vertical. Thus, different air masses may be characterised as cold, moist and unstable or warm, dry and stable, or by other combinations of these properties, with the terms referring generally to the lower layers of the air mass. Since the atmosphere is both heated and derives its moisture from underlying surfaces, it follows that air masses inherit their basic properties by long-period residence (a week or more) over surfaces, with uniform thermal and moisture characteristics, which are known as source regions.

Clearly, the essential requirement of a source region, apart from homogeneity of surface, is an anticyclonic regime, which alone allows sufficiently long stagnation for the slow impression of properties from surface to air. The major source regions are thus the permanent and semi-permanent high pressure regions of the global circulation. These include the polar highs·over snow- and ice-bound surfaces, the subtropical highs, whether over warm seas or the arid lands of similar latitudes and the thermal highs that develop over the continental interiors, when they are mantled by winter snow. The net outflow from the base of these anticyclones ensures a persistent spread of the associated air masses towards regions of lower pressure. From the point of view of air-mass climatology, there are 'donor regions' and 'receptor regions' and it is in the latter that the frontal zones of the globe must be sought.

Air-mass properties may be more or less significantly modified by passage of the air over surfaces different in character from the source region. Since all air masses originate in high pressure regimes, they tend to be initially stable with a well-marked inversion layer. A cold air mass moving equatorwards will be warmed from below, with lapse-rate steepening (section 3.2), eventually to become unstable in its lower layers. If its movement is over water, it will also acquire moisture by evaporation, the more readily since its temperature is also increasing. On the other hand, a warm air mass moving polewards will be progressively cooled from below, with a temperature inversion developing to reinforce its initial stability. The weather of a particular air mass in a 'receptor region' may be very different from that in its source region and air-mass identity is often destroyed due to a complex history.

The Norwegians originally classified air masses according to their source region and subsequent paths (in order to make some allowance for modification). Air from cold sources was termed *polar*, whether or not it originated near the Poles: air from warm sources was *tropical*. A second element depended on moisture characteristics: thus air might be *maritime* or *continental* in origin. The basic classification was therefore fourfold: *polar maritime* (Pm) originating over cold seas like the Arctic Ocean; *polar continental* (Pc) born in the winter thermal highs of the Eurasian and North American continents; *tropical maritime* (Tm) from the subtropical oceans under the great high pressure cells and *tropical continental* (Tc) from the desert regions of the same latitudes. Air moving over warmer surfaces received the suffix K (*kalt*), implying warming from below and loss of stability (e.g. PmK). Air moving over colder surfaces was suffixed W (*warm*) implying cooling from below and a reinforcement of its stability (e.g. TmW).

This broad classification proved inadequate when applied on a regional scale, and various other terms are in use in particular parts of the world. Sometimes the terminology departs from the logic of the original scheme. For example, the term *equatorial air* is sometimes used to describe air of various origins and histories which has become stagnant over equatorial waters and has acquired excessive warmth and moisture. When such air is drawn into the summer monsoon circulation of Southeast Asia, it is often called *equatorial monsoon air* (Em). The air-mass climatology of three contrasted continents will be considered in sections 12.4–12.6, together with the related frontal conditions.

12.2 Air-mass boundaries and frontogenesis

The juxtaposition of air masses at a front has to be understood in a three-dimensional model (Fig. 12.1) and there may be confusion due to the use of terms. The front is a transitional layer of the order of 1 km in thickness, sloping very gently (with a gradient usually between 1 : 25 and 1 : 300) up towards the cold air side and within this layer, there is turbulent mixing between the two air masses. The term 'frontal surface' is sometimes (loosely) applied to this layer and where this 'surface' intersects the ground is also termed the front. However, because of the gentle frontal slope, this is in reality one edge of a belt of considerable width (up to 200 km depending on the gradient), the *transition* or *mixing zone* being conventionally placed in the cold air. For convenience, these intersections are represented as lines on the weather map. The diagrammatic tem-

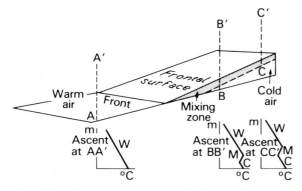

Fig. 12.1 Schematic structure of a front. The frontal slope is much exaggerated. The graphs represent idealised radio-sonde ascents at various points (W – warm air; M – mixing zone; C – cold air).

perature–height graphs drawn from points A, B and C (Fig. 12.1) illustrate the composite nature of the structure. At these points, the mixing zone is represented by a *frontal inversion* or by an isothermal layer or one of small lapse rate, if the air-mass temperatures are less contrasted.

Such a frontal structure does not necessarily denote rainy weather. There are *inactive* fronts, at which the two air masses flow in parallel streams and no more than a little cloud development betrays the frontal presence. A front becomes *active* when a convergent flow pattern results in the encroachment of one air mass upon the other, with the consequent ascent of the warm air. The frontal cloud and weather thus depend largely on the moisture and equilibrium properties of the warm air mass (section 3.4). If this happens to be dry and stable, which is not common in middle latitudes, the uplift has little effect and the front is weak, marked only by a belt of shallow cloud. It is more usual for the warm air to be potentially unstable and moist throughout much of its depth. Then, the result of frontal uplift is a thick sheet of rain-giving cloud extending upwards into cirrus levels or sometimes a multi-layered cloud denoting a variable moisture content in different layers. Sometimes, the warm air is conditionally unstable (see Fig. 3.5a) in which case vigorous ascent may be unleashed, with cumulonimbus embedded in the layer cloud and a risk of frontal thunderstorms.

The process by which fronts form, or are intensified, is termed *frontogenesis* and is common in the low pressure regions (convergence zones) of the general circulation. On the other hand, exist-ing fronts weaken if the air-flow pattern becomes divergent, with associated subsidence of air from higher levels. These processes accompany anticyclonic development and fronts become inactive and eventually disappear (*frontolysis*) when pressure rises.

The familiar terms 'warm front' and 'cold front' refer to movement of the entire boundary zone, which is not to be confused with the relative motion between the two air masses already referred to. If the movement is such that warm air replaces cold on the ground (i.e. in the direction A–C in Fig. 12.1), a warm front is specified. In the case of a cold front, the cold air wedge advances (direction C–A in Fig. 12.1). A front without movement is neither warm nor cold but *quasistationary*. Fronts may change their direction of movement and thus their specification and it is also quite common for a front to pivot about a stationary point, being warm to one side and cold to the other. Warm fronts and cold fronts might be expected to have a mirror-image relationship, differing only in direction of movement, but this is not quite the case. In fact, there are a variety of frontal models, to which many actual frontal structures approximate more or less, and there are also examples which hardly conform to any model.

The passage of a front is marked by more or less pronounced changes of pressure and wind. Reference to weather maps will show that moving fronts are mostly located along the axes of troughs of low pressure, so that pressure falls ahead of an advancing front and rises or at least steadies behind it. The surface wind veers (changes direction in a clockwise sense) as the front passes. The isobars on the synoptic chart similarly change direction and are conventionally drawn at an angle across the front. Cold fronts move at the speed appropriate to the pressure gradient imposed on them (ascertained by placing a geostrophic wind scale along the front and reading off, as described in section 8.2). However, warm fronts move at about three-quarters the geostrophic speed (section 13.3) and quasi-stationary fronts lie more or less parallel to the isobars.

12.3 Frontal models

The main differences between warm- and cold-front characteristics stem from the different slope of the frontal surfaces, cold fronts being generally much steeper. This is partly due to the frictional drag exerted by the earth's surface on the lower-air layers, which flattens the frontal slope in the case of the warm front but necessarily steepens it in the case of the cold front. As a consequence, the ascent of warm air and the related cloud formation are usually of more limited horizontal extent in the case of the cold front than in that of the warm. Warm-front cloud sheets measured from the high cirrus, which leads the sequence, to the thick nimbostratus near the surface front itself may reach 1000 km in horizontal extent.

Something like one-third of this cloud mass is likely to be thick enough to give precipitation on the ground. Frontal movement varies from near-stationary to something like 25 m s^{-1}. However, assuming a more usual speed of 15 m s^{-1}, this means that the entire warm-front sequence will take some 18 hours to pass a given spot and that frontal rain will fall for some 6 hours. With active cold fronts, the steeper frontal slope often results in a more vigorous up-thrust of the warm air which

leads to more vertical (cumuliform) cloud development (section 5.3). Furthermore, the cold dense air behind the front tends to subside, thus limiting cloud in that direction. Consequently, all these features combine to give a horizontal dimension often only half that of the warm-front sequence. Cold-front cloud and precipitation are therefore commonly more short-lived but these generalisations, however, mask a great variety in frontal behaviour.

An elaboration of frontal models was first suggested by Bergeron of the Norwegian school and further developed by Sansom (1951) for cold fronts over the British Isles. This divided fronts into *ana-fronts-* and *kata-fronts*, the prefixes meaning respectively 'ascending' and 'descending' (compare anabatic and katabatic winds, Ch. 14). An ana-front is active at all levels in the sense, earlier described, of encroachment of one air mass upon the other at all levels so that there is general uplift of the warm air and extensive frontal cloud

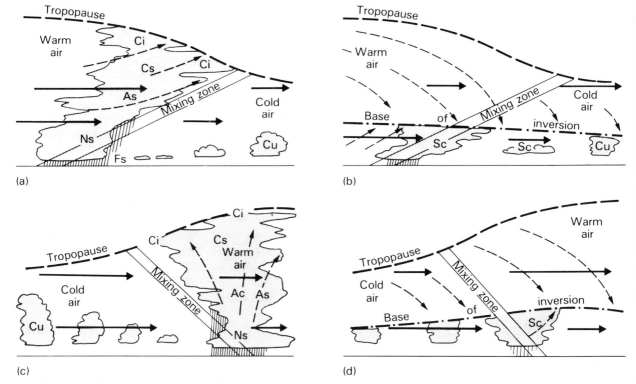

Fig. 12.2 Diagrammatic cross-sections through ana- and kata-fronts. Thick horizontal arrows represent relative motion of the two air masses (length proportional to speed), broken arrows represent vertical motions: (a) ana-warm front; (b) kata-warm front; (c) ana-cold front; (d) kata-cold front. Common cloud types indicated – Ci (cirrus); Cs (cirrostratus); As (altostratus); Ac (altocumulus); Ns (nimbostratus); Fs (fractostratus); Cu (cumulus); Sc (stratocumulus). (Adapted from Pedgley 1962)

often to the tropopause. With a kata-front, the necessary convergent motion is present only in the lowest few kilometres and warm air ascent is confined to this shallow layer. Above this, a divergent or frontolytic situation obtains and the stable warm air here actually sinks and dries out, with a subsidence inversion marking the base of the subsided air.

Figure 12.2 depicts these models diagrammatically. In the *ana-warm front*, the warm air is shown as overtaking the cold air (convergence) at all levels, resulting in warm air ascent throughout the troposphere (Fig. 12.2(a)). In the *ana-cold front* shown, the cold air is encroaching at all levels, but the result is very similar, except that the frontal slope is steeper (Fig. 12.2(c)). In this example, potentially unstable warm air is assumed, giving a thick sheet of cloud, rather than the more cumuliform development associated with many ana-cold fronts. In the kata-fronts, both warm and cold (Fig. 12.2(b) and (d)), air-mass encroachment is restricted to the lowest layers, while above this general subsidence occurs. The frontal cloud may be only 3 or 4 km thick, no more than thick stratocumulus, yielding a little light rain or drizzle and sometimes no precipitation at all. Clearly the ana-fronts are the main rain-bringers. In relation to the surface front, warm-front cloud and rain are largely pre-frontal and cold-front weather is largely post-frontal.

In western Europe, ana-warm fronts are more common than kata-warm fronts and kata-cold fronts more common than ana-cold fronts, though all types occur. In general, ana-fronts represent greater activity and probably reflect an earlier and more vigorous stage of frontal development, while kata-fronts belong to later, dying stages. The lack of frontal activity does not reduce the significance of a kata-front as an air-mass transition zone. While fronts of whatever type may be considered as isolated synoptic features with their associated weather sequences, their involvement in the development of middle-latitude disturbances adds to their significance as will be shown in sections 13.2 and 13.3. Two more frontal models, the so-called warm and cold occlusions, will be introduced at that stage, since they can best be understood in the context of the life-cycle of frontal depressions.

12.4 Air masses and fronts in Europe

The variety of weather and climate to be found within Europe can largely be explained by the alternating residence of different air masses, the fronts that separate them and the frontal disturbances that develop over the continent or, more often, invade it from outside. All the basic types of the Norwegian air-mass classification are represented here and various sub-types may be added for a more complete description in particular regions. Western Europe experiences a highly changeable climate, being most exposed to air masses of Atlantic origin, while still receiving occasional visitations from continental sources, and also experiences Atlantic frontal systems in still vigorous condition. Central and eastern Europe, while not immune to them, are further from oceanic influences and their more continental regime is characterised by seasonal air-mass dominance. With the 'grain' of relief running east–west, Mediterranean Europe is to some extent sheltered from these influences and displays distinctive features which also owes much to the proximity of North Africa.

Figure 12.3 gives a broad indication of the air-mass regions and principal frontal zones of the northern hemisphere. In these maps, the air-mass regions include the source regions under the semi-permanent and seasonal anticyclonic cells of the global circulation shown in Fig. 11.1. They also include the fringe areas near the major frontal zones (mainly the polar front), where the air masses are furthest from their sources and therefore most modified and also where air-mass changes most frequently occur. Some climatologists (e.g. Crowe 1971) prefer to distinguish these areas as 'secondary air masses', from the 'primary' air-mass source regions, the belt broadly between latitudes 40° and 60° being designated 'mid-latitude air'.

The major characteristics of European air masses are shown diagrammatically in Fig. 12.4. The cold air masses originate mainly over the Arctic Ocean and neighbouring land under the influence of the weak North Polar High. Alternatively, they develop over the snow-bound Eurasian interior when in the grip of the winter high pressure. The former source region generates Arctic air (A) which can directly 'attack' regions of northern and western Europe, but much of this air is modified over the North Atlantic and reaches the continent as Pm. Both A and Pm are typically K air masses, losing stability by travel over warmer waters: they are respectively fresher and less fresh variants of the same type. The potency of modification is well illustrated by Pm returning air, which, though Arctic

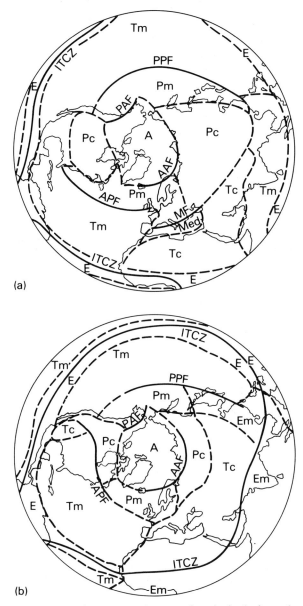

(a)

(b)

Fig. 12.3 Air-mass regions and principal frontal zones in: (a) winter, and (b) summer in the northern hemisphere. AAF, PAF – Atlantic and Pacific Arctic fronts; APF, PPF – Atlantic and Pacific polar fronts; MF – Mediterranean front. The ITCZ is also shown. (Adapted from Petterssen 1956; Crowe 1971 and others)

in source, has warmed up considerably over the waters west of the Iberian peninsula and 'returns' northwards as a stabilising type. Polar Continental is the only truly seasonal air mass, spreading sub-freezing temperatures and clear skies over much

of Europe and occasionally to the British Isles. Here, passage over the North Sea may destroy the initial stability and bring snow showers to east coasts.

The warm air masses originate in the subtropical highs and become W types on approaching Europe. Tropical maritime betrays its oceanic origin by bringing marine (advection) fog (section 5.2) or low stratus cloud, while Tc transports scorching dust-laden Saharan skies. However, both air masses may be substantially modified during their movement. For example, Tm generally loses its stability away from fog-bound windward coasts on summer days and may stagnate to give heat waves culminating in thunderstorms. A long track over the Mediterranean Sea adds moisture to dry, 'thirsty' Tc air which, with daytime heating, may likewise lead to strong convection and thundery activity. In summer, the Eurasian interior generates a thermal low and, while strictly this cannot be a true source region, air drawn in from various tends to stagnate and become modified to something akin to Tc in temperature and equilibrium properties.

Most frontal disturbances affecting Europe originate on the Atlantic polar front (and some on the weaker Arctic front), but the Mediterranean Basin is distinctive in harbouring a detached portion of the polar front (which is separately identified as the *Mediterranean front*) during the winter season. The air-mass contrast here is between polar air (of either Atlantic or continental origin) and either Tc from North Africa or so-called Mediterranean air (much modified air warmed by long residence over the sea). The bursts of cold air through gaps in the mountain wall (the Mistral of southern France and the Bora of the Balkans) are part of the unique climate of the Mediterranean Basin, not paralleled in other regions of so-called Mediterranean regime. In summer, high pressure dominates and the front disappears, so that the region enjoys hot dry weather with much of its air of North African origin (sometimes as the hot, desiccating *sirocco* or *khamsin*).

12.5 Air masses and fronts in North America

Although North America experiences basically the same air-mass types as Europe, obvious contrasts in the grain of relief lead to differences in their incidence and properties. The Rocky Mountains

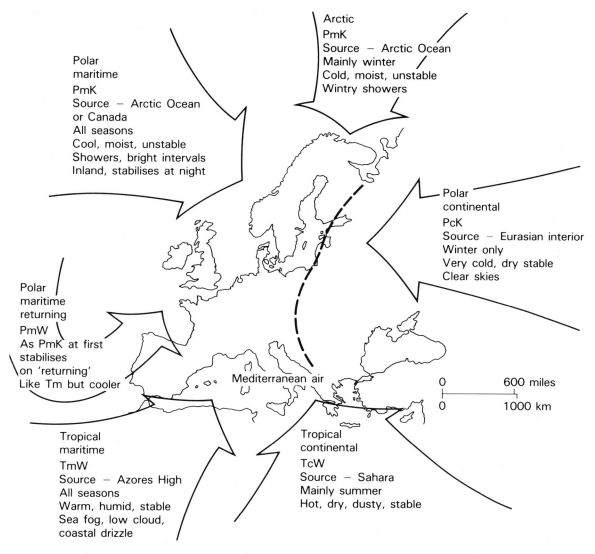

Fig. 12.4 Air masses affecting Europe. The arrows indicate the main directions of approach. The broken line indicates where the incidence of maritime and continental air masses is equal, according to Berg (1940).

confine Pacific influences to a narrow west coast zone and the interior of the continent lies open to invasion from both polar and tropical sources. As Fig. 12.5 shows, the winter cold is supplied largely by Canadian Pc air (with some Arctic, with little difference at that season), which on occasion penetrates far to the south, giving sub-freezing temperatures on the coast of the Gulf of Mexico. This air can lose its stability by passage over the Great Lakes and may then yield snow showers in the Lake Peninsula, or, with orographic uplift, over the Appalachian Mountains. In summer, the Canadian Shield, now snow-free but often water-

logged in places, generates a very mild variety of Pc which, however, when it advances southwards, gives welcome relief from the Tm air prevalent in that season.

North America receives maritime air from two oceans. Polar Pacific air is very much like its European counterpart in winter, bringing showers to the coastal areas and snowfall to the high ground. The inherent stability of tropical Pacific air, which affects the coastal area mainly south of about latitude 40 °N, is strongly reinforced by passage over the cold California current offshore, and coastal fog (section 5.2) is common in that region. This

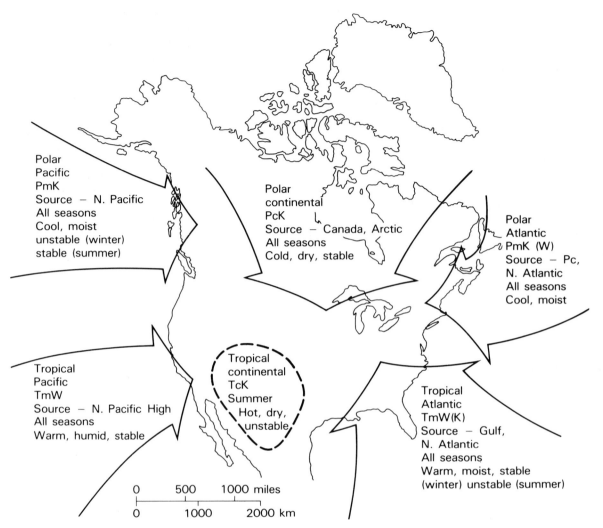

Fig. 12.5 Air masses affecting North America. The arrows indicate the main directions of approach. (Mainly after Trewartha 1968)

tendency to stability applies also in summer, even with polar Pacific air, since the offshore waters at that season are cooler than the neighbouring land along most of the Pacific coast. Passage of Pacific maritime air over the Rockies modifies the initial properties to give something resembling a dry stable Pc (further north) or dry stable Tc (further south).

Polar Atlantic air is actually Canadian Pc in origin and is kept cool and cloudy by passage over the cold Labrador current. Tropical Atlantic is typical TmW in winter giving low cloud or advection fog (section 5.2) as it moves northwards over the cold continent. In summer it dominates most

of the United States east of the Rockies. Losing its initial stability over heated surfaces inland, this very moist oppressive air is capable of unleashing heavy showers and thunderstorms, especially when lifted in frontal disturbances. The Atlantic polar front has its mean position in summer over the heart of the continent (Fig. 12.3(b)). Marked contrasts between Tm and Pc occur along highly active cold fronts in spring over the east-central States, when not only thunderstorms but also destructive tornadoes may develop (section 14.5). A relatively small region of the southwestern USA is described as a source region for Tc air in summer, although as already suggested, this is largely heated and desiccated air of tropical Pacific origin.

12.6 Australian air masses

Some interesting contrasts are offered by examination of the air masses of Australia, all of which (except Tasmania) lies north of 40 °S. There is no Pc air, and even Pm (SPm on Fig. 12.6) is much modified by travel over middle-latitude waters. Gentilli (1971) has suggested the classification shown in the diagram. There is not a great deal of difference in properties between Tm from both the Indian and Pacific oceanic sources and the equatorial air that invades the extreme north in summer. Other authors do not differentiate, as Gentilli does, between the tropical and subtropical varieties of Indian Ocean and Pacific Ocean air: the latter is simply a cooler version of the former. All these maritime air masses are moisture-laden and yield substantial rainfall when lifted by high relief (especially along the eastern seaboard, with a rapid decrease westwards) or in frontal situations. Tropical continental air dominates most of the interior, especially in winter when cooler land surfaces reinforce the prevailing anticyclonic regime and the penetration of the peripheral air masses is frequently blocked. Gentilli points out that air-mass analysis is not used now for practical forecasting purposes in Australia, for reasons noted in the following section.

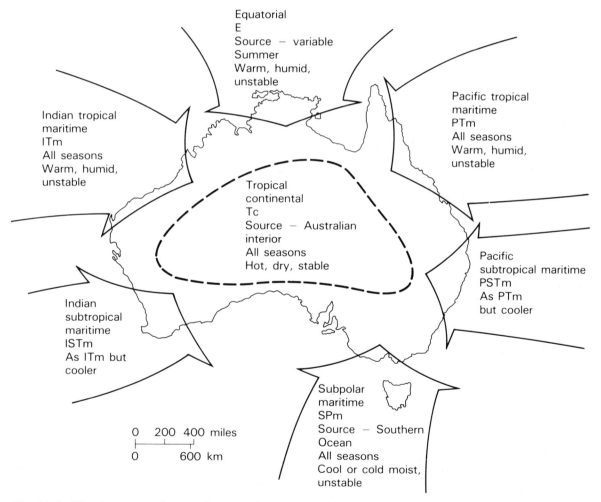

Fig. 12.6 The air masses of Australia, according to Gentilli (1971). Within the broken line, the frequency of Tc air is at least 75%.

12.7 Failure of air-mass and frontal concepts in the tropics

The impetus given to synoptic studies by the Norwegian concepts and their value in illuminating many aspects of middle- and high-latitude weather and climate led, for a while, to the assumption that these notions could be applied with equal success to tropical latitudes. However, forecasting experience in these regions (particularly during the Second World War) showed that this was an unwarranted assumption and, with hindsight, the reasons are not far to seek.

It is evident that in the generally unstable low pressure belt around the equator, no source region in the accepted sense can exist and it is, if anything, a receptor region for air spreading from other sources. In fact, air arrives from the subtropical highs of both hemispheres as either land-derived or ocean-derived trade winds, which might initially be described in air-mass terms as Tm or Tc in origin. However, studies have shown that, whatever its earlier history, all incoming air is forced to stagnate for a while over warm equatorial waters or 'steaming' land. Consequently, it eventually becomes homogenised into a very warm, moist, conditionally unstable air-mass type. This much-modified air is often termed equatorial (E) (Fig. 12.3) though, in some air-mass maps, the historical terminology may be preferred.

For a while, it was thought that the confrontation of the trade wind airstreams from either hemisphere must be marked by a distinct front. Consequently, circulation maps in some publications (even as late as 1960) include a so-called intertropical front (ITF) near the axis of the equatorial trough. However, it is now accepted that with uniformly 'tropicalised' air, no temperature or density contrasts exist and hence no fronts occur which are comparable to those of higher latitudes. Uncertainty as to nomenclature remains and the convergence feature, that must exist between the two trade wind belts, has been variously described as a broad intertropical convergence zone (ITCZ) or as a double northern and southern convergence zone (ITCN and ITCS) or as the *intertropical confluence* (ITC). Whatever it is called, this cannot be a continuous feature around the globe since it is interrupted by regions of divergence or of stagnant air. However, there are some situations where maritime equatorial air is brought against (warmer) Tc air, when the use of the term 'front' in the middle-latitude sense has some justification.

Yet another distinctive feature of weather analysis in low latitudes is the abandonment of the isobaric pattern as the basis of the synoptic chart. Within about 10° of the equator, the Coriolis parameter is so weak that the actual winds are no longer represented by the geostrophic direction and speed. Instead, the tropical air flow is represented by *streamline* patterns which indicate the direction of motion at a given instant. Where a geostrophic balance can prevail, isobars are in fact streamlines. However, if both isobars and streamlines are drawn in near the equator, it can be seen that the latter intersect the former at large angles.

12.8 Retrospect

Air masses and fronts are components of large-scale circulation systems and the maps of Fig. 12.3 should be seen in relation to the mean pressure distributions of Fig. 11.1 and the idealised planetary circulation scheme of Fig. 11.2. The genesis of air masses over source regions in anticyclonic regimes and their modification during passage elsewhere, illustrate very clearly the basic processes of heat and moisture transfer between surface and air. The convergence of air masses of different densities in middle and high latitudes leads to large-scale ascent of the warmer air and the consequent formation of thick frontal cloud and abundant precipitation.

No longer regarded as an analytical tool in meteorological science, the air mass nevertheless remains a useful descriptive concept, except in tropical regions. To speak of a 'polar maritime day' is to use a convenient shorthand to describe a complex of characteristic weather events and conditions. Inland, it implies a chilly but bright start to the morning, then the formation of cumulus clouds as surfaces warm. This is followed by more or less frequent blustery showers as the clouds build to sufficient depth, separated by sunny intervals. Finally, the convection clouds decay during the evening to give a return to fine, clear conditions. On the other hand, a 'tropical maritime day' suggests a quite different weather sequence and 'feel' to the air.

Because of their value in describing the total atmospheric environment, the frequency of occurrence of different air masses in particular regions has been used as the basis of an air-mass climatology. For example, it undoubtedly adds to an appreciation of the climate of Britain to know that cool or cold maritime air (Pm or A) is in residence for 30–40% of the year depending on location (Belasco 1952). On the other hand, the value of this approach is limited, because of days of indeterminate air and the problems inherent in modification, including those due to dynamic processes such as uplift or subsidence. For this reason, a classification of air-flow patterns by Lamb (1950) based on general direction of approach and pressure regime (cyclonic or anticyclonic), which may cut across air-mass types, is often preferred.

Fronts can also be regarded as elements of everyday weather experience, heralding the advent of warmer or colder air, as well as bringing much of the rain of middle latitudes. During the first flush of success of the Norwegian air-mass frontal model, much forecasting began with extrapolation of the movement and behaviour of fronts, according to guide-lines owing a good deal to experience. The more objective methods of today use giant computers to predict the pressure-fields at various levels and the fronts are inserted at appropriate locations almost as afterthoughts. This does not lessen their reality or diminish their

contribution to middle-latitude weather, particularly when involved in the frontal disturbances on which attention is focused in the next chapter.

Key topics

1. The nature of air masses and fronts.
2. The contrasts between warm and cold fronts.
3. Topography as a controlling factor of air-mass and frontal differences between Europe and North America.
4. Air characteristics in the tropics as compared with those of higher latitudes.

Further reading

Barry, R. G. and Perry, A. H. (1973) *Synoptic Climatology*. Methuen; London and New York.

Pedgley, D. E. (1962) *A Course in Elementary Meteorology*. HMSO; London.

Chapter 13 Secondary circulations and disturbed weather

Apart from convectional and orographic precipitation, which depend largely on air-mass properties, most of the disturbed weather of the globe is associated with organised travelling circulation systems which are capable of lifting deep layers of moisture-laden air. These systems are known by a variety of names such as lows, disturbances, perturbations, storms, cyclones or depressions (with or without waves), and their general characteristics were considered in Chapter 10. The present chapter is concerned with a more detailed examination of middle-latitude depressions and tropical cyclones, which are sometimes referred to as secondary or synoptic-scale circulations. They have dimensions considerably less than those of the primary or global circulation but greater than those of the tertiary or mesoscale features which are associated with terrain differences (Ch. 14).

13.1 The changing models of the middle-latitude depression

Weather maps were produced on a regular basis by the national meteorological services of both the UK and the USA from the early 1870s. Before that, the relationship between weather type and barometric pressure was rather dimly perceived, although there was much interest in tropical cyclones as travelling vortices, rather like large-scale whirlwinds. Indeed, both Admiral FitzRoy (the first director of the Meteorological Office of Great Britain) and the German Dove in the 1860s had suggested that middle-latitude storms were somehow connected with the meeting of different airstreams, thus anticipating later notions. The new synoptic charts, however, were basically pressure maps which clearly illustrated the association of cloudy, rainy weather with isobaric 'lows' and of dry, quiet weather with isobaric 'highs'. Many thought that the lows were simple thermal systems in origin.

A well-known 'cyclonic' model of that period (Abercromby and Marriott 1883) showed a simple, roughly concentric pattern of isobars (Fig. 13.1(a)). Furthermore, it also included descriptive labels, the fruit of careful observations, which suggested no such symmetry about the centre. In 1911, Shaw (1933) plotted air trajectories into depressions and picked up clues planted earlier by FitzRoy and Dove. He found no smooth circulation of air around the centre of the system, but rather distinct discontinuities between a band of warm air from the south and cool or cold air both to the north and to the west. His original sketch (Fig. 13.1 (b)) also contained comments on rainfall.

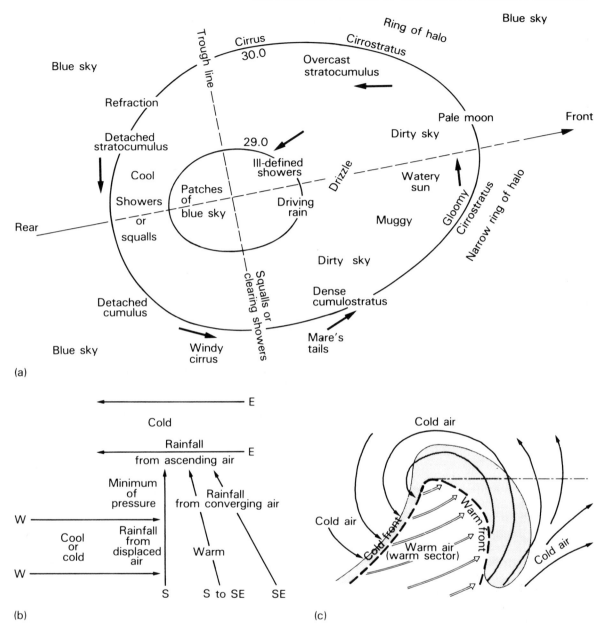

Fig. 13.1 Models of the middle-latitude depression: (a) Abercromby and Marriott (1883); (b) Shaw (1911); and (c) the Norwegian model of Bjerknes and Solberg (1922). In (c), the shaded area denotes cloud.

However, it was left to the Norwegians (particularly V. Bjerknes and J. Bjerknes, father and son) to complete the evolution of thought. They superimposed on the model a pattern of warm and cold fronts, stemming from the centre of the depression. The Norwegians thus succeeded in explaining many of the surface weather features embodied in the earlier models and in integrating their air-mass, frontal and pressure development notions. The original Bjerknes model (Fig. 13.1 (c)) was described by Shaw as 'the most widely known illustration of modern meteorology'. In the same work (Shaw 1933), with typical elegance, he headed the section on the new approach with the title 'The Norwegian duet for polar and tropical air with cyclonic accompaniment'.

115

The Norwegian model of the middle-latitude depression has stood the test of time remarkably well, as a glance at almost any weather map will confirm but, like all models, it represents an idealisation and a simplification. Many actual depressions resemble the model more or less, though there are some alternative models which serve reality better in some circumstances and of course (section 10.3), there are also non-frontal depressions.

13.2 Cyclogenesis in middle latitudes

The Norwegians formulated their theory of the development of depressions (*cyclogenesis*) without the benefit of more than rudimentary upper-air data, relying almost entirely on careful analysis of surface synoptic charts. Their point of departure was an undisturbed section of the polar front, the gently sloping frontal 'surface'. This intersected the ground in a straight line and represented an air-mass boundary between warm air to the south overlying cold air to the north. The 'birth' of a series of frontal depressions was seen as the development of undulations or waves on the frontal surface in much the same way as wind-waves are raised on the surface of the sea. In the latter case, the waves arise because of a difference in motion between two immiscible fluids, the movement of air being greater than that of water.

A similar discontinuity of motion or *wind shear* could be said to exist across the frontal surface, provided that the warm air mass moved faster than the underlying cold air. Bearing in mind the general direction of flow in these latitudes, this meant in effect that the warm air must have a faster westerly motion relative to the cold. The intersection of the undulating frontal surface with the ground necessarily assumed a wave-like form, with a salient of warm air (the so-called *warm sector*) tongueing into the cold, with its leading edge forming the warm front and its rear edge the cold front (Fig. 13.1 (c)). The term 'frontal wave' has by usage become transferred to this two-dimensional synoptic chart view of a three-dimensional structure.

In the Norwegian scheme, further development of the frontal wave depended on its dimensions. Some waves were described as stable in the sense that, having reached a certain amplitude, they simply travelled unchanged as a distortion along the line of the polar front. Furthermore, they trav-

elled at a speed approximately that of the geostrophic wind speed in the warm sector, bringing a sequence of warm-front weather followed by cold-front weather to areas under their path. This applied particularly to shallow (small-amplitude) waves of large wavelength (more than about 3000 km) and also to those of very small wavelength (less than 600 km). Both theoretical considerations and experience showed that most waves of intermediate wavelength were unstable. They developed further, with an increasing amplitude and a fall of pressure at the tip of the wave. Eventually, the wave character was replaced by a true cyclonic circulation, depicted by a closed pattern of isobars on the synoptic chart. At this point, the system was described as a *mature* frontal depression, though with much of its life-cycle yet to come.

Despite the considerable attractions of the Norwegian theory of cyclogenesis, it left some questions less than adequately answered, particularly that of the origin of the pressure fall at the tip of the warm sector of an unstable wave. A satisfactory answer to this question necessarily awaited the establishment of a realistic picture of upper-air behaviour. As this information became available, it was first considered that the surface processes outlined above were the basic ones and that related patterns of middle- and upper-tropospheric air flow were imposed on those levels from below. However, the post-Second World War developments of regular drawing of contour and thickness charts, and the recognition of Rossby waves and frontal jet streams, brought about a change in viewpoint.

As was explained in Chapter 10, the pattern of Rossby waves creates divergences and convergences at upper levels which induce the opposite conditions at lower levels. In particular, the forward limb of a trough in the wave pattern favours cyclonic development in the region below, as is illustrated in Fig. 13.2. This shows a trough on the 500 mb contour chart (a) and the corresponding surface chart (b) with two frontal depressions under the forward limb and an anticyclone (H) under the rear limb of the trough. As a further illustration of the significance of this particular contour pattern for the weather experienced below, Fig. 13.2 (c) shows the mean 500 mb contour pattern over western Europe during the month of April 1983. This was an unusually wet April over the eastern UK especially, with some areas receiving between twice and three times the average monthly rainfall. It can be seen that the month

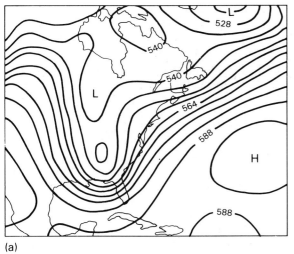

(a)

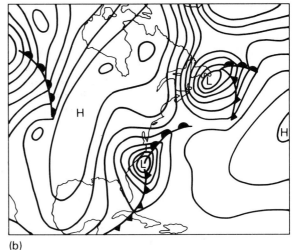

(b)

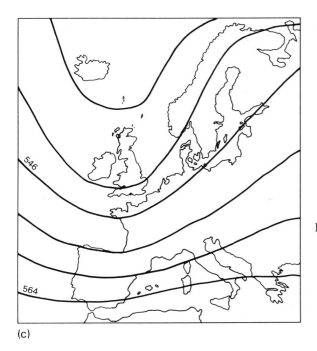

(c)

Fig. 13.2 Relation of Rossby wave and surface synoptic development: (a) 500 mb contour chart for 12 November 1968, 0000 GMT; (b) surface chart for same date and time; and (c) mean 500 mb contour chart for April 1983 over western Europe (contour heights in decametres). ((a) and (b) adapted from Petterssen and Smebye 1971)

was dominated by a persistent 500 mb trough with its axis over the western UK, implying frequent cyclogenesis under the forward limb which persisted over eastern areas.

It would seem that while conditions astride the polar front are ripe for cyclogenesis probably throughout much of its length, the actual areas of development are dictated by the suitability of the upper-air pattern. Only under an upper divergence will there be the mass ascent of air necessary for the mechanism of the low pressure system

to come into operation. The wave analogy then has limited application and it may be as useful to regard the vicinity of the surface polar front as a belt generally conducive to the formation of horizontal eddies, because of the cyclonic wind shear across the front. In much the same way, to use another analogy, wind swirls (made visible by raised dust and leaves) form on pavement surfaces because of the wind shear between the open roadway and the proximity of walls and hedges.

The arrival aloft of the forward limb of a trough

in the moving Rossby wave pattern provides another essential ingredient in the required mixture, and the release of latent heat, as air ascends, adds to the energy supply. Finally, the problem of removal of air from the upper reaches of a depression (section 10.1), one that has exercised the minds of so many meteorologists, now seems less intractable when the presence and role of the frontal jet stream are considered. Sir Napier Shaw (1919–31) had already guessed that an upper-air current acted 'the part of scavenger' in this respect. Middle-latitude frontal depressions are invariably associated with jet streams which, through their function of rapid air removal, are said to steer their movement. Yet, the jet streams exist because of the juxtaposition of warm and cold air at fronts (section 9.4), an example among others of the difficulty of distinguishing cause and effect in meteorological processes.

13.3 Life-history of a model frontal depression

If the original Norwegian conception of cyclogenesis has been modified by subsequent knowledge, there is no doubt that their depiction of the life-cycle of a frontal depression remains valid as a model which enhances understanding of actual weather events. Inspection of consecutive synoptic charts of the North Atlantic, for almost any week of the year, will reveal a sequence broadly similar to that shortly to be described. It is illustrated in Fig. 13.3, in both map and cross-section form, as a sequence along the same length of front from the earliest stage (left) to the end-stages (right), which is a convenient way of representing a model life-cycle. However, although successive stages of development may sometimes be seen on the weather map related in this way, actual patterns are often more complex.

The initial stage (A) is a quasi-stationary portion of a major front such as the polar front of the North Atlantic. The warm air south of the front is shown as having a westerly flow, the cold air to the north an easterly flow. However, this is not the only pattern providing the required cyclonic wind shear since both air masses may have westerly flow, as long as the warm air moves faster than the cold. The next stage (B) shows the beginnings of an eddy superimposed on the front, which is the 'birth' stage now differentiated into a warm and a cold front. The central pressure begins to fall at this time and one or two closed isobars may now be drawn on the synoptic chart to denote the incipient depression. In (C), the system has deepened and the amplitude of the wave-like form on the ground has increased to give a substantial warm sector (Plate 13.1). This is the stage of the

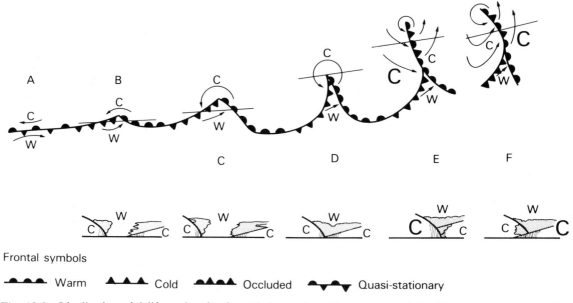

Frontal symbols

▬●●● Warm   ▲▲▲ Cold   ▲●▲● Occluded   ●▼●▼ Quasi-stationary

Fig. 13.3 Idealised model life-cycle of a frontal depression. Stages are explained in text: w – warm air; c – cold air; C – colder air; cloud masses shaded.

Plate 13.1   Satellite view (NOAA5) of a wave depression, North Atlantic, 23 September 1976.

stable 'frontal wave', already described, which may develop no further.

The cross-section diagrams depict the composite character of the system, so that an observer stationed on the line of cross-section will experience, successively, the warm-front weather sequence, the warm sector, the cold-front sequence and finally the return of the cold air. With further development, which is more usual, the amplitude of the warm sector increases and the low deepens until the mature stage is reached (Fig. 13.1 (c)). During this time, the low has not only developed but it has also progressed eastwards or northeastwards in the general southwesterly flow of these latitudes. In the North Atlantic, it is common for such a low to appear first off the northeast coast of North America and to have reached mid-Atlantic at the mature stage. Its further progress towards the European continent is usually attended by the process of elimination of the warm sector, which is known as *occlusion*.

Occlusion occurs because the cold front of such a system has an inherent tendency to move faster than the warm front and is destined, under most

119

circumstances, eventually to overtake it. With an active warm front, the warm air mass moves faster than the cold, encroaching upon it and rising up over it. Meanwhile, at the following cold front, the cold air mass moves faster than the warm, undercutting it. It follows that the cold front must catch up with the warm front (Fig. 13.3, stage D), a process which begins at the tip of the warm sector and works its way along, progressively forcing the warm air off the surface. Eventually, two limbs of the cold air mass have joined forces and, in a fully occluded depression, the warm air is found only aloft (E, F).

It was mentioned briefly in Chapter 10 that the juxtaposition of warm, light air and cold, heavy air at a front represents a high concentration of potential energy. The occlusion process clearly ensures that denser air underlies less dense air and, as this occurs, the potential energy becomes available for conversion into kinetic energy, which is manifested by increasing vigour of the circulation. The stage of near-complete occlusion is also that of maximum deepening and strongest winds. From this point on, the disturbance begins to 'die' as surface friction takes its toll of the energy of circulation.

It might appear that the forcing aloft of the warm-sector air removes all frontal characteristics from the system, at least as far as the experience of a ground-based observer and appearances on the weather map are concerned. That this is rarely so is because the cold air mass, which can be assumed to be homogeneous at stages A and B of Fig. 13.3, has become differentiated into two portions by stages E and F. The imposition of a circulation directs the air into different paths and ensures different modifications. Temperature and therefore density contrasts appear within the once-uniform cold air which result in the formation of a new frontal surface extending to ground level.

This is the *occluded front* (or occlusion). Reference to Fig. 13.3 shows two possibilities, but the most common circulation of the cold air is as in E. In the rear of the system, the cold air arriving from a northwesterly direction has been freshened by its recent path and is colder than the air ahead of the warm front (suggested by large C and small c in the diagram). As a result, the colder air undercuts its own leading portion, giving a so-called *cold occlusion*. Alternatively, the air behind the cold front is directed south into the circulation of a trough and returns (usually as Pmr air) northwards. The rear limb is then warmer than the forward limb (F) and the result is a *warm occlusion*.

The weather associated with an occluded depression depends partly on the type of occlusion. Although there is now only one surface front, rain may fall for a while from the lifted warm air cloud (the *upper front*) and occluded lows of whatever type may give protracted rain, especially if slow-moving. Eventually, the upper front loses its moisture supply and the weather pattern depends on the surface front so that, for example, the cold occlusion becomes indistinguishable from a cold front. A fully occluded depression is near the end of its life-cycle. The basic mechanism fails, the latent heat supply is exhausted, the low 'fills' and the winds slacken. Satellite photography reveals how the occluded front becomes tightly coiled round the original centre. The cloud pattern, initially so strongly organised, becomes diffuse and frontal rain is replaced by occasional showers from patchy cumulus cloud. Ultimately, the low is undetectable on the synoptic chart. Many depressions are partially or wholly occluded by the time they reach western Europe: these may have travelled 5000 km across the Atlantic in perhaps three days.

13.4 Complex frontal lows

Close acquaintance with weather maps over a long period reveals an almost bewildering number of variations on the simple theme of the preceding section. A common occurrence is the development of a *secondary low* within the circulation of a parent or primary low. This may remain the minor partner in the system or may deepen to the same extent as the primary, giving a complex area of low pressure with the two centres circling each other in a cyclonic direction. Sometimes, the secondary appears at the tip of what remains of the warm sector of a partially occluded primary – the so-called *triple point* where cold, warm and occluded fronts meet. On some occasions, this secondary develops and moves away rapidly eastwards as a *breakaway depression*. On others, the secondary deepens *in situ* to become the main centre, while the primary degenerates into a trough of low pressure containing the occluded front. This feature may be carried round in the circulation of the dominant centre to follow behind the cold front as a *back-bent occlusion*, thus adding another facet to the weather sequence.

Secondaries often form on the trailing cold front of an occluded depression. These *cold-front waves* are sometimes stable, progressing as a ripple along the cold front for a while before fading. Often, they are unstable, developing into a system as vigorous as the primary and undergoing a similar life-cycle. Repeated development of this kind gives rise to a so-called *family of depressions* often seen strung out along the Atlantic polar front. It typifies the high-index mode of circulation (section 11.4) and provides characteristic unsettled weather for regions such as the British Isles, which may be visited by each member of the family (which may number three, four or five) in turn. However, successive lows in the sequence tend to track further south than their predecessors and eventually the polar front trails into subtropical latitudes where it frontolyses under the influence of the prevailing high pressure. Under these circumstances, polar air finds its way into tropical circulations, one of many ways in which heat exchanges are effected across the lines of latitude.

Depressions may possess a double-frontal structure. Often, this takes the form of a *secondary cold front* following the main cold front 100–300 km behind and usually the more vigorous of the two. A depression passing eastwards in the North Atlantic may draw very cold air from the Greenland ice-cap into its circulation. Frontogenesis now exists between this fresh Arctic continental air and the more modified Pm air already present in the disturbance and creates the secondary cold front. Sometimes a large Atlantic low may involve both the polar and the Arctic fronts so that the system boasts a complete double frontal pattern, incorporating three air masses. North American synoptic analyses sometimes reveal triple-frontal structures involving four air masses, and some of these complexities are shown in Fig. 11.5(b) and 11.6(b).

13.5 The tropical hurricane

Intense tropical low pressure systems are given different regional names – *hurricanes* in the western North Atlantic, *cyclones* in the Indian Ocean, *typhoons* in the west Pacific and *willy-willies* in northern Australia. However, for convenience, they will all be referred to as hurricanes in the following discussion. They are extremely vigorous disturbances, with winds regularly achieving hurricane force (that is at least 33 m s^{-1}), a speed only occasionally matched in middle-latitude systems. By no means all tropical rain-giving disturbances are hurricanes, since there are also easterly waves and incipient hurricanes that never fully develop. The number of mature hurricanes is relatively small and, according to Simpson and Riehl (1981), they number between 75 and 100 in any year, though their impact is considerable.

They tend on average to be about one-third of the size of middle-latitude systems, although the actual storm diameter varies from 100 to 800 km, taking the entire cloud envelope into account. The central pressure averages about 950 mb, no lower than in some deep temperate lows, but it can be in the range 860–900 mb. On the synoptic chart, the isobars have a tight concentric pattern (Fig. 13.5) and the wind speeds they represent can sometimes exceed 75–100 m s^{-1}. On satellite photographs, the mature hurricane shows up as a roughly circular cloud mass with spiral cloud 'streets' attached. Furthermore, in the centre of the thickest cloud, a conspicuous dot proclaims the so-called *eye* (Plate 13.2).

The cloud structure in a hurricane is cylindrical in form, extending from a low base often to the tropopause. This cloud cylinder, which widens out in its upper reaches, consists mainly of cumulonimbus towers massed together. Also, it is characterised by spiralling bands of stratocumulus which enter at low levels and cirrus and cirrostratus which spread out at high levels. The eye is a central well or tube of largely calm, cloud-free air (except sometimes for a little broken low cloud) extending from top to bottom of the system, with a diameter of some tens of kilometres. Most of the energy of a large hurricane is concentrated in a ring within 100 km radius of the centre. In this zone, the winds attain maximum force, the cloud is thickest and most precipitation falls (at rates that can exceed 500 mm day^{-1}).

As mentioned in section 10.3, hurricanes are of composite origin. They are in part thermal systems since nearly all form over oceans with surface temperatures of 27 °C or more, especially in the late summer–early autumn season when the ocean waters are at their warmest. While such temperatures are necessary for the lapse rate steepening and deep free convection (sections 3.1 and 3.2) inherent in any hurricane, they alone cannot explain why relatively few such systems form and why only in certain preferred locations. Satellite imagery, which has spectacularly furthered the monitoring of these disturbances, shows that many begin life as close clusters of large cumulonimbus towers. These are less likely to be found over the eastern parts of the oceans where the circulation

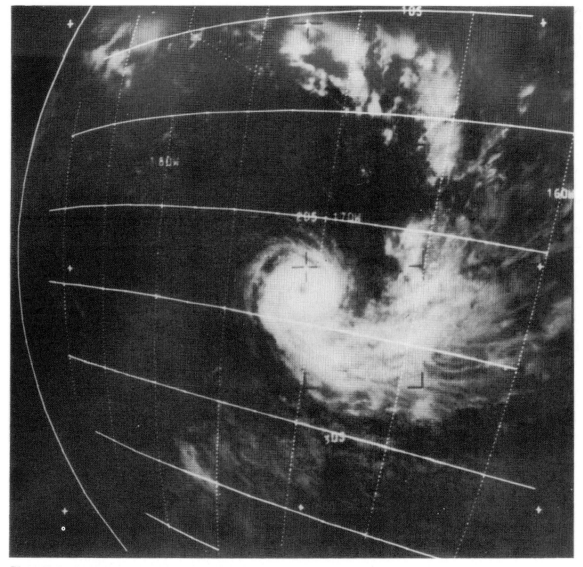

Plate 13.2 Satellite view (ESSA3) of a hurricane, South Pacific 12 February 1968.

of the subtropical highs, together with cold surface waters, combine to produce a marked low inversion, than the western parts where the inversion is absent (section 3.2). Sometimes, these cumulonimbus clusters organise themselves into disturbances which are not yet hurricanes (in the sense that they do not display hurricane dimensions, structure or vigour) but are sometimes called 'hurricane seedlings'. This stage of development requires a location at least 5–6° from the equator (where the Coriolis parameter is sufficient to sustain a circulation) and a pre-existing low pressure with attendant cyclonic shear and convergence.

This requirement may be satisfied by location on the equatorial trough (when well away from the equator) or by the presence of an old trailing cold front from higher latitudes or by the contribution of an easterly wave. The existence of low-level convergence, of whatever origin, must also be matched by a divergent flow at very high levels (above about 12 km or the 200 mb pressure level).

Easterly waves are shallow non-frontal troughs of low pressure, first noted in the tropical North Atlantic and later in the Pacific. They move westwards at speeds of 5–7 m s^{-1} in the deep trade wind flow of the southern limb of the Azores or

North Pacific High. Furthermore, they are recognised on the weather map by the poleward tongueing of the isobars or streamlines. Ahead of the trough-line, the trade wind inversion is particularly low and the weather is fine. Behind it, the inversion is much higher and free convection can take place here with large cumulonimbus building in the deep, unstable, moist air and depositing heavy rain. The origin of these waves is not clear but some of them generate hurricane seedlings. However, only about 10% of seedlings survive to become mature hurricanes.

Unlike the middle-latitude depression, in which the provision of energy rests initially on the confrontation of warm and cold air, hurricanes form in homogeneous air. However, the preconditions for hurricane development already outlined seem to provide maximum opportunity for the creation of a persistent and vigorous convective system, which requires a chimney or core of rising air warmer than its surroundings. Plainly, this depends largely on an abundant release of latent heat as condensation occurs into already warm air, richly supplied with moisture from underlying warm ocean waters, including the plentiful spray thrown up by strong winds. This assessment is complicated by the presence of the eye. The eye has been interpreted as the result of subsidence and adiabatic warming of high-level air drawn down into the low pressure at the heart of the vortex. This subsiding air is cloud-free but, because the subsidence usually ceases about 1500 m above the surface, there may be some thin cloud in the more moist air below this level.

Whatever the origin of the eye, the centre of the system appears to have a double chimney structure. The inner chimney contains warm descending air and the surrounding ring contains rising air warmed by latent (and some sensible) heat release from the surface. The establishment of the warm core, however constituted, marks the mature stage of the hurricane. The essential structure of the mature hurricane is depicted in Fig. 13.4. The inflow into the system is usually concentrated in the lowest levels, below about 2000 m (or the 800 mb pressure level). This feeds a vortex which is strongly cyclonic up to about 7–8 km (400 mb), above which the rotation lessens and finally reverses to give strong anticyclonic outflow above about 12 km (200 mb). Compensating downcurrents of cold high-level air are assumed to occur around the edges of the storm, beyond the confines of the diagram.

The tracks of most hurricanes are dictated by the circulation of the subtropical highs in which they are embedded, although any one hurricane is apt to pursue a highly erratic path. Initially, the systems are steered from east to west in the trade wind flow. This, incidentally, introduces an asymmetry in the wind pattern, with the winds north

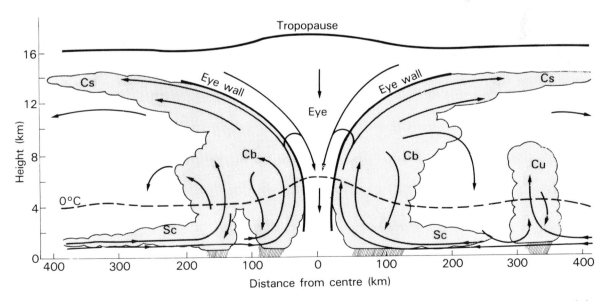

Fig. 13.4 Suggested vertical structure of a mature hurricane. Air-flow patterns, main cloud types, precipitation and freezing-level are indicated. Cloud-type key as for Fig. 12.2, Cb (cumulonimbus).

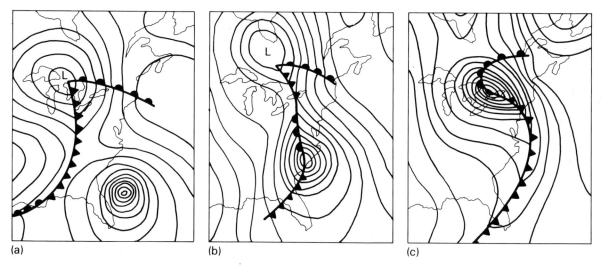

(a)   (b)   (c)

Fig. 13.5 The transformation of Hurricane Hazel into an extratropical depression. Surface charts for: (a) 15 October 1954, 03.00 GMT; (b) 15 October 1954, 15.00 GMT; and (a) 16 October 1954, 03.00 GMT. (Adapted from Palmén and Newton 1969)

of the centre (i.e. those blowing with the general flow) stronger than those to the south. In contrast to the winds within it, the movement of the whole system is slow, usually 5–7 m s^{-1}.

Eventually, the hurricane tracks curve polewards around the western edge of the subtropical high and the systems now decay as they move over the colder waters of high latitudes or over land, where the latent heat supply is finally cut off. Despite a general tendency to remain over water, many hurricanes move over coastal regions and sometimes over a long land track, creating considerable havoc. Some Atlantic systems cross Central America and are regenerated in the Pacific. Hurricanes decay from the base up, with the upper vortex sometimes lingering at high levels. Simpson and Riehl (1981) cite the case-history of Camille, an Atlantic hurricane of the 1969 season, which developed when an easterly wave moved under the detached anticyclonic outflow of a previous hurricane.

A number of hurricanes which would normally end their life-cycle having moved over colder surfaces in middle latitudes actually take on a new lease of life by transformation into frontal depressions. A famous example is that of hurricane Hazel, part of the unusually severe hurricane season of autumn 1954 in the western Atlantic. The sequence of events is portrayed in Fig. 13.5, where the three charts are 12 hours apart. Chart (a) shows Hazel off the Florida coast, while a middle-latitude frontal low is centred over the Great

Lakes. In (b), Hazel has moved northwards and incorporated the trailing cold front into its circulation. In (c), it has amalgamated with the pre-existing centre and its regeneration as an extratropical depression is complete. Such rejuvenated systems usually cross the Atlantic behaving as typical (but vigorous) frontal lows and giving little hint of their exotic origin.

13.6 Retrospect

There are important differences and similarities between temperate lows and tropical hurricanes. An obvious difference is that the former are, in general, the beneficial rain-bringers of middle latitudes, while the excessive rains, the violent winds and the coastal flooding associated with hurricanes bring hazards and disasters to certain tropical regions. The Gulf coastlands of the USA and the Bay of Bengal are particularly vulnerable to *surges* (excess tidal height above computed astronomical level) raised by hurricanes or cyclones. These surges are in part due to the change of pressure (sea-level rises by 1 cm approximately for every 1 mb fall in pressure), but much more so to the stress of the wind impelling the water against a coastline (see Perry and Walker 1977 and Lockwood 1974).

It should be added that very deep temperate depressions can also cause storm surges, the serious flooding on both sides of the southern North Sea in February 1953 being a case in point (Barnes and King 1953). The Thames barrage below London, completed 30 years later, is intended as a protection against such events. The harmful aspects of the hurricane are hardly offset by other benefits. Most of the useful rainfall in low latitudes derives from less intense and destructive disturbances and much hurricane rain is lost in rapid runoff and flooding (section 7.3).

In energy terms, both systems rely on the potential energy inherent in rising warm air, aided by latent heat release, which is eventually converted into the kinetic energy of vigorous circulation. Whereas the mechanism of the middle-latitude low is triggered by the juxtaposition of warm and cold air at the surface, the tropical hurricane must organise a warm core through the processes earlier outlined. The energy supply of the middle-latitude low is largely shut off when occlusion is complete: thereafter surface friction and upper-air changes hasten the decay. The hurricane 'mechanism' breaks down when the latent heat supply is cut off through passage over cold water or land.

The hurricane is in essence an intense, free convective system with ascending air in the centre and descending air around, whatever the complications introduced by the eye and, for that matter, precipitation downdraughts within the cumulo-nimbus clouds (Ch. 14). It is less easy to visualise the middle-latitude low in the same light, with its marked asymmetry and the part played by the frontal jet stream as an outlet for air removal.

Yet, if the frontal low is considered together with the cold high that usually follows close behind, the elements of a forced convective system are there. Ascent is the function of the low, organised broadly along the frontal belts, vigorous in the early stages when both fronts are likely to be of the ana variety. The high is an area of descent, of subsidence in the cold settling air, a tendency which often extends into the structure of the rear of the low as the cold front transforms into a kata-front.

Key topics

1. The problems of modelling mid-latitude depressions.
2. The controls on cyclogenesis.
3. The occlusion process and associated energy transformations.
4. The genesis and behaviour of hurricanes.

Further reading

Barry, R. G. and Chorley, R. J. (1982) *Atmosphere, Weather and Climate*, Methuen; London and New York.

Lockwood, J. G. (1974) *World Climatology. An Environmental Approach.* Edward Arnold; London.

Pedgley, D. E. (1962) *A Course in Elementary Meteorology.* HMSO; London.

Chapter 14 Mesoscale circulations

There is no general agreement concerning the scale limits of the so-called *tertiary* or *mesoscale* (or simply *local*) circulations. The problems of scale delimitation in space and time are discussed by Oke (1978) and Barry (1970), among others. Horizontal distances of 1–100 km are sometimes specified but, since these circulations are often linked to terrain features such as mountains and valleys, their dimensions obviously vary widely. Usually, quite a shallow layer of air is involved but, on the other hand, thunderstorm cells can extend vertically to the tropopause. Land and sea breezes and mountain and valley winds are parts of thermal circulations which can develop only when general gradients are weak, i.e. in anticyclonic conditions. Other circulations, such as that involving the föhn wind, are mechanically forced and depend on particular relationships of underlying relief and synoptic situations. Whatever their scale or origin, such circulations may be highly significant to human beings and sometimes represent the difference between the regional weather forecast and the actual weather experienced in a given locality.

14.1 Thermal circulations – coastal breezes

Land and sea breezes are the familiar components of thermal circulations set up in coastal areas, due to the differential response of land and water to daytime heating and nocturnal cooling processes. Surface water temperatures show little diurnal variation, compared with considerable land surface fluctuations, because of the high heat capacity of water, its mobility, the depth of penetration of incoming radiation and evaporation at the surface. Under suitable conditions, a sufficient thermal contrast develops between adjacent land and water strips which generates pressure differences and a consequent circulation (section 10.3), which necessarily reverses in direction as between day and night. The daytime sea breeze, directed from relatively cool sea to relatively warm land, may reach speeds of up to $5-7$ m s^{-1}. The oppositely directed land breeze at night, when less energy is available, rarely achieves half such speeds and may often be reinforced by slope effects.

A few experiments with balloons or kites have confirmed the presence of return flows at height (usually about 500 m in middle latitudes, but often twice that in the tropics) to complete the circulation. Other investigations have shown that the sea breeze mechanism is rather more complex than is suggested by the idea of a static thermal cell astride the land–sea boundary. The arrival of a sea breeze at a meteorological station can often be demonstrated by a fall in temperature, an increase in relative humidity and a change in wind direction and speed and, if autographic records are available, the timing of this event is easily established.

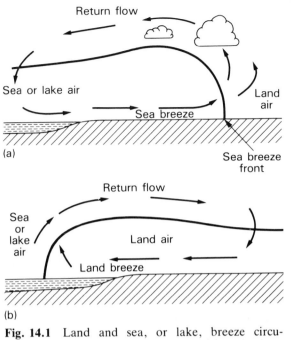

(a)

(b)

Fig. 14.1 Land and sea, or lake, breeze circulations: (a) daytime; and (b) night-time. (Adapted from Oke 1978)

Such studies have shown that the sea breeze, having sprung up at coastal sites around mid-morning, then penetrates inland as a shallow wedge of cooler air behind an aptly named *sea breeze front*. This encourages uplift in air which is already unstable and the development of cumulus clouds which move seawards in the return flow (Fig. 14.1). They do not survive in the subsiding air over the cooler sea and thus a line of cumulus during the day often betrays the presence of the coastline beneath it. Maps of *isochrones* (equal time lines) of particular situations in southern England have shown the sea breeze to be still detectable as a weak feature 80 km inland (Simpson 1964). The cooling effect of the sea breeze is welcome on some warm coastlands: the so-called Fremantle 'doctor' of Western Australia is a well-known example.

14.2 Thermal circulations – slope winds and along-valley winds

In the very different conditions of mountainous topography, aspect and relief nevertheless combine to create a regime of diurnally reversing wind circulations which resemble land and sea breezes in certain respects. Differential receipt of solar radiation during the day, among slopes of different aspects, creates marked horizontal temperature gradients (much depending on the form and orientation of ridges and valleys) which in turn give rise to pressure gradients. The air above a sunlit valley slope will be warmer than air at the same level above the valley floor. The pressure gradients set up will move air at ridge level from valley side to above the valley floor and surface air will move from the floor up the valley sides as a lateral upslope or *anabatic* wind: air will also sink above the valley floor to close the circulation (Fig. 14.2(a)). Cumulus clouds often form at ridge levels in the buoyant upslope currents. Sometimes, if these clouds grow sufficiently, their tops

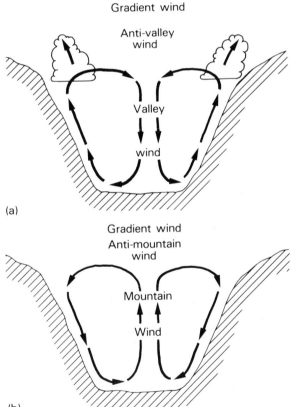

(a)

(b)

Fig. 14.2 Mountain and valley winds. The view is up-valley. Arrows denote the slope winds: (a) daytime – anabatic slope winds, valley wind into, and anti-valley wind out of, the page; (b) night-time – katabatic slope winds, mountain wind out of, and anti-mountain wind into, the page. (Adapted from Oke 1978)

127

may be seen to distort as they are blown by the (necessarily light) prevailing gradient wind. These lateral circulations are reinforced by a longitudinal circulation of *valley wind* (blowing up-valley from the neighbouring plain) and return *anti-valley wind* at heights, which forms for the same reason. This is associated with the temperature difference between warmed air above the valley head and cooler air at the same elevation above the plain.

At night, the valley sides cool first and temperature gradients are reversed. Chilled air drains downslope by gravity as a katabatic flow (section 5.2) and, to maintain continuity, air rises above the valley floor (Fig. 14.2(b)). The related longitudinal circulation comprises the so-called *mountain wind* and the return *anti-mountain* flow at ridge levels. These circulations are typical of clear nights, and related long-wave radiation loss, with little general wind. The tendency of cold air to gravitate to the lowest levels (which will happen with quite modest relief) is responsible for the phenomena of *valley inversions* and *frost* and *fog pockets* (section 5.2). So prevalent are these inversions in certain valleys, that there is a favoured *'thermal belt'* between the more exposed upper slopes and the usual top of the cold air 'lake', which finds a response in the siting of settlements and orchards.

These winds are light (rarely exceeding 5 m s^{-1}) but are significant for human occupance. The above outline generalises a wide variety of cases. For example, the *glacier wind* (cold air fed off ice or snow surfaces) is a very shallow gravity flow which reinforces the mountain wind at night and may persist during the day, undercutting the warmer up-valley wind. On the other hand, in some regions where all slopes are snow-covered in winter, these diurnal circulation systems tend to be insignificant. Fuller discussions will be found in Defant (1951) and Barry (1981).

14.3 Thermal circulations – urban winds

Urban areas tend to be warmer than their rural surroundings, especially at night in the summer season, for a number of reasons. The most notable one is associated with the capacity of urban materials to retain heat gained during the day from the absorption of solar radiation (section 3.1). However, secondary heat sources are also important and include the 'anthropogenic' heat released

from a variety of combustion processes and the 'blanketing' role of smoke in reducing nocturnal radiation heat loss, with its contribution to the so-called 'greenhouse effect' (section 2.2). On calm and clear nights, this well-known heat-island effect may reach 6–8 °C and it would be expected that such temperature gradients would generate thermal circulations comparable to those already discussed. Some evidence that this is so is provided by studies establishing the reality of convergent air flow into city centres on clear calm nights. Examples of this *'country breeze'* (following the convention that winds are named from their source) come from Frankfurt, Munich, London, Toronto and Asahikawa, Japan. Evidence of up-currents over city centres come from balloon investigations in New York and St Louis in the USA and various examples of convectional rainstorms localised over cities. It must be remembered that only a city set in a featureless plain can display purely urban climatic characteristics and, in most cases investigated, other mesoscale influences are inevitably also involved (Chandler 1965; Findlay and Hirt 1969; Landsberg 1981).

14.4 Thunderstorm circulations

Every large cumulus cloud betrays a thermal cell, triggered by heating from below, lapse-rate steepening and free convection (sections 3.1–3.2). Furthermore, whereas up-currents are concentrated within the cloud, slow compensating down-sinking is more widely spread in the surrounding clear air, and low-level inflow and high-level outflow are found to complete the circulation. Under highly unstable conditions (sometimes aided by frontal or orographic lift), the rising air may become organised into a large warm air 'chimney' (often some kilometres in diameter), with the energy of ascent boosted by latent heat release at both condensation and freezing levels. This is an incipient thundercloud or cumulonimbus, from which (as described in section 5.3) heavy precipitation, often with hail, and electrical activity may be expected. Less obvious accompaniments may escape the casual observer but they are detected by recording instruments at the time of, or sometimes preceding, the onset of heavy rain. They include a marked fall in air temperature, a rise in relative humidity, an upward 'jump' on the barograph and sudden gusting winds, with a more or less marked change of direction.

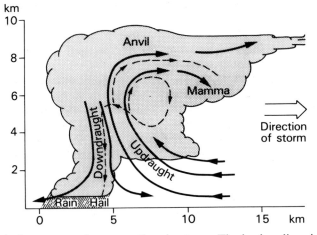

Fig. 14.3 Suggested vertical structure of a severe thunderstorm. The broken lines indicate the paths of hail-stones. (Adapted from Ludlam 1961)

These features are explained by a model of thunderstorm development produced by Byers and associates (Byers and Braham 1949), which comprises three stages. The first or *cumulus stage* is as described above, with *updraughts* throughout the cloud, commonly reaching 10 m s⁻¹ (and often twice that). They are capable of holding in suspension the ice crystals already forming through the Bergeron process. Eventually, some of these grow sufficiently in weight to overcome the terminal velocity (section 5.3) and fall through the cloud as precipitation and, in the process, drag down cold air by friction. This cold air *downdraught*, kept cold by evaporation of raindrops and associated latent heat use below the cloud base, reaches the ground with the precipitation. Here, it spreads out, often against the prevailing surface wind, to give the changes referred to above. The downdraught air often spreads beyond the area receiving precipitation to give 'ghost' effects, sometimes many kilometres from the actual storm. This stage in the sequence, in which both updraughts and downdraughts coexist within the cloud, is the so-called *mature stage*. Downdraughts then become established throughout the cloud, which rains itself out in the final or *dissipating stage*. According to Byers, a typical life-cycle takes about an hour.

Frequent experience of very severe and much more protracted thunderstorms has thrown doubt on the adequacy of the Byers model. Attention has focused on the downdraught, which is obviously an essential feature of thunderstorm development. However, it appears to play a self-quenching or negative feedback role in that the spread of cold air at ground level cuts the warm air supply to the 'chimney'. Detailed studies of several intense storms, including one at Wokingham, England in July 1959 (Browning and Ludlam 1962) have led to the conclusion that the updraught and downdraught are in fact separated. The updraught is back-tilted while the downdraught of cold heavy air is more nearly vertical. The forward surge of downdraught air at ground level then assists in the thrusting up of incoming air, so that the storm is self-fed rather than self-destructive (Fig. 14.3). This diagram also shows typical circuits within the cloud that allow the growth of large hailstones (section 5.3).

The requirement of an inclined updraught points to a strong wind shear with height and this suggests that a frontal situation is particularly favourable to intense thunderstorm development. To return to an earlier analogy, a frontal jet stream aloft provides ideal conditions for the warm air 'chimney' to 'draw well'. Lines of such thunderstorm cells, known as *squall lines*, are often observed advancing ahead of a cold front. They are thought to be triggered by downdraught scoops from initial thundery outbreaks at the front itself, which are particularly common in the central plains of the USA.

14.5 Tornadoes

Tornadoes are further accompaniments of severe thunderstorms. They are very intense, rotating mesoscale systems commonly 100 m, but

129

occasionally up to 1 km in diameter, in which wind speeds may approach 100 m s^{-1}. The vortex is visible as a writhing *funnel cloud*, which apparently originates in a cumulonimbus base and builds down towards the ground. The cloud evidently represents saturation as a result of adiabatic cooling (section 3.3) due to the marked pressure drop (which may be 50–100 mb) in the centre of the vortex. It is this abrupt fall of pressure, as well as the high wind speeds, that explains the destructive power of the tornado, capable of uprooting trees, de-roofing buildings and lifting vehicles, albeit in a somewhat narrow swathe.

The general conditions for tornado development are those that favour thunderstorms and squall lines, usually a particularly active cold front lifting warm moist unstable air, aided by the jet stream aloft. In such conditions, some 150 of these 'twisters' are reported on average annually in the Mississippi lowlands of the USA, mainly in spring and early summer with 'peak' thermal contrasts between Pc and Tc air along active cold fronts. The precise mechanism of formation and the sources of the prodigious energy involved are by no means understood. It has been suggested that there are preferred areas below the edge of the cumulonimbus base where interplay between outspreading downdraught air and inblowing air would concentrate the cyclonic vorticity already present in the system. Equivalent systems over the sea are called *waterspouts*, which are able to suck sea spray into the base of the pendant cloud.

14.6 Forced circulations

Irregularities of terrain are capable of imposing circulations on the flow of air which owe nothing to thermal influences. With all but light winds, an abrupt obstacle such as an isolated hill (or, for that matter, an isolated building) will cause marked separation of flow. This establishes a turbulent eddy about a horizontal axis on the lee side of the obstacle and such turbulence sometimes also develops on the windward side. In the case of tall slab buildings, such eddying motions often create nuisances by causing zones of strong gusty winds at pedestrian level or by bringing down pollution to the ground. However, these circulations are on a smaller than mesoscale. In hilly terrain, similar eddies form on the lee side of hills and other breaks of slope like cliffs or escarpments. In a narrow valley, a regional wind blowing along the val-

ley axis may be accelerated as a *funnelled wind*, but a cross-valley wind is likely to produce a lee eddy below the windward ridge. On a smaller scale, the same effects may be observed in a typical 'canyon' street in a city.

In more mountainous terrain, forced circulations on a larger horizontal scale occur frequently. A study by Förchgott (the original 1949 paper is in Czech, but a summary will be found in Corby 1954) relates the behaviour of air flow over a simple elongated ridge (perpendicular to the wind direction) to the wind strength and vertical profile, under stable or neutral conditions. With very light winds changing little with height, the air flow is smoothly adjusted to the shape of the ground and vertical motion is minimal (with laminar streaming). With stronger winds, increasing somewhat with height, the air descends on the lee side of the ridge in a *standing eddy*. With still stronger winds, the lee eddy disappears and the air is sent into a series of oscillations downwind of the ridge. These are known as *lee waves* and have a wavelength (crest to crest) comparable to the width of the mountain ridge. Trains of up to

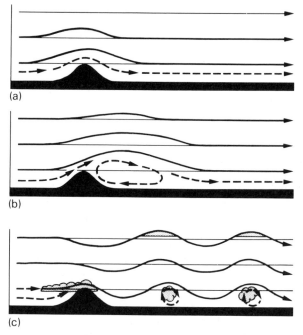

Fig. 14.4 Some types of air flow over ridges: (a) undisturbed (laminar) streaming; (b) standing eddy streaming – one stationary lee eddy; (c) wave streaming – standing waves with rotors: lenticular clouds, crest cloud and rotor (bar) clouds. (Adopted from Corby 1954)

Plate 14.1 Altocumulus lenticularis cloud near Palmerston North, New Zealand.

six waves have been observed over distances of 60 km and sometimes, large turbulent eddies known as *rotors* form under individual wave crests (Fig. 14.4).

The presence of these wave features is well known to glider pilots, but is also often made visible by the clouds that form if the air being lifted into the wave crests is not far from saturation. Under these conditions and assuming stable stratification, each wave is marked by a characteristic *lenticular* (lens-shaped) cloud (Plate 14.1). These clouds are stationary while the meteorological situation remains unchanged and betray the *standing* waves tied to a particular mountain feature. One such cloud – the so-called *cap* or *crest* cloud – is seen on or above the mountain ridge itself (depending on the lifting condensation level of the air involved). This is somewhat different from the *banner cloud* which forms on the lee side of isolated peaks. Here, the separation of flow causes a pressure reduction which encourages upward air movement. The resultant adiabatic cooling gives rise to condensation and a cloud tied to, but streaming downwind from, the mountain peaks.

Tiers of wave clouds up to cirrus levels are sometimes observed, betraying various moist layers in the air mass and showing that the waving process may disturb much of the troposphere. Rotors are also often marked by clouds, but *rotor* or *bar clouds* have a ragged round appearance, quite

unlike the lenticular wave clouds. In the northern UK, the Crossfell Edge escarpment overlooking the Eden Valley to the southwest offers excellent examples of hill wave features whenever the wind is strong from the northeast: the 'Helm cloud' caps the escarpment and the 'Helm bar' sits under the first lee wave downwind (Manley 1945).

14.7 Föhn winds

The föhn (*foehn*) wind of the Alps, the Chinook of the Rockies, the *Santa Ana* of southern California and the *Zonda* of Argentina are some of the regional names given to warm, dry winds descending in the lee of high ground (Plate 14.2). Such winds have been known to raise temperatures by 25 °C in a hour, reduce relative humidity to below 10% and gust to gale force. Their impact on the lives of valley dwellers in these mountain regions has long been respected, since they have frequently caused avalanches and the destruction of entire villages by fire. However, the föhn mechanism is by no means understood and the term has been used to denote rather wide-ranging phenomena. Sometimes, the föhn is classed as a 'fall wind', which makes it an unlikely companion of the cold 'bora' or the katabatics, which are true gravity winds.

Plate 14.2 Föhn bank over Mt Sefton, New Zealand.

The classic explanation of the föhn dates back to Hann (1866) and has already been outlined (section 3.3) in connection with forced uplift and adiabatic temperature changes. In essence, it assumes saturated-adiabatic ascent on the windward side of a mountain and the dry-adiabatic descent of stable air on the lee side. This explanation necessitates precipitation on the windward side, which often happens, and is further supported by the föhn wall or föhn bank (Plate 14.2). The gain in temperature at the foot of the lee slope is thus the latent heat released on condensation. A number of difficulties prevents a complete acceptance of the classical mechanism to cover all föhn occurrences (Brinkmann 1971). Some of the temperature increases recorded are too great to be explained by that process and perfectly good föhns occur without precipitation on the windward side. Furthermore, Crowe (1971) questions why warmed air should descend on the lee side, displacing colder, denser air.

It seems inevitable to regard the föhn as basically part of a forced circulation, i.e. as a lee eddy or a lee wave wind, like the Helm wind discussed earlier which descends the escarpment slope of Cross Fell before rising into the first downwind wave. The rise in temperature may indeed derive from the often-quoted classical mechanism, but could occur equally if air of high potential temperature at summit level (i.e. air warm for that elevation) were brought down, warming dry-adiabatically in its forced descent. On this view, air ascent up the windward slope is not an essential part of the story. In the Rockies, instances have been identified of Pacific air being advected above a shallow pool of polar air before being brought down as a warm Chinook.

Confusion, however, still exists. Some writers refer to an 'anticyclonic föhn', with an obvious reference to anticyclonic subsidence. Whatever the mechanism, although the föhn is effective at the mesoscale, such a cross-mountain wind can blow only in response to a suitable pressure gradient, i.e. in certain synoptic-scale situations. In Alpine valleys, the most dramatic temperature rises occur with the south föhn, when pressure is high to the south and low to the north, and Mediterranean air is involved. The reverse situation produces the so-called north föhn, in which case temperature increases are minimal. A full discussion of the föhn and all aspects of mountain weather will be found in Barry (1981).

14.8 Retrospect

The circulations and phenomena outlined in this chapter are of undeniable significance to living organisms. However, they are on a scale that often allows them to slip unnoticed through the lacunae in a network of meteorological stations dedicated to recording what is representative of the region. But they can be studied by special investigations and inferred from personal observation. Land and sea breezes and mountain and valley winds derive their energy from thermal gradients: so basically do thunderstorm cells. Forced circulations in mountainous regions represent diversions of existing kinetic energy into different channels. The thermal circulations may be more or less closed off from the regional or synoptic-scale circulations.

However, they are still basically dependent on them in the sense that they are favoured in locations and during periods in which anticyclonic conditions prevail. Forced circulations are more closely linked to synoptic events.

Key topics

1. The characteristics of diurnally reversing thermal circulations.
2. The influence of topography on local air flow.
3. The modification of air flow by urbanisation.
4. The mechanisms involved in the development of thunderstorms.

Further reading

Barry, R. G. (1981) *Mountain Weather and Climate*. Methuen; London and New York.

Landsberg, H. E. (1981) *The Urban Climate*. Academic Press; New York and London.

Oke, T. R. (1978) *Boundary Layer Climates*. Methuen, London; John Wiley and Sons, New York.

Part III
Sediment transport systems
Chapters 15–23

Geological forces lead to the differential uplift of the earth's crust but, even during this uplift, the action of air, water and ice initiates the transfer of the earth's materials to lower levels. In the following nine chapters, the characteristics of earth movements, rock types, weathering and earth surface materials are discussed, although emphasis is placed on the operation of the above sediment transfer systems. Because of the complexity of their operation both in time and space, an understanding of these systems remains far from complete. Nevertheless, in recent years field, laboratory and theoretical studies have provided a firm framework for their investigation.

Glacier sediment transfers – the Athabasca Glacier, Canadian Rockies.

Chapter 15 Introduction to sediment transport systems: the influence of endogenic processes and base-level changes

The relief of the earth is continuously under attack by erosive forces, which transport debris derived from rocks towards the sea. The inputs to this *sediment transport system* are numerous, with contributions from weathering, vegetation, hydrological and solar energy systems. The weathering system is important, because it leads to great reductions in the resistance of materials to erosion (Ch. 16). Vegetation exerts its influence in many ways, but it primarily acts to dampen inputs, and hence reduces erosion rates by its presence. The hydrologic cycle (section 2.1) is intimately involved with many components of the debris system not only through its role as the medium of transfer, especially in river channels (Ch. 20) and glaciers (Ch. 22), but also because of its effect on the strength of the materials being eroded (Ch. 18.). The solar energy system (section 2.2) provides inputs indirectly through the hydrologic system and through the movement of air in aeolian transport (Ch. 21), due to differential receipts of energy across the globe (Ch. 11).

The sediment transport system can be divided into a number of subsystems, each of which is a cascade system in its own right: in each cascade, inputs of energy and matter are transformed leading to the output of sediment, which is ultimately delivered to the oceans. In the slope debris system, materials are usually moved downslope to the stream channel system which provides a local *base-level*. The most important control of the slope system is its *potential energy* which results from its height above the stream at its base. Potential energy (E_p) is simply defined by the following equation:

$$E_p = mgh = Wh \qquad [15.1]$$

where m is the mass of the material, g is the acceleration caused by gravity and W is weight which equals mg.

The slope debris cascade delivers sediment to the stream channel (or glacier) system by a number of routes which are analysed in Chapters 18, 19 and 23. The operation of the stream channel (or glacier) system is itself controlled by the potential energy provided by its height above the next junction with a larger stream, which in turn has potential energy as a consequence of its height and so on until the base-level provided by the sea is reached. So far as the land areas are concerned, sea-level provides the ultimate base-level at any given time but, as discussed in Chapter 33, subsequent transport of materials can take place down to lower parts of the ocean floor. For aeolian systems, the influence of potential energy and base-levels is less important since materials can be transported in the air from depressions and are deposited on higher ground. Although this section emphasises the processes of sediment transport, it is important to stress that the land surface is not simply being passively denuded. The operation of these processes leads to the creation of landforms which themselves affect the operation of processes through numerous feedback loops. Chapter 20, for example, examines the many ways in which stream channels adjust their form to inputs of sediment and in turn affect the output of sediment.

Fundamental to all aspects of the sediment transport system is the existence of relief, from which debris can be transported. The creation of this relief is a consequence of processes operating beneath the surface, which produce differential uplift. These *endogenic systems* operating at a global scale form the subject of section 15.1. The influence of base-level provided by the sea is not constant since considerable fluctuations in sea-level have occurred in recent geological times, in a process called *eustasy*. The earth's crust also responds to changes in weight due, for example, to additions of ice or removal of materials by erosion

and this response is known as *isostasy*. Eustasy and isostasy are analysed in section 15.2 and the final section examines the effects of endogenic movements at more local scales, through the creation of folds and faults.

15.1 Endogenic changes

Estimates of current rates of erosion indicate that the earth's continents would be reduced to sea-level in about 25 million years (Ollier 1981). Since this is a relatively small time-span in geological terms (Table 15.1), then uplift of the earth's crust must therefore take place either continuously or periodically. Table 15.2 shows one set of estimates of contemporary erosion and uplift suggesting that

Table 15.1 The geologic time-scale

Era	Period	Epoch	Million years ago
Cenozoic	Quaternary	Holocene or Recent	0.01
		Pleistocene	
			1.9
	Tertiary	Pliocene	5
		Miocene	23
		Oligocene	38
		Eocene	54
		Palaeocene	
			65
Mesozoic	Cretaceous		135
	Jurassic		195
	Triassic		225
Palaeozoic	Permian		280
	Carboniferous		345
	Devonian		395
	Silurian		435
	Ordovician		500
	Cambrian		570
Precambrian			
	Formation of earth		4700

Table 15.2 Estimated rates of erosion and uplift. (Adapted from Ollier 1981)

	mm per 1000 yr
Erosion of lowlands	50
Erosion of highlands	500
Uplift rates	1 000
Lateral movements	10 000

we are currently in a period of net *mountain building*. It is now believed, from many lines of evidence, that the Quaternary (Table 15.1) is a period of considerable *tectonic* activity. An understanding of the forces leading to creation of the earth's relief requires knowledge of the geology far below the surface and Fig. 15.1 shows the normally accepted subdivision of the earth's near-surface materials.

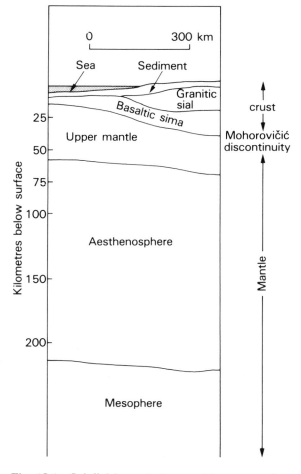

Fig. 15.1 Subdivision of the earth's near-surface materials. (After Ollier 1981)

The continents are composed primarily of granitic materials called *sial*, after its two most abundant minerals *si*lica and *al*umina, although surface sediments mask most of this granitic material. Beneath the ocean, predominantly basaltic materials called *sima* are found, after its most abundant materials *si*lica and *ma*gnesium. The sima also extends beneath the continents with the lighter sial (density about 2.7) 'floating' on top of the sima, with a density of roughly 2.9. It is now widely accepted that the relative positions of the continents have changed, and continue to change, although there remains considerable doubt about the precise mechanisms involved (Davies and Runcorn 1980). The current configuration of the earth's land areas is believed to result from the splitting up of a single continent called *Pangea*, which started some 200 million years ago in the Triassic (Table 15.1). Moreover, the process of *continental drift* occurs through the movement of relatively rigid plates and Fig. 33.4 shows the movement of an oceanic plate moving below the continental plate, as the two plates collide.

The concepts of *plate tectonics* and continental drift are now generally accepted but other explanations for the relative movements of continental land masses have been proposed, including that of an expanding earth (Carey 1976), a contracting earth and constant earth with shrinking continents (Chevallier and Cailleux 1959). Given the relative movements of the continents which appear to have taken place, some of the possibilities for mountain building appear fairly obvious. Collision of continental plates can lead to one of them being pushed up over the other, as appears to have happened in the Himalayas. Alternatively, movement of a continental plate over an oceanic plate can lead to the former being forced up, as is believed to be the case for the Andes, and further possibilities are discussed in Chapter 33. These relative movements cause folding and faulting and the whole process is termed *orogenesis*. Gravity can be an additional cause of folding since the piling up of large rock masses by lateral movements may be followed, over long periods of time, by a downward pull associated with gravitational sliding. This process is believed to be responsible for much of the intense folding in the Alps, instead of being a result of compressive forces (Rutten 1969).

Originally, it was thought that the mechanism of lateral movements and resultant folding was the direct cause of mountain building and hence was given the name orogenesis. However, there is now evidence to suggest that subsequent to folding, vertical uplift may often take place to create mountains, as in the case of the Appalachians. Hence, orogenesis as defined above is not necessarily sufficient for mountain building to occur (Ollier 1981). Horizontal forces are undoubtedly important in controlling the configuration of rocks found within mountains but the generation of relief can occur without them. Plateaux are often underlain by unfolded rocks and the suggested mechanisms for plateau formation include the addition of low-density material to the base of the crust from the mantle and the volumetric increases in the crust and upper mantle by heating or phase changes (Hobbs *et al.* 1976).

King (1962) proposed the theory of cymatogeny to explain the relief of Africa, particularly the structural warping which produced *up-doming* and *rift valleys*. In this theory, materials eroded from the continent are deposited in marginal positions and, due to the weight of these sediments, the crust is depressed allowing further deposition. The removal of materials from the continent reduces the weight of the crust, which therefore rises isostatically (section 15.2). A further mechanism for creating relief is through the action of volcanoes which brings molten materials to the surface from deep within the earth. Active volcanoes are only found in a few discrete linear zones on the earth's surface, almost all being at the margins of the Pacific, in the Atlas–Alpine–Himalayan belt, the East African rift system and the mid-Atlantic Ridge (section 33.4). The resultant forms and eruptive characteristics are a direct function of the acidity of the lava.

Basaltic lavas are basic (i.e. have low acidity) and can flow large distances, typically producing broad convex shaped domes. The best contemporary example is Mauna Loa in the Hawaiian islands which rises approximately 40 000 m above the Pacific floor, though only 12 500 m are exposed above sea-level. On continental land masses, extensive plateaux can be created by successive lava flows infilling existing topography, as was the case in the Deccan Plateau of India and the Columbia River Plateau in the northwestern USA. The basaltic lavas of the Columbia River Plateau originated in the Miocene epoch (Table 15.1). However, contemporary eruptions at the Pacific margins are very different in character, due to the greater acidity and hence increased viscosity of the lava, and their explosive nature was dramatically exemplified in the eruption of Mt St Helens, USA, in May 1982. During explosive outbursts, mixtures of solid particles in gas can move rapidly downslope like a fluid and, after multiple eruptions, the resultant deposits can eventually produce *ignimbrite plateaux*. Because of the evacuation of magma from below the surface, subsequent collapse of volcanoes can take place producing *caldera* and hence negating the previous relief-producing activity.

15.2 Eustasy and isostasy

Changes in base-level arise not only from vertical movements of the earth's crust but can also occur independently as a consequence of changes in sea-level and evidence for such change is abundant, in the form of old shorelines well above present sea-levels. These changes are termed *eustatic* and originally were thought to be manifested by parallel changes in sea-level throughout the world. However, more recent work has shown that because of variations in gravitational pull, sea-level varies substantially across the globe at any given time (Mörner 1983). The principal cause of eustatic changes in sea-level in the recent geological past is glaciation, and specifically the amount of water in frozen form which was stored in ice sheets. During the Quaternary, sea-level was probably lowered to approximately 150 m below current sea-level during the Riss glaciation, whereas during the last glaciation it probably reached approximately 120 m. These falls in sea-level resulted in large areas of continental shelf being exposed, whereas a complete melting of the Antarctic and Greenland ice sheets would lead to a rise in sea-level of about 65 m.

Some Quaternary regressions may result not from glaciation but from gravitational changes, due to factors such as changes in the earth's rate of rotation, distribution of relief and latitudinal gravitational waves (Mörner 1983). Observations in the Mediterranean region in particular suggest that earlier sea-levels in the Quaternary were at much higher levels than at present and that, superimposed on the fluctuations caused primarily by glaciation, there has been an overall fall in sea-level. On this basis, other mechanisms apart from simple glacial eustasy need to be invoked to explain sea-level changes (Goudie 1977). One possibility may be associated with eustatic changes caused by tectonic movements and erosional processes which arise from alterations in the overall distribution of the earth's materials. Erosion of land areas will ultimately cause deposition in the oceans which will displace the water and hence lead to a rise in sea-level. Consequently, during periods of mountain building, there will be a net renewal of materials from the oceans causing an overall fall in sea-level.

The weight of ice sheets causes downward flexing of the earth's crust, a process termed *glacial isostasy*, and removal of the ice during deglaciation leads to a rise in the surface. In Sweden, near the Gulf of Bothnia, and Hudson Bay, Canada, the post-glacial rise in land level has been up to 300 m and because of the very viscous character of the earth's rocks, the upward response has continued long after the ice has disappeared. *Isostatic* readjustment also takes place as a result of the erosion of continental land masses and, as discussed in the last section, is the basis of the theory of cymatogeny. Furthermore, the lowering of sea-levels during the Quaternary by eustatic change probably led to isostatic readjustment, due to the removal of the weight of water from the exposed areas.

15.3 Effects of rock structure on stratified rocks

The deformations of the earth's crust (discussed in the previous section) have a profound impact on landforms when the rocks are exposed at the surface and this is especially apparent for stratified (i.e. layered) rocks, where the resistance to erosion of different layers differs substantially. Stratiform rocks are usually sedimentary but they may be metamorphic and, in the case of volcanic lavas, they can even be igneous. Rocks which have a single angle of inclination are described as uniclinal. They result in a landform which is called a *cuesta* (Fig. 15.2(a)) because the less resistant rocks are more readily eroded and the more resistant rocks therefore form the steeper escarpment slope, since they are transported less rapidly. It should be noted that the precise mechanisms of transport do not need to be known since all that is required is a differential resistance to erosion between the two layers.

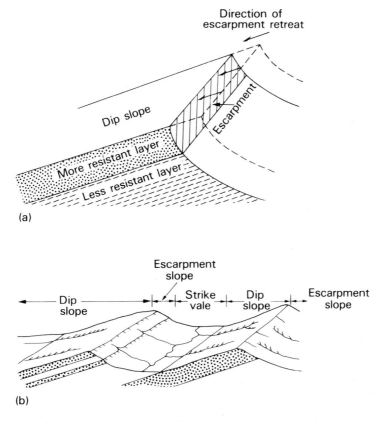

Fig. 15.2 Relation between dipping rocks and landforms: (a) elements of a simple cuesta; (b) cuesta and strike vale topography.

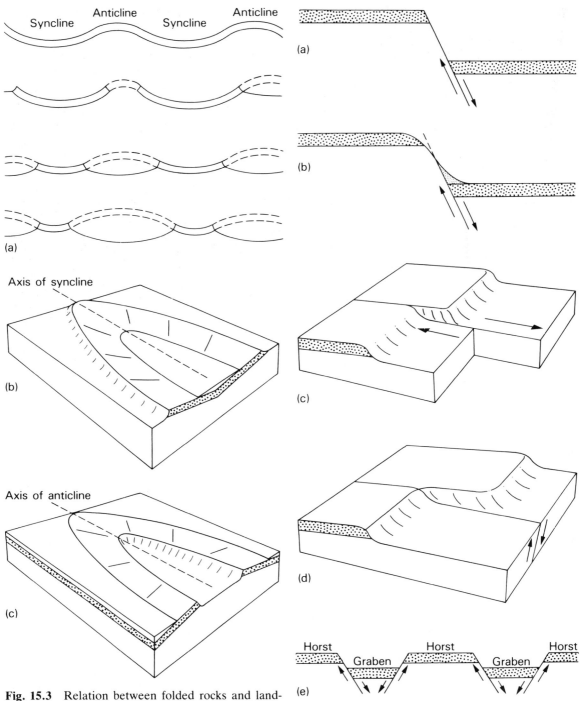

Fig. 15.3 Relation between folded rocks and landforms: (a) simplified sequence of erosion of folded rocks leading to relief reversal; (b) relationship between landform and underlying rocks in a typical plunging syncline; (c) relationship between landform and underlying rocks in a typical plunging anticline

Fig. 15.4 Landforms produced by faulting: (a) fault scarp; (b) fault-line scarp; (c) escarpment displacement by tear faulting; (d) escarpment displacement by faulting with movement at right angles to escarpment; (e) horst and graben produced by faulting.

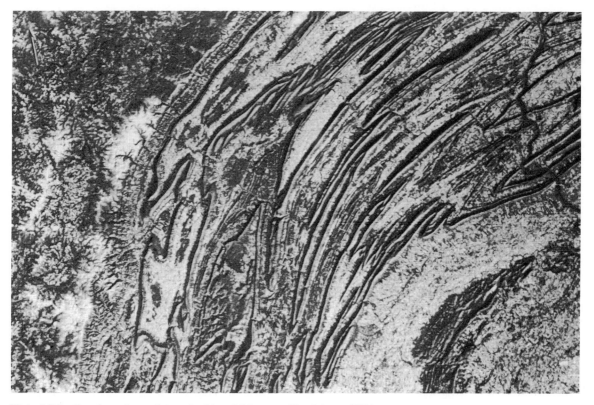

Plate 15.1   Landsat view of Appalachian folds, near Harrisburg, USA.

A series of alternating resistant and less resistant rocks can give rise either to escarpment slopes with *structural benches* or a cuesta and strike vale topography (Fig. 15.2(b)) where the strike is the direction in plan view at right angles to the dip of the rocks. In general, for a rock of a given resistance, the greater the angle of dip the lower will be the resultant relief. Thus, a symmetrical *hog's back* formed by steeply dipping rocks will be lower than a cuesta in the same rocks dipping at a lower angle.

In the case of *folded rocks*, it might be expected that the upfolds of anticlines would form the higher ground and the downfolds of synclines the lower ground (Fig. 15.3(a)). In fact, such a relationship is not common, although the Jura Mountains of France contain some of the best examples in Europe, and the more usual relationship between folded structures and relief is shown in Fig. 15.3(b) (c). Explanation of this relationship may be sought in the fact that the crest of the anticline is the first part exposed to erosion. Once cut through, it produces in effect a pair of cuestas whose escarpment slopes are eroded more rapidly than their dip slopes, due to the exposure of the

underlying less resistant rock. Hence, the core of the anticline is exposed and the axis of the syncline typically forms the higher ground. Thus, anticlines are usually topographically found as a pair of cuestas with their escarpments facing each other and synclines as a pair of cuestas with their escarpments facing outwards. Where the axes of the folds plunge, the limbs of the folds converge (Fig. 15.3). Plate 15.1 shows a satellite image of part of the Appalachians in the USA containing several plunging folds although, because of the steepness of the dip of the rocks, the escarpment and dip slopes are almost symmetrical. Consequently, it is difficult to distinguish between anticlines and synclines.

Faulting in rocks is often manifested at the surface by scarp slopes (Fig. 15.4(a)), although the fault plane itself is often obscured by erosion and deposition. These processes may well start during the actual faulting when it occurs at the surface and continued erosion may lower the ground surface sufficiently, so that all signs of the original fault scarp have been removed. However, if the rocks on either side of the fault-line have differential resistance, a scarp will still be produced, which is called a *fault-line scarp*, although it may point in the opposite direction of the original faulting (Fig. 15.4(b)). Horizontal *tear faults* give rise to horizontal displacement of scarps (Fig. 15.4(c)) and an apparently similar displacement can arise from near vertical faulting if it affects dipping rocks (Fig. 15.4(d)). Frequently, sets of parallel faults are arranged giving rise to areas of uplift called *horsts* and zones of downfaulting known as *grabens*, with elongated grabens known as rift valleys (Fig. 15.4(e)).

15.4 Retrospect

This introduction to sediment transport systems has stressed their intimate coupling with other surface terrestrial systems and has demonstrated the importance of understanding subsurface geological phenomena. The latter are difficult to observe directly but the resultant altitudinal changes in the earth's surface appear to be broadly comparable with those produced by sediment transport systems. It seems likely in future years that the interplay between tectonic systems and sediment transport systems will be increasingly recognised as having a crucial significance for geomorphological investigations and the synthesis by Ollier (1981) provides an excellent basis for future investigations in this subject. At the scale used in this chapter, human influences are relatively small although, more locally, the effects of mining, petroleum extraction and groundwater pumping can cause major land subsidence. For example, in both the San Joaquin Valley in California and the Houston–Galveston region of Texas, USA, an area of more than 12 000 km^2 has been affected by this subsidence, with damage estimated to have cost \$100 million in California and more than \$1000 million in Texas (Coates 1983).

Key topics

1. The interrelationships between sediment transfer systems and other terrestrial systems.
2. The importance of tectonic processes in creating potential energy by changing base-levels.
3. The significance of vertical and lateral earth movements in mountain building and in the mechanics of relief creation.
4. The mechanisms of land/sea-level changes.

Further reading

Barnes, C. W. (1980) *Earth Time and Life*. John Wiley and Sons; New York and London (Ch. 7).

Brown, G. C. and Mussett, A. E. (1981) *The Inaccessible Earth*. George Allen and Unwin; London and Boston (Chs 7–9).

Chappell, J. (1983) Aspects of sea-levels, tectonics and isostasy since the Cretaceous, in Gardner, R. and Scoging, H. (eds) *Mega-geomorphology*. Clarendon Press; Oxford, pp. 56–72.

Strahler, A. N. and Strahler, A. H. (1978) *Modern Physical Geography*. John Wiley and Sons; New York and London (Chs 24–26).

Windley, B. F. (1977) *The Evolving Continents*. John Wiley and Sons; New York and London (Chs 14 and 15).

Zumberge, J. H. and Nelson, C. A. (1977) *Elements of Physical Geology*. John Wiley and Sons; New York and London (Chs 5–7, 9).

Chapter 16 Parent rocks, weathering and weathering products

Sediment transport depends on the detachment and exposure of surface particles. This results from *weathering*, which can be thought of as a cascading system in which the soil and mantle are subsystems. Heat, water, chemical solutions and clastic debris cascade through these subsystems to alter their size, composition, structure and location. The operation of the system is accompanied by the physical disintegration of the bedrock or raw debris and by the release of minerals from their original chemical bonds. It is conditioned by the stability of the rocks and their component minerals and by the processes, mainly derived from climate, which cause them to be transformed into mobile products. The weathering system thus forms a part of the environmental *super-system* in which rock materials are broken down and cycled through phases of transportation, deposition and consolidation into new rocks.

16.1 Parent rocks

Parent rocks can be classified in many different ways although their behaviour under weathering and in forming surface soils is governed mainly by surface topography and by their internal structure and chemistry. Furthermore, since both of these derive from their mode of origin, it is most convenient to subdivide them first genetically and then chemically. *Igneous* and *metamorphic* rocks compose about 95% of the earth's crust but only cover about 25% of its surface. They are generally crystalline and grains can vary in size from the large crystals found in *pegmatites* (coarse-grained rocks with the same general composition as granites) and *porphyries* (rocks with large crystals in a finer-grained groundmass) to the fine mineral particles in relatively amorphous rocks, such as *basalts*. Their main significant differences from the point of view of soil formation and sediment transport are the sizes of their constituent crystals and the amount of silica (SiO_2) in their bulk chemical composition.

In general, *plutonic rocks* (formed by the solidification of magma deep within the earth's crust) are coarsely crystalline, *hypabyssal rocks* (which form dykes, sills and veins) are intermediate and *volcanic rocks* (formed by solidification at the earth's surface) are fine grained. Acid materials tend to be relatively more common in plutonic and basic materials in volcanic rocks. Furthermore, while the disintegration of rocks with large crystals is relatively more rapid than that of rocks with fine crystals, the decomposition of the individual crystals is more rapid in the latter. The chemical composition of some common minerals in igneous rocks is shown in Table 16.1.

Minerals vary in stability with the nature of their chemical composition and crystal structure and with the amount of energy change involved

Table 16.1 Approximate contents of silica (SiO_2) in some important minerals

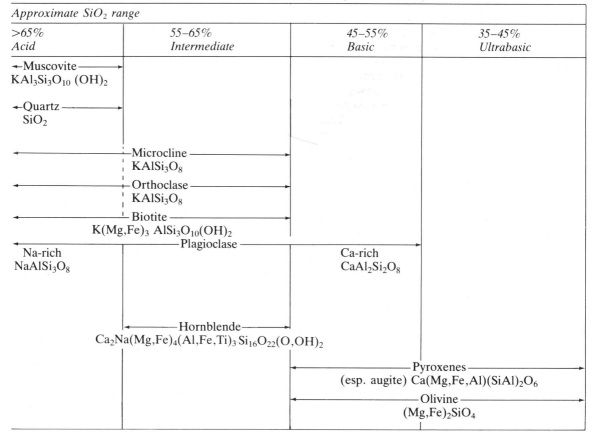

Approximate SiO_2 range			
>65% Acid	55–65% Intermediate	45–55% Basic	35–45% Ultrabasic

In the table (reading across the silica-content bands):

- ◄—Muscovite——► $KAl_3Si_3O_{10}(OH)_2$
- ◄—Quartz——► SiO_2
- ◄——————Microcline——————► $KAlSi_3O_8$
- ◄——————Orthoclase——————► $KAlSi_3O_8$
- ◄——————Biotite——————► $K(Mg,Fe)_3 AlSi_3O_{10}(OH)_2$
- ◄——————Plagioclase——————► Na-rich $NaAlSi_3O_8$ ... Ca-rich $CaAl_2Si_2O_8$
- ◄——————Hornblende——————► $Ca_2Na(Mg,Fe)_4(Al,Fe,Ti)_3 Si_{16}O_{22}(O,OH)_2$
- ◄——————Pyroxenes——————► (esp. augite) $Ca(Mg,Fe,Al)(SiAl)_2O_6$
- ◄——————Olivine——————► $(Mg,Fe)_2SiO_4$

in weathering reactions. Where crystals are individually vulnerable and weakly bonded and where the surrounding reactant has substantial free energy, conditions are maximised for rock weathering. Basic minerals (especially those of a ferromagnesian type, which tend to be dark in colour) are relatively the most vulnerable to weathering so that the order from left to right in Table 16.1 is also one of increasing speed of weathering. Acidic rocks weather to permeable gravelly, sandy and silty soils whereas basic rocks weather mainly to clays. Some minerals (such as zircon and tourmaline) are so resistant that they will survive through several weathering cycles and can be used as an index in calculating the rate of loss of accompanying minerals.

Sedimentary rocks occupy only 5% of the earth's crust but cover 75% of its surface. In geology, they are often classified according to whether their mode of deposition was detrital, chemical, organic or pyroclastic. However, in considering weathering, it is preferable to subdivide them broadly according to chemical composition because this is a good indication of the types and sizes of product into which they will be converted. The commonest rocks, forming perhaps three-quarters of the earth's crust, are those formed of silicate and aluminosilicate minerals. In sediments, these form the shales and mudstones which tend to be poorly cemented and to break down into soils rich in clay (particles smaller than 0.002 mm). Next commonest are the siliceous rocks which (including those of igneous origin) constitute some 13% of the earth's crust. The main sedimentary types are conglomerates (formed dominantly of rounded stones and gravel), sandstones (formed mainly of sands) and quartzites, which are metamorphic rocks of almost pure silica. These yield mainly gravels and sands which are coarser than the products of silicate and aluminosilicate minerals. Calcareous rocks (such as limestone, chalk and dolomite) are slightly soluble and are removed in solution, leaving lime-rich soils and a concentration of their insoluble residues (such as clay or flint) after weathering. Recent deposits are normally poorly consolidated or weathered and their constituent particles are relatively unaltered. Sediments formed of granite or sandstone particles, for instance, may contain relatively intact crystals or grains which have been inherited from the original rocks.

16.2 Weathering processes

Rocks may be prepared for weathering by physical *unloading*. This is the process in which the removal of overlying rocks or sediment reduces pressure on the freshly exposed rock, permitting it to expand and produce intergranular dislocations which may lead to cracks and fissures. Organic activity also leads to rock breakdown through the action of plant roots in widening cracks, by the flaking of rock surfaces associated with the wetting and drying of organic materials (in intimate contact with the surface) and by the churning and moving effect of earthworms and other soil organisms. Most weathering of rock surfaces, however, is by physical disintegration and chemical decomposition.

(a) Physical disintegration

Physical disintegration mainly results from a combination of changes in temperature (*thermoclasty*) and moisture conditions (*hydroclasty*). Temperature changes cause intergranular stresses in juxtaposed mineral fragments with different coefficients of expansion, which weakens their cohesion. The effect of moisture depends on the extent to which it is combined with temperature changes since heat speeds reactions. A repeated alternation of wetting and drying attacks rocks and wide, rapid temperature changes cause water to swell and contract, especially when this includes repeated freezing and thawing. The freezing of water in a confined space generates an outward force of about 1500 t m^{-2} and as this mainly acts near the surface of rocks, it sets up pressures between outer and inner layers which can lead to *exfoliation* (i.e. the splitting and peeling-off of outer rock layers). The results of such breakdown can be seen in the rubble and talus slopes formed in mountain areas subject to frequent freezing.

Salt weathering (*haloclasty*), due to the crystallisation of salts in confined spaces, may also generate destructive pressures. Since the solubility of common minerals increases with pressure, crystallisation can occur against pressure only in a supersaturated solution. This happens, for instance, in deserts where surface heating and drying winds cause excessive evaporation. Repeated solution and recrystallisation of such supersatured water may have a cumulative effect. There is also some evidence that salt crystallisation, acting in combination with freeze–thaw, is more destructive than the aggregate of their separate effects. Physical weathering is dominant in soils in early stages of development and in desert, Arctic or Alpine conditions where free groundwater is lacking and organic processes are undeveloped. Such soils have little internal differentiation and have a large proportion of coarse materials.

(b) Chemical weathering

Chemical weathering is the decomposition of rocks resulting from attack by the chemically active constituents of the soil and atmosphere. The more complete the fragmentation of the rocks, the more intense are the chemical processes because of the greater specific surface exposed. There are two main phases of alteration: firstly, the destruction of certain minerals and secondly, the formation of secondary products. These are illustrated in Fig. 16.1. The chemical destruction of minerals depends on five main types of reaction, namely

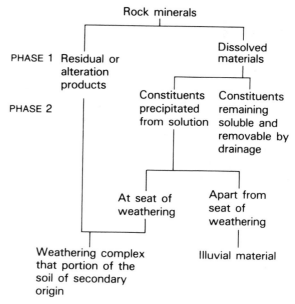

Fig. 16.1 Processes in the weathering of rocks. (After Robinson 1949)

solution, hydration, hydrolysis, carbonation and *oxidation/reduction*. These seldom act alone but occur in various combinations with each other. They all increase with temperature and so are most effective in tropical climates.

Simple solution affects all rocks but it is mainly significant on soluble materials such as halite (NaCl) and the sulphates and carbonates of calcium and magnesium. Table 16.2 shows some of the wide contrasts in solubility between different minerals. When rocks are in contact with the soil solution, some of the water will be added to their molecular structures and this is known as hydration. It is especially marked in the feldspar, pyroxene, amphibole and mica groups which contain iron oxides, and one of the most important reactions of this group is as follows:

$$2Fe_2O_3 + 3H_2O = 2Fe_2O_3.3H_2O \quad [16.1]$$
$$\text{haematite} \qquad\qquad \text{limonite}$$

Under arid climates, the reaction goes into reverse with the dehydration of yellow limonite to red haematite being especially noticeable in mature desert sands. Anhydrite may be converted to gypsum by the following reaction:

$$CaSO_4 + 2H_2O = CaSO_4.2H_2O \quad [16.2]$$
$$\text{anhydrite} \qquad\qquad \text{gypsum}$$

This is accompanied by a 60% increase in volume and a softening of the rock material and the reaction may be reversed when gypsum is deposited from solutions rich in NaCl or KCl.

Minerals which are exposed to the soil solution also become detached and ionise at a rate proportional to their solubility levels (Table 16.2). At the same time, a small proportion of water molecules dissociate into H^+ and OH^- *ions* (electrically charged particles) which react chemically with other ions and compounds. This process is known as *hydrolysis* and is the most important form of chemical weathering, since it is the basis for the conversion of silicate minerals (such as feldspars and micas) into soil clays. It is rendered many times more effective in humid climates

Table 16.2 Solubility of some common minerals in water (mg l^{-1})

SiO_2 (quartz)	Very slight
Fe_2O_3	Very slight
$CaCO_3$	14 (25 °C)
$MgCO_3$	100 (18 °C)
$CaSO_4.2H_2O$ (gypsum) or	
$\quad CaSO_4$ (anhydrite)	2 080 (25 °C)
NaCl (halite)	264 000 (20 °C)
$NaNO_3$	468 000 (20 °C)

where vegetation yields organic acids (notably H_2CO_3) from the following reaction:

$$CO_2 + H_2O \rightarrow H_2CO_3 \quad [16.3]$$

This acid attacks rocks, causing carbonation (i.e. their conversion to soluble carbonates), and limestone, for instance, experiences the following reaction:

$$CaCO_3 + H_2CO_3 = Ca(HCO_3)_2 \quad [16.4]$$

The soluble calcium bicarbonate ($Ca(HCO_3)_2$) is subject to removal in the drainage water.

The hydrolysis of silicate minerals to form soil clays is commonly a double process involving *desilicification* and *dealkalisation*. Desilicification releases a silicic acid and an alkali by reactions such as:

$$2KAlSi_3O_8 + 11H_2O =$$
$$\text{orthoclase}$$
$$Al_2Si_2O_5(OH)_4 + 4H_4SiO_4 + 2KOH \quad [16.5]$$
$$\text{kaolinite} \qquad\qquad \text{orthosilicic acid}$$

Dealkalisation then converts the alkali into a soluble salt, which can be removed in drainage following carbonation:

$$2KOH + CO_2 = H_2O + K_2CO_3 \quad [16.6]$$

Oxidation is the change of state resulting from the loss of an electron and most commonly occurs in soils through the action of dissolved oxygen. The stability of minerals is thus governed by the amount of oxygen present. The element most commonly oxidised is iron and this usually occurs in the divalent (double-charged) ferrous (Fe^{2+}) state in its sulphide, carbonate and silicate mineral compounds. Reaction with oxygen changes it to the trivalent (treble-charged) ferric (Fe^{3+}) state, which commonly requires two stages, the first yielding ferrous *cations* (positively charged metallic ions) and the second ferric oxide. The weathering of ferrous orthosilicate in the presence of carbonic acid provides an example with first hydrolysis, as follows:

$$Fe_2SiO_4 + 2H_2CO_3 + 2H_2O =$$
$$\text{ferrous}$$
$$\text{orthosilicate}$$
$$2Fe^{2+} + 2HCO_3^- + H_4SiO_4 + 2OH^- \quad [16.7]$$
$$\text{ferrous}$$
$$\text{cations}$$

and then oxidation, as follows:

$$2Fe^{2+} + 4HCO_3^- + \tfrac{1}{2}O_2 + 2H_2O =$$
$$\qquad\qquad \text{oxygen dissolved}$$
$$\qquad\qquad \text{in water}$$
$$\qquad Fe_2O_3 \downarrow + 4H_2CO_3 \quad [16.8]$$
$$\qquad \text{precipitated}$$
$$\qquad \text{ferric oxide}$$

The reverse reduction process occurs under water-logged conditions called anaerobic because of the absence of oxygen. *Anaerobic bacteria* (Ch. 25) must obtain their oxygen from chemical compounds rather than air. All soils have a tendency towards one process or the other, depending on their *oxidation–reduction* (or *redox*) *potential*, expressed by the symbol *Eh*. This is determined electrometrically by inserting a platinum electrode and measuring the difference against another electrode of known potential. The condition at which gaseous hydrogen loses an electron and ionises is arbitrarily fixed at a value of 0.0 V. A positive *Eh* value indicates oxidising whereas a negative *Eh* value represents reducing conditions, and soils generally range between extremes of about $+800$ and -1000 mV. The redox potential is especially important in indicating the mobility of iron and manganese cations. These can occur as reduced, relatively mobile divalent cations (Fe^{2+}, Mn^{2+}) or as oxidised polyvalent cations (Fe^{3+}, Mn^{3+}, Mn^{4+}) depending on the *Eh* of the solution. At low redox potentials, ferric oxide is reduced to ferrous oxide; nitrates to nitrites, ammonia and nitrogen gas; sulphates to sulphides (sometimes involving the evolution of the malodorous gas H_2S) and carbohydrates to hydrocarbons such as CH_4 (methane or 'marsh gas').

16.3 Rates of weathering

The rate of weathering is the product of all the component individual reactions at the rock–soil interface. It is considered to be in a *steady state* when the controlling variables are in *dynamic equilibrium*, that is, when inputs of energy and materials are balanced by outputs. When subjected to drastic change (such as when the surface soil is removed by erosion), the soil system will re-establish a new equilibrium. Weathering at the rock–soil interface will tend to replace the lost material and, in a steady state system, rates of weathering and erosion will again attain equal values.

Weathering rates depend mainly on the nature of the materials, the annual mean temperature and the amount of percolating water. An indication of the maturity of a soil can sometimes be obtained from the proportion of silt it contains. This is because both the sand and the clay fractions are chemically stable, while the silt fraction tends to weather to clay. However, as little clay

is formed below 5 °C and not much below 10 °C, this is most characteristic of warmer climates. A number of attempts have been made to measure rates of weathering, some of which have been reviewed by Ollier (1975) and Trudgill (1976). The methods have included assessing the rate of destruction of tombstones, Cleopatra's Needle and buildings of known age and the laboratory testing of rock samples in simulated field conditions. Rates of soil formation also indicate rates of weathering and studies have compared quantitatively the physical and chemical characteristics of soils of unknown age with those of nearby soils, which have developed since a known starting date. Examples have included soils which have developed on volcanic ash, moraines, sand dunes or coastal polders (whose dates of original deposition are known) and podsols which have developed under forests of known age. Although results are very variable, they seem to indicate that parent materials can be weathered into mature soils in the humid temperate zone in 100–500 years and in the tropics in a few decades.

Another promising quantitative approach is that of measuring the amount of cations in drainage water (Perrin 1965). This can be applied to all rocks but especially to those (such as limestones) which are subject to loss by solution. Values of 14.6 m^3 km^{-2} yr^{-1} were obtained for sandstones, 3.3 m^3 km^{-2} yr^{-1} for granites and 0.9 m^3 km^{-2} yr^{-1} for metaquartzites in New Mexico by Miller (1961) and impermeable greywackes in Wales showed solution losses of all dissolved solids of 1.55–2.05 m^3 km^{-2} yr^{-1} (Oxley 1974). Considerable research has been devoted to the measurement of calcium in drainage waters from limestone areas. This has been supported by studies which have traced the water to its surface origin, measured its flow across weirs and statistically calculated its changes through time. Values obtained lie mainly between 50 and 102 m^3 km^{-2} yr^{-1} of $CaCO_3$ lost (Smith and Newson 1974). Recent studies have emphasised the need for supporting such calculations with studies of the regional morphometry, the chemistry of the soil water and petrological differences within limestones (Douglas 1976; Smith and Atkinson 1976).

16.4 Weathering products

The products of weathering consist of particles of all sizes from boulders to clay. In general, physical

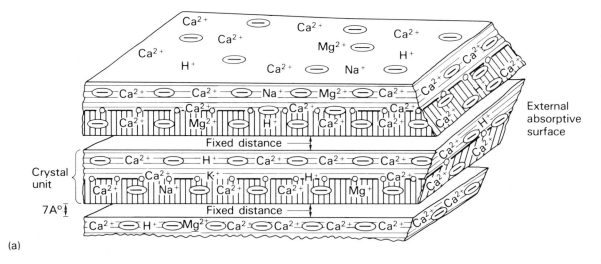

(a)

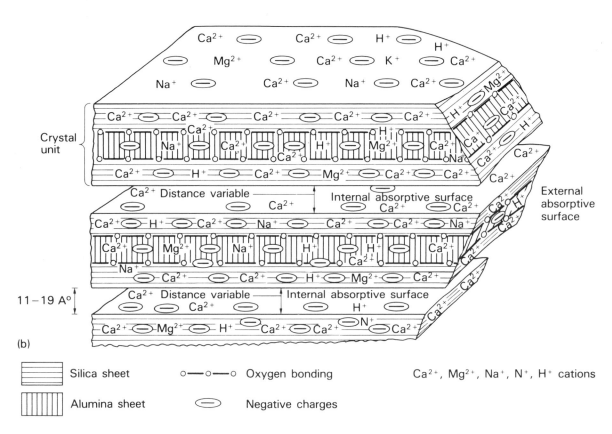

(b)

▤ Silica sheet	○—○—○ Oxygen bonding	Ca^{2+}, Mg^{2+}, Na^+, N^+, H^+ cations
▥ Alumina sheet	⊖ Negative charges	

Fig. 16.2 Diagrammatic representation of clay micelles with their sheet-like structure, negatively charged surface and adsorbed cations: (a) kaolinite; (b) montmorillonite. Illite has the same general structure organization as montmorillonite except that K atoms add further structural connections between the sheets, supplementing the oxygen bonding.

weathering yields the coarsest materials, namely stones, gravel, sand and silt, whereas chemical weathering produces the colloidal particles. The coarser fractions are relatively inert but provide the skeleton of the soil, supporting the plants and facilitating the infiltration of water (section 6.1).

The colloids consist mainly of silicate clays, hydrous micas and iron oxides and are softer than the parent rocks, although they occupy a greater volume. Furthermore, they show more subdued colours, which tend to be reddish or yellowish from their contained iron compounds. Perhaps the most important products are the silicate clays which play a key role in determining soil properties and processes. Individual clay particles are known as *micelles*. They are platy in form and have diameters of less than 2 μm, with most of them smaller than half or even a quarter of this size. Each particle has a negative charge and acts somewhat like a large complex *anion* (negatively charged ion) in attracting positively charged cations.

There are three main types of clay minerals (depending on the chemical composition of the material) which are known generally by their commonest forms as *kaolinite, montmorillonite* and *hydrous micas* (Fig. 16.2). Kaolinite clays are characterised by a 1 : 1 crystal lattice in which a sheet of Al_2O_3 is bonded by oxygen atoms to a sheet of SiO_2, with a fixed inter-sheet spacing of 7° Angstroms (Å). Micelles are relatively large (i.e. 0.1–5 μm) and water and solutes can only approach their outer edges. This restricts the effective surface and thus the amount of swelling when wetted and contraction and cracking when dried, making them relatively stable as can be seen, for instance, in their suitability for ceramics.

Montmorillonite clays, by contrast, are composed of smaller particles than kaolinite (0.01 μm). They have a 2 : 1 expanding crystal lattice with two sheets of SiO_2 sandwiching one of Al_2O_3, forming layers with a spacing of about 14 Å, although this can vary from 11 to 19 Å depending on the amount of water and cations present. Because of the smaller size of the particles and the fact that ions can penetrate between the layers of the lattice, montmorillonites have a much greater surface area. They can thus become more plastic and cohesive and, by absorbing more water, swell on wetting and shrink and crack on drying. They also have a much larger cation exchange capacity (see Ch. 26). Often associated with montmorillonite clays are the hydrous micas, of which illite is a

common type. They have a 2 : 1 non-expanding crystal lattice and differ from montmorillonites in that the particles are larger. Furthermore, about 15% of the Si atoms are replaced by Al and the negative charges are largely satisfied by K atoms. The clay is thus less subject to expansion and contraction and, both because of the size (0.1–2.0 μm) and the chemical charge of the particles, it has an intermediate exchange capacity between the other two types of clay.

The formation of the different types of clay mineral depends on the environment. In general, clay formation is accelerated by high temperatures and montmorillonite tends to be formed in the presence of high pH, a good base supply and poor drainage whereas kaolinite occurs in strongly leached, acid, low-base soils. Illites are intermediate between the two and are especially common where the soil solution is rich in potassium. For these reasons, illite and montmorillonite tend to be relatively more frequent in arid areas whereas kaolinite is more common in humid regions.

16.5 Retrospect

This chapter has discussed weathering in relation to sediment transport systems and has suggested some of the theoretical approaches being taken in understanding it. The movement of surface materials is limited by their rate of detachment from the parent rock. This detachment is caused by the physico-chemical change of the rock surface which is induced by the action of weathering agents, of which water and its dissolved ions are the chief factors. The resulting processes operate at rates which have been measured, especially in drainage waters from soluble rocks such as limestones. The products of weathering are mainly mineral particles ranging in size from stones to clay, the latter being especially important because of its role in determining soil properties and processes.

Key topics

1. The significance of weathering in sediment transfer systems.

2. The susceptibility of different rock types to weathering.
3. The processes and rates of weathering.
4. The characteristics of the products of weathering.

Further reading

Birkeland, P. (1977) *Pedology, Weathering, and Geomorphological Research*. Oxford University Press; London and New York.

Buol, S. W., Hole, F. D. and McCracken, R. J. (1989) *Soil Genesis and Classification*. Iowa State University Press; Ames.

Curtis, C. D. (1976) Stability of minerals in surface weathering reactions, *Earth Surface Processes*, **1**, 63–70.

Derbyshire, E. (ed.) (1976) *Geomorphology and Climate*. John Wiley and Sons; New York and London.

Duchaufour, P. (1982) *Pedology*. George Allen and Unwin; London and Boston.

Trudgill, S. T. (1977) *Soil and Vegetation Systems*. Oxford University Press; London and New York.

Chapter 17 Strength of materials and their stress–strain behaviour

In order to understand the debris cascade (section 15.1), it is essential to obtain a basic knowledge of the factors controlling the resistance of materials to movement. This inevitably involves use of some elementary concepts of physics, which will be familiar to many readers with a scientific background, and the most fundamental of these relate to the forces controlling the movement of a block of material sitting on an inclined plane. At first sight, this may seem to be a very crude model of a slope system but, as later chapters will show, it is a surprisingly accurate one in many situations. The internal cohesion of materials is also dealt with (to be discussed further in Ch. 18) and particular emphasis is placed on the crucial role of moisture in controlling the strength of materials. It is apparent that materials respond to applied stresses in several different ways: for example, some may shatter, others break along a single fracture line whereas others may bend or flow. Consequently, this chapter concludes with a consideration of stress–strain relationships.

Those readers without a mathematical or physics background may find some of the material to be rather difficult. In this case, such readers are strongly encouraged not just to read between the equations, but to try and understand the chapter in its entirety since it forms the basis of our understanding of sediment transport systems and hence of geomorphology as a whole. However, it is worth noting that the inclusion of basic equations will increasingly be a trend throughout physical geography in the future.

17.1 Introductory ideas of force and resistance

The most pervasive and important force acting on the earth's surface materials is the pull of gravity and, in large part, the study of sediment transport systems is the study of the gradual failure of these materials under the gravitational pull. Gravitational force, like all others, is measured by the acceleration which it gives to a *mass* and the equation defining *force* (F) is simply:

$$F = mA \qquad [17.1]$$

where m is mass measured in grams (g) and A is acceleration measured in metres per second (m s^{-1}). *Weight* needs to be distinguished from mass since the former is the force acting on an object as a result of gravity, whereas the mass of an object is an inherent property which indicates its resistance to movement by force. For example, a boulder will have the same mass on the earth's or moon's surface but because of the very different gravitational pull exerted by the two bodies, its weight on the moon will be only one-sixth of that on earth.

It is useful to consider a boulder or pebble resting on a horizontal flat surface (Fig. 17.1(a)) where, for the sake of simplicity, the particle is block-shaped. A force is exerted on the flat surface by the pebble's weight and assuming the particle is stationary, there must be an equal upward force (R_w) exerted by the surface opposing this weight. If a horizontal force is exerted on the particle, then the forces acting are those shown in Fig. 17.1(b)). The horizontal pushing force could be exerted by flowing water or wind and assuming the particle does not move with this applied force, then there must be an equal and opposite resisting force (R_f). This force is due to friction acting between the particle and the surface. Figure 17.1(b) is already becoming somewhat confused and for the sake of clarity, it is usual therefore to represent this and other similar force diagrams as shown in Fig. 17.1(c). In terms of measuring and relating the forces involved, it does not matter that R_w is shown emanating from the top of the particle, or that the pushing force is represented by a pulling force.

When F is gradually increased, a force is eventually reached at which the particle begins to move, and this is termed the *critical force*. At greater forces, the resisting force will be less than the pushing force and F and R_w will no longer be balanced. Furthermore, in practice, it is found that once motion has been initiated, the frictional resistance usually drops slightly. If the weight of the particle is increased (for example, by adding weights to its upper surface), then a larger critical force (F_c) will be needed to initiate motion. However, the ratio of F_c to the weight (W) does remain constant and, as a result, for a particle on a horizontal plane and with a horizontally applied force, the relationship can be expressed as follows:

$$\frac{F_c}{W} = \text{constant} = \mu \qquad \qquad [17.2]$$

where μ is known as the *coefficient of friction* and since there is a force on the top and bottom of the expression, it is dimensionless.

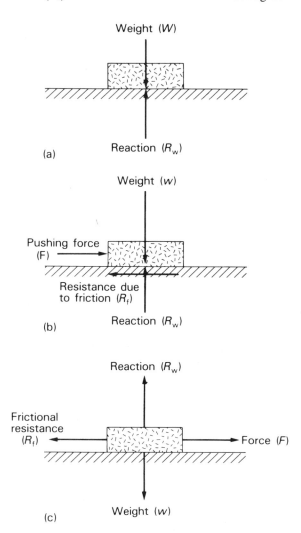

(a)

(b)

(c)

Fig. 17.1 Forces acting on a pebble on a horizontal surface with: (a) no horizontal forces acting; (b) horizontal forces acting; and (c) usual diagrammatic representation of (b).

17.2 Force and resistance on a slope

Most surfaces across which debris moves are of course not horizontal and it is therefore necessary to examine how the considerations of the previous section have to be modified for an inclined surface. The weight (W) and reaction (R_w) continue to operate in the same directions (Fig. 17.2(a)) but interest now focuses on understanding how the force of gravity acts on the particle in order to move it downslope, as determined by the forces shown in Fig. 17.2(b). The *normal reaction* is the force resulting from the pebble's weight, which acts at right angles (or is normal) to the surface. For a horizontal surface, this is clearly equal to the weight (W) whereas for an inclined surface, this is clearly not the case since there is a component of this weight acting down the slope.

Determination of the components shown in Fig. 17.2(b) can be obtained by use of a parallelogram of forces since the vertical force exerted

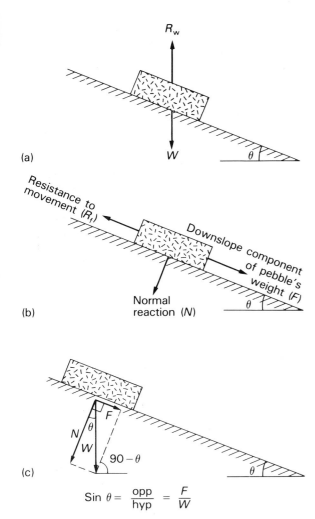

$$\text{Sin } \theta = \frac{\text{opp}}{\text{hyp}} = \frac{F}{W}$$

Fig. 17.2 Forces acting on a pebble on an inclined surface at angle θ (see text for explanation of other symbols).

by the particle's weight can be represented by components operating at an angle to the vertical. Furthermore, the component forces can be geometrically calculated since their magnitude is equal to the length of the sides of the parallelogram as shown in Fig. 17.2(c). This parallelogram has right angles at its corners and, as shown in Fig. 17.2(c), the downslope and normal components are equal to the weight (W) multiplied by the sine (sin) and cosine (cos) of the slope angle respectively, expressed as:

$$\begin{aligned} F &= W \sin \theta \\ N &= W \cos \theta \end{aligned} \qquad [\textbf{17.3}]$$

where F is the downslope component and N is the normal component.

If the particle is stationary on the slope, it follows that R_f equals F. However, if the angle (θ) is progressively increased, then F will increase until it reaches a critical threshold value (F_c) when sliding downslope will commence. This value is called the *critical angle* (θ_c) and at this angle, the frictional resisting forces will be at their maximum (R_{fc}) and are equal to F_c. Thus, equation [17.3] can be rewritten as follows:

$$R_{fc} = W \sin \theta_c \qquad [\textbf{17.4}]$$

and the normal reaction at this angle will be:

$$N = W \cos \theta_c \qquad [\textbf{17.5}]$$

As in the case of the particle resting on a horizontal plane, it is apparent that as the weight of the pebble is increased, then the force required for movement increases. It is found experimentally that the ratio between F_c and N (the corresponding normal reaction) is a constant, as follows:

$$\frac{F_c}{N} = \text{constant} = \mu \qquad [\textbf{17.6}]$$

where μ is again the coefficient of friction. This coefficient has a direct connection with the critical angle (θ_c) since it is in fact equal to its tangent (adjacent/hypotenuse in a right-angle triangle). This can be simply proved by substituting equations [17.4] and [17.5] into equation [17.6]:

$$\mu = \frac{F_c}{N} = \frac{W \sin \theta_c}{W \cos \theta_c} = \frac{\sin \theta_c}{\cos \theta_c} = \tan \theta_c \qquad [\textbf{17.7}]$$

The latter follows from:

$$\left(\frac{\sin}{\cos} \right) = \left(\frac{\text{opp/adj}}{\text{hyp/hyp}} \right) = \left(\frac{\text{opp}}{\text{adj}} \right) = \tan$$

Thus, the coefficient of friction (μ) equals the tangent of the slope angle when sliding of the particle begins. From the above, it follows that for a slope and pebble composed of the same materials, μ is independent of the pebble's weight. Conventionally, the critical slope angle θ_c, is given the symbol ϕ and, since at the moment at which the pebble slides $F_c = R_{fc}$, equation [17.6] can be rearranged to obtain the following expression:

$$R_{fc} = F_c = \mu N = \tan \phi \, N \qquad [\textbf{17.8}]$$

In other words, the maximum frictional resistance equals the tangent of the angle at which sliding commences, multiplied by the normal reaction to the slope. For slope angles less than θ_c, no movement occurs and hence the resistance to movement equals the downslope force and equation [17.8] becomes:

$$R = F = \tan \phi \, N \qquad [17.9]$$

where $\tan \phi$ is called the *coefficient of planar friction*.

It should be noted that all of the above equations are expressed in terms of force. However, since *areas* of slopes are of most common concern, it is usual to express the equations in terms of force per unit area, the name for which is *stress*. Equation [17.9] can therefore be rewritten as follows:

$$S = \tau = \sigma \tan \phi \qquad [17.10]$$

where σ is the normal stress rather than the normal reaction, S represents the resistance of a pebble or any other non-cohesive particle to downslope movement and τ is the downslope stress, caused by gravity. Equation [17.10] also indicates the approximate stresses exerted on the materials within a slope, which consist entirely of non-cohesive particles, rather than one such particle resting on a solid surface. However, the angle at which sliding commences will be higher than that for a particle resting on a slope, because of interlocking between the particles within the slope's materials. Symbolically, this change is indicated by $\tan \phi_i$, representing the *coefficient of internal friction*, instead of $\tan \phi$ which represents the coefficient of planar friction. For a slope of non-cohesive materials, S is the total resistance of the material to shear stresses and hence represents their *shear strength*. Thus, equation [17.10] is rewritten as follows:

$$S = \tau = \sigma \tan \phi_i \qquad [17.11]$$

Particulate materials, which have been densely packed due to the presence of overlying rocks before erosion, have a *peak resistance* to shear which declines after initial failure to a residual strength.

17.3 Influence of cohesion and moisture

The previous section has analysed the forces acting on surface materials although it is a simplification in many respects, particularly since it ignores cohesion and the influence of water. In the case of unweathered rocks, the cohesion of small hand samples may be extremely high whereas for larger rock masses, the presence of joints will significantly lower such values. In rocks and soils, cohesion may be associated with carbonates, sil-

ica, alumina, iron oxide and organic compounds. In addition, clays may also possess significant cohesion due to physico-chemical forces (Ch. 16) acting between the particles. As discussed below, the presence of water can also add cohesion due to capillary action (section 6.1). The effects of cohesion can be included in equation [17.11] by a simple additive quantity to give:

$$S = C + \sigma \tan \phi_i \qquad [17.12]$$

where C is cohesion expressed in units of stress. This equation is called the *Coulomb equation* although, in this form, it fails to take full account of the effects of water. In particular, water occupying the pores between particles changes the balance of forces dramatically, because the water supports part of the weight of the particles. Such reduction in weight is due to the effects of buoyancy, which is experienced by all bodies that are immersed in water. In the context of slope materials, this effect is known as *pore-water pressure* (u). The normal stress (σ) will therefore be reduced by the pore-water pressure to give the *effective normal stress* (σ') as follows:

$$\sigma' = \sigma - u \qquad [17.13]$$

For saturated materials, equation [17.12] is modified to show that the internal shear strength is reduced under such conditions. The cohesion is also modified by the presence of water and is given the symbol C' in the following equation:

$$S_c + C' + (\sigma - u) \tan \phi_i \qquad [17.14]$$

Following heavy rainfall, the pore-water pressure isolines may well be approximately parallel to the surface. In this situation, the shear stress will not be reduced to compensate for the reduction in shear strength evident in equation [17.14] and thus slope failure is very likely to occur. This explanation, rather than spurious ones involving lubrication or the additional weight of water present, accounts for the geomorphic significance of heavy storms on rapid mass movements (Statham 1977). If slope materials are unsaturated, then a negative pore-water pressure will be present due to capillary effects, yielding an increased effective normal stress.

The likelihood of slope failure is determined by the factor of safety (FS), as follows:

$$FS = \frac{\text{Shear strength}}{\text{Shear stress}} = \frac{S}{\tau} \qquad [17.15]$$

For example, if $FS < 1.0$, the slope is stable whereas at values of 1.0 or just above, slope fail-

Table 17.1 Geomorphic Rock Mass Strength Classification and Ratings (After Selby, 1980)

Parameter	1 Very Strong	2 Strong	3 Moderate	4 Weak	5 Very Weak
Intact rock strength (N-type Schmidt Hammer 'R')	100–60 r:20	60–50 r:18	50–40 r:14	40–35 r:10	35–10 r:5
Weathering	unweathered r:10	slightly weathered r:9	moderately weathered r:7	highly weathered r:5	completely weathered r:3
Spacing of joints	>3 m r:30	3–1 m r:28	1–0.3 m r:21	300–50 mm r:15	<50 mm r:8
Joint orientations	Very favourable. Steep dips into slope, cross joints interlock r:20	Favourable. Moderate dips into slope r:18	Fair. Horizontal dips, or nearly vertical (hard rocks only) r:14	Unfavourable. Moderate dips out of slope r:9	Very unfavourable. Steep dips out of slope r:5
Width of joints	<0.1 mm r:7	0.1–1 mm r:6	1–5 mm r:5	5–20 mm r:4	>20 mm r:2
Continuity of joints	none continuous r:7	few continuous r:6	continuous, no infill r:5	continuous, thin infill r:4	continuous, thick infill r:1
Outflow of groundwater	none r:6	trace r:5	slight <25l/min/10 m² r:4	moderate 25–125l/min/ 10 m² r:3	great >125l/min/10 m² r:1
Total rating	100–91	90–71	70–51	50–26	<26

ure is about to take place. Selby (1980; 1982) has proposed a useful empirical guide to the estimation of rock strength for geomorphic purposes (Table 17.1). The characteristics include weathering, the spacing orientation, width and continuity of joints and the outflow of groundwater. The ratings applied are based on empirical observations and for a given rock are summed to give the overall rock strength.

17.4 Stress–strain behaviour of materials

In a sense, all sediment transport systems (and indeed the whole of geomorphology) could be described as the study of stress–strain behaviour of materials since *strain* describes the resultant deformation of surface materials under applied stress. In this section, the principal types of be-

haviour will be outlined as a basis for Chapters 18–23, which give an account of particular types of transport systems.

The earlier analysis of a block resting on a planar surface or inclined plane (Figs 17.1 and 17.2) corresponds to the first type of stress–strain behaviour. The materials possessing the characteristic are known as *rigid plastics*, a term which is somewhat confusing given the everyday use of the term 'plastic'. In the present context, the term 'rigid plastic' refers to materials which show no strain (or deformation) below a critical threshold of stress (Fig. 17.3(a)). When this threshold or yield point is reached, failure suddenly occurs and continues infinitely if the same stress is applied. If such changes occur within materials, the movement is termed *plastic flow*. Materials displaying considerable plastic flow are called *ductile* and those displaying little or no flow are called *brittle*.

In contrast to this type of stress–strain behaviour, *elastic deformation* involves a strain directly

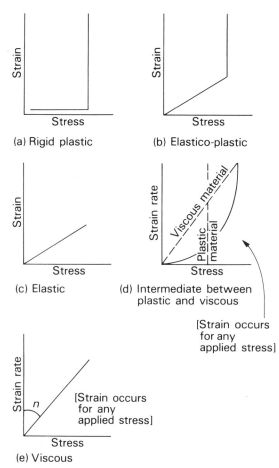

(a) Rigid plastic

(b) Elastico-plastic

(c) Elastic

(d) Intermediate between plastic and viscous

[Strain occurs for any applied stress]

(e) Viscous

Fig. 17.3 Stress–strain relationships for model materials.

the behaviour of a *Newtonian fluid* and is expressed as:

$$\tau = n \ (d\varepsilon/dt) \qquad [17.16]$$

where n (the reciprocal of the steepness of the line in Fig. 17.3(e)) is the *viscosity*. However, in *non-Newtonian fluids*, non-linear relationships between stress and strain rates are found.

Another type of behaviour, which is intermediate between plastic and viscous behaviour (Fig. 17.3(d)), is represented by *Glen's Law*. It is used to analyse ice movement, in which the strain rate (dε/dt) has a simple power relationship with applied stress, as follows:

$$\frac{d\varepsilon}{dt} = A\tau^n \qquad [17.17]$$

where A is a coefficient dependent upon temperature and n is constant for a given confining pressure. Increasing moisture content can have a profound effect on the behaviour of soils, as described by the *Atterberg limits* (Fig. 17.4).

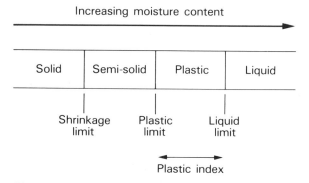

Fig. 17.4 Atterberg limits for soils showing influence of moisture.

proportional to applied stress (Fig. 17.3(c)). An ideal elastic material recovers its original form when the stress is removed and if the elastic limit is exceeded, then failure occurs. Those materials possessing cohesion will display limited elastic behaviour, at least until the applied stress exceeds the yield point. Perfect elastic behaviour, followed by sudden failure, is modelled by elastico-plastic or 'St Venant' materials (Fig. 17.3(b)).

In complete contrast to elastic and plastic behaviour, *viscous deformation* involves strain at all stresses and is a characteristic of fluids. The amount of deformation or strain is therefore independent of the applied stress, although the time during which stress occurs does affect the magnitude of the strain. Thus, in Fig. 17.3(e), the strain rate (dε/dt) is shown to be linearly proportional to the shear stress (τ). This relationship describes

17.5 Fluid motion and effects of applied stress

The previous section has emphasised the distinctive nature of fluids in terms of their stress–strain characteristics, compared with other materials. The behaviour of flowing water is of particular importance in sediment transport systems because of the large amounts of material which are directly carried by it. Water tends to flow in one of two quite distinctive ways, although there is an intermediate condition as well, and these are termed laminar and turbulent flow (Fig. 17.5). In *laminar*

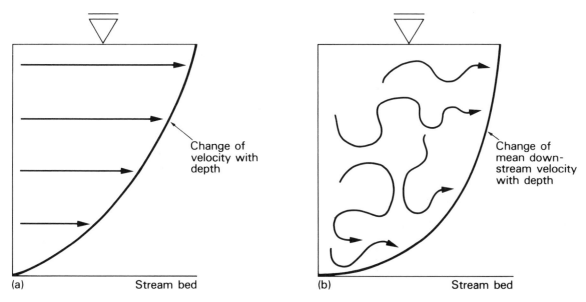

Fig. 17.5 Laminar and turbulent flow. Note that for turbulent flow, there is a very rapid increase in velocity above the bed: (a) laminar flow, no mixing between layers; (b) turbulent flow, random fluctuations inherent part of flow.

flow, lines of flow within the water are essentially parallel with little or no mixing (Fig. 17.5). In *turbulent flow*, a more rapid increase in velocity occurs from the bed upwards but, more importantly, there is very strong mixing of the fluid with numerous vertical eddies. Hence, the flow is inherently more erosive than laminar flow.

The occurrence of laminar and turbulent flow is defined by the Reynolds number (*Re*) expressed as:

$$Re = \frac{\rho_w VL}{\nu} \qquad [17.18]$$

where ρ_w is the density of water, V is the velocity, ν is the dynamic viscosity and L is a characteristic linear measurement. For overland flow or wide channels, L equals the depth of flow whereas for narrow, relatively deep channels, it equals the hydraulic radius. This is equal to the cross-sectional area (A), divided by the perimeter (p) (Fig. 17.6). Laminar flow is found for values of *Re* which are less than 500 whereas fully turbulent flow occurs with values greater than approximately 2000. For stream channels, flow is invariably turbulent, except very close to the channel bed and sides. However, on hillslopes, laminar flow and intermediate types of flow (i.e. when *Re* is more than 500 but less than 2000) are much more common.

The question of the forces exerted by flow can be approached by considering the situation which

exists when there is *uniform flow*. This means that there are no changes in velocity with time within the flow. A reasonable assumption concerning the force exerted by the water per unit area which is resisting flow (i.e. the tractive stress τ_0) is that it varies with the square of the velocity (V) as follows:

$$\text{Resisting stress} = \tau_0 = kV^2 \qquad [17.19]$$

where k is a coefficient describing channel roughness.

Key:
A = cross-sectional area
p = wetted perimeter
L_c = length of channel reach
R = hydraulic radius = A/p

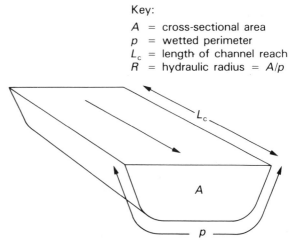

Fig. 17.6 Channel reach dimensions and definition of hydraulic radius.

159

For a channel reach (L_c), the total resistance exerted to the water flow is the product of the resisting stress (τ_0) and the total bed area over which it exerts, including the channel sides (Fig. 17.6), i.e. $\tau_0 p L_c$. Furthermore, the force (F) exerted on the bed by the water can be derived from equation [17.3] (namely $F = W \sin \theta$). For the channel reach in Fig. 17.6, the weight of water is equal to the product of the water's volume (AL_c), density (ℓ_w) and the force of gravity (g):

$$W = AL_c \rho_w g \qquad [17.20]$$

For uniform flow, the force exerted by the water ($W \sin \theta$) equals the total resistance ($\tau_0 p L_c$). Substituting for W, as in equation [17.20], we obtain:

$$\tau_0 p L = AL\rho_w g \sin \theta \qquad [17.21]$$

Rearranging equation [17.21], the resisting stress can be expressed as follows:

$$\tau_0 = \rho_w g (A/p) \sin \theta \qquad [17.22]$$

This expression can be simplified in the following ways: channel area (A) divided by the wetted perimeter (p) is called the hydraulic radius (R), and the unit weight of water (γ_w) is the weight of water per unit volume which equals $\rho_w g$. Moreover, for small slope angles (as in almost all river channels), $\sin \theta$ approximately equals $\tan \theta$, the latter being the gradient S. Thus, the resisting stresses (and under balanced conditions, the tractive stresses operating) can be expressed in the following simplified version of equation [17.22]:

$$\tau_0 = \gamma_w RS \qquad [17.23]$$

A relationship between stream velocity, channel dimensions and gradient can be derived by combining equations [17.19] and [17.23], which both give an expression for τ_0:

$$kV^2 = \gamma_w RS \qquad [17.24]$$

Rearranging equation [17.24], and combining k and γ_w into one coefficient (C), the *Chezy formula* is obtained, which shows how velocity changes as a function of channel dimensions and slope:

$$V = C\sqrt{RS} \qquad [17.25]$$

where C is a measure of channel resistance to flow. A very similar equation, which has been found empirically to be more applicable to actual channels, is the *Manning equation*, in which velocity is a function of the hydraulic radius and stream gradient, but with a slightly different coefficient for the latter, as follows:

$$V = \frac{R^{2/3} S^{1/2}}{n} \qquad [17.26]$$

where n is the Manning roughness coefficient. It should be noted that n will vary according to factors such as the coarseness of the bed material, the type of vegetation present, the channel cross-section and plan form. Although this formula gives a first-order understanding of the relationships between channel form and water flow, it is valid only for uniform flow and without the incorporation of sediment into the flow. The transporting ability of channelled flow will be further examined in Chapter 20.

17.6 Retrospect

Surface materials vary markedly in terms of their mineral composition, size distribution of particles, degree of interlocking between particles, pore size, organic content and moisture. Furthermore, the variation of these factors in three-dimensions means that an application of the simple relationships discussed in this chapter will provide less than perfect predictions of the behaviour of surface materials, especially as inputs to the debris cascade. Nevertheless, they do provide a basic introduction to the factors which have to be considered if sediment transport systems are to be fully understood. However, lack of space has prevented the inclusion of the techniques used to measure several of the properties mentioned, especially shear strength. These techniques are adequately covered in the further reading and in the works of Goudie (1981) and Selby (1982).

Key topics

1. The influence of force and resistance on earth surface materials of horizontal and inclined surfaces.
2. The effects of pore-water pressure and cohesion on shear stresses.
3. The characteristics of water flow and associated stresses.
4. The relationship of velocity to channel resistance and form.

Further reading

Carson, M. A and Kirkby, M. J. (1972) *Hillslope Form and Process*. Cambridge University Press; Cambridge (Chs 3 and 4).

Costa, J. E. and Baker, V. R. (1981) *Surficial Geology*. John Wiley and Sons; New York and London (Chs 5 and 8).

Finlayson, B. and Statham, I. (1980) *Hillslope Analysis*. Butterworths; London and Washington (Chs 3 and 4).

Richards, K. (1982) *Rivers Form and Process in Alluvial Channels*. Methuen; London and New York (Ch. 3).

Chapter 18 Sediment transport by mass movements

Gravity exerts a downslope stress on all the surface materials of the earth and the response of the materials to this stress is manifested in a variety of ways. Locally, very rapid transport may take place in the form of landslides and, much more pervasively, slow movements (commonly called creep) gradually transport material downslope, usually to the fluvial system. The general name given to all these transport mechanisms, which do not involve incorporation by ice, water or air, is *mass movement* or *mass wasting*. Any attempted classification of mass movements inevitably involves considerable simplification, since more than one process is nearly always found to be operating at any given site. Nevertheless, it is useful to distinguish between three basic types of movement (Carson and Kirkby 1972), namely *slides, flows* and *heaves*.

Slides involve failure along a slide plane, without any internal shear within the transported material and once started, the resistance to movement drops sharply. In flow movements, no sharply defined shear zone occurs and the failure is spread throughout the mass, with maximum shear at the base (strictly speaking, rivers could be included in this class). Heaves represent the third type of movement and involve expansion and contraction relative to the slope surface.

These simple types are usually not found in practice, but can be used to define the corners of a triangular diagram, with actual movements lying in intermediate positions (Fig. 18.1). The relative speed of transfer varies according to the time-scale considered and when a landslide occurs, it is obviously much quicker than the slow processes of creep. However, the contrast in rates will be much less over the longer term, since the former occurs very infrequently whereas the former may operate more or less continuously.

18.1 Shallow translational slides

As the name implies, this type of failure occurs close to the surface and the slides are characterised by a shear plane parallel to the surface (Fig. 18.2). They typically occur in soils with low cohesion and in the absence of moisture, failure will take place at an angle of ϕ (section 17.2), where the coefficient of internal friction (μ) equals tan ϕ. The friction between the particles is thus the primary factor giving the material its shear strength. However, as pointed out in section 17.3, the presence of water can reduce the normal stress (σ) thus decreasing the shear strength (s), which equals σ tan ϕ (equation [17.10]), without a corresponding decrease in shear stress. Consequently slopes, whose form is a result of failure by shallow sliding, will have slope angles at a much lower slope angle than ϕ. Translational slides tend to

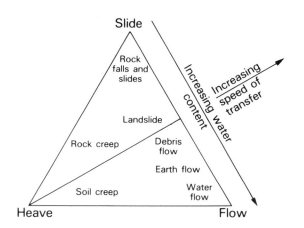

Fig. 18.1 Examples of mass movements based on the three basic movements of slide, flow and heave. (After Carson and Kirkby 1972)

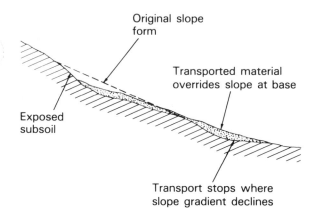

Fig. 18.2 Typical features of a shallow translational slide. Note that the transported material has a somewhat different shape compared with its original form as indicated by the dashed line. This indicates that internal deformation has taken place and that, as is usually the case, a simple slide as defined in the introduction has not occurred.

be shallow, because failure takes place through the weakest material and that is usually close to the surface, where the highest degree of weathering (Ch. 16) has taken place.

Intuitively, this type of sediment transport might be expected to be limited to materials with high proportions of sand-sized particles, with low cohesion. However, it is worth noting that even small amounts of silt–clay particles in sandy materials can significantly increase shear strength, largely through the additional moisture which is

retained, leading to a resultant negative pore-water pressure (section 17.2). Furthermore, in some clays shallow failures may be common, even in those with measurable cohesion in the laboratory. For example, in the strongly fissured London Clay of southern England, it was found that the strength is at a peak (ϕ_p) immediately after a slope is cut. Then, the shear strength progressively drops to a residual level (ϕ_r) and, as a consequence, artificial slopes may fail up to 100 years after they were initially cut.

The mechanism by which ϕ_p declines to ϕ_r has been associated with the penetration of water down the fissures which induces weathering (Skempton 1964). It is the intact material between fissures within the slope which prevents failure, and progressive weathering between the fissures reduces the cohesion until failure occurs. Materials which have been subject to previous failure are also found to have a much lower residual strength (ϕ_r) than the original peak strength (ϕ_p) before failure.

Work by Chandler (1982) on clay slopes in the English East Midlands, with angles between 6 and 15°, indicated that landsliding was very common and often occurred as shallow movements on the lower slope segments. Moreover, the residual strength (ϕ_r) depended strongly on the size of the landslide, with ϕ_r falling as the size of the landslide declined and hence as the effective stresses increase. The implication of this is that at least for clay slopes, there may be no unique threshold slope inclination; for the slopes considered by Chandler (1982), the range of resultant slope angles was between 6 and 9°.

18.2 Deep-seated slides

Slope materials (whether comprised of rocks or soils) possess a strength which is independent of the inter-particle frictional forces and with increasing depth, the weight (W) exerted by the overlying material progressively increases. In Fig. 18.3, the downslope stresses acting on a block of material of unit width within the slope are represented and the forces and stresses are directly comparable with those shown in Fig. 17.2 and equation [17.3]. As depicted in Fig. 18.3, the shear stresses (τ) are directly proportional to the depth (z) below the surface and can be described by the following equation:

$$\tau = \gamma z \cos \theta \sin \theta \qquad \text{[18.1]}$$

163

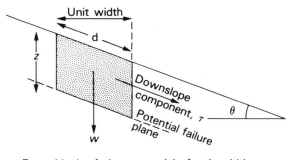

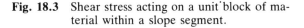

For a block of slope material of unit width:
$$W = \gamma z \cos \theta$$
∴ Downslope component of weight, shear stress (τ):

$$\tau = W \cdot \sin \theta$$
$$= \gamma z \cos \theta \sin \theta$$

Fig. 18.3 Shear stress acting on a unit block of material within a slope segment.

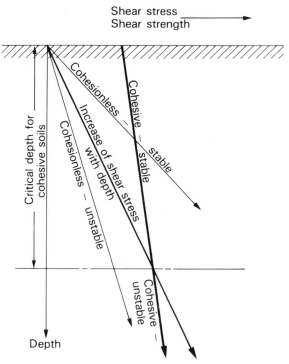

Fig. 18.4 Changes of shear stress and shear strength with depth. (Adapted from Carson and Kirkby 1972; Statham 1977)

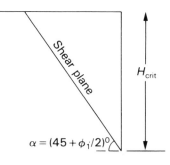

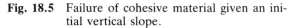

α = angle of shear plane
ϕ_i = angle of internal resistance

Fig. 18.5 Failure of cohesive material given an initial vertical slope.

The shear strength (s) of the slope material can be expressed by means of the Coulomb equation (equation [17.13]). For a unit width of slope, the normal reaction (σ) can be rewritten as $\gamma z \cos^2 \theta$, and therefore the Coulomb equation can be written as:

$$s = c' + (\gamma z \cos^2 \theta - u) \tan \phi_i \qquad [18.2]$$

Thus, the shear strength (s) also increases with depth (z) but less rapidly than the shear stress.

In Fig. 18.4, the shear stress and shear strength are both plotted against increasing depth. Failure occurs if shear strength is exceeded by the shear stress (i.e. once the factor of safety exceeds 1.0) and therefore, there will be a critical depth (z_c) before instability occurs. This means that in cohesive materials, the height of the slope as well as slope angle will control slope failure and hence control the downslope transport rate of materials by this mechanism. For cohesionless materials, the shear strength is zero at the surface and is a straight line above or below the line of shear stress. Depending on the water content, the orientation of the shear strength line will vary substantially for materials with or without cohesion.

In the case of a vertical face in homogeneous materials (caused for example by vertical stream incision), the form of the failure is relatively simple (Fig. 18.5). The angle of the shear plane (α) is directly related to the angle of internal shearing resistance (ϕ_i) with the height also being determined by the cohesion. in the following equation:

$$\alpha = (45 + \phi_i/2)° \qquad [18.3]$$

The height of the face necessary before failure (H_{crit}) occurs is given by the following equation:

$$H_{crit} = \frac{2c}{\gamma} \tan \alpha \qquad [18.4]$$

where c is the cohesion and γ is the unit weight. Full derivation of the above equation is beyond the scope of the present account, but it is worth noting here that an understanding of forces involved in deep-seated movements depends on the concept of *principal stresses*, (i.e. the maximum, minimum and intermediate stress intensities along each of three mutually perpendicular axes in a given material).

Within a slope, a cube of materials with arbitrary orientation can be imagined and, on each of its faces, the stresses can be resolved into those acting parallel to the face (i.e. shear forces) and those normal to the face. It is possible to choose an orientation of a cube such that the only stresses acting are those that lie normal to the faces. These stresses are called the major (σ_1), intermediate (σ_2) and minor (σ_3) principal stresses and the corresponding planes are also termed major, intermediate and minor. It can be shown (experimentally and theoretically) that the shear plane will be at an angle $\alpha = (45 + \phi_i/2)°$ to the major principal plane. Thus if the major principal stress is vertical, then the resultant failure will be along a linear plane at the angle α.

Except at coastlines and near incised river channels, vertical initial slopes are uncommon al-

though in such situations, the direction of the major principal stress is curved towards the lower ground. However, the angle (α) between this direction and the failure plane will still be at $(45 + \phi_i/2)°$ (Fig. 18.6), which results in a curved failure plane (Fig. 18.6). Hence, in such situations, a rotational slip will be produced with the characteristic features shown in Fig. 18.6. This outline explanation of the cause of deep-seated failures shows that they can occur in homogeneous materials and that their occurrence is not dependent on less permeable rock layers underlying more permeable ones, although slopes where these geological conditions are found are especially likely to be unstable (Brunsden and Jones 1976).

18.3 Rock falls and rock avalanches

The analysis of sediment transport by shallow and deep-seated failures outlined in the two previous sections assumed that the materials are nearly homogeneous. Hence, this analysis tends to be most applicable for well-weathered materials and for rocks without marked jointing patterns, as is found within some sedimentary rocks. However, in many rocks the orientation, length and density of joints have a major role in controlling the mode of slope failure. In particular, the angle (α) of the joints in relation to hillslope angle and the frictional angle (ϕ_i) of the rock mass (equation [17.8]) are of major significance. Where the major joints or bedding planes dip downslope, failure is more likely to occur and the resultant slope angles are less than where joints are horizontal or dip into the slope.

The shear strength of the rock is not a simple function of small intact specimens, but is controlled by the proportion of intact rock along potential failure planes and the frictional and cohesive forces of joints along such planes (Terzaghi 1962). Because jointing is so common, the latter is likely to be of prime importance and failure will occur once weathering has sufficiently extended its way along these lines of weakness. The form of the resultant *rock slides* and *avalanches* is consequently strongly affected by the vertical and horizontal arrangement of rocks and especially the arrangement of joints within them (Plate 18.1).

Where bare rock surfaces are exposed, slope material may occur by the detachment of individual blocks or fragments which result in *rock falls* (Plate 18.2). Detachment is by processes of weather-

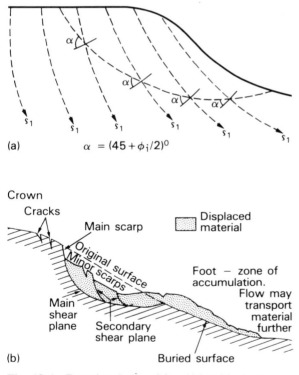

(a) $\alpha = (45 + \phi_i/2)°$

(b)

Fig. 18.6 Rotational slip: (a) relationship between angle of shear plane (α) of rotational slip and principal stress (s_1); (b) main components of rotational slip within cohesive material.

Plate 18.1 Failure in Quaternary sands, Basilicata, southern Italy.

ing immediately behind the face and in extra-tropical areas, processes of freeze-thaw are particularly important (section 16.2). By contrast, if large joints occur parallel to a steep face, then slab failure may occur. Such failure can occur in mechanically weak rocks where jointing is rare (Plate 18.1). If these are strongly undercut, horizontal pressure caused by the weight of the material can open up tension cracks. The plane of failure at the base of the slab fracture is likely to be close to $(45 + \phi_i/2)°$, as discussed in section 18.2. Slab failure also characteristically occurs in massive igneous rocks, where jointing develops with the release of pressure due to the erosion of the surrounding rock.

At the other extreme, especially in mechanically weak rocks, individual grains may be detached as a result of solution of the cement which binds particles, drying out of the surface which reduces capillary binding forces and by freeze-thaw. These grains may then accumulate at the base of the slope following *granular detachment* associated with subaerial weathering processes. In cold climates, the mechanism for mechanical weathering and hence for rock falls and granular

detachment, is generally ascribed to freeze-thaw activity, with the expansion of ice just below freezing-point exerting stresses within rocks (section 16.2a). Both White (1976) and Fahey (1983), among others, indicate that hydration (the adsorption of water) may exert considerable internal pressures within rocks and has comparable importance with freeze-thaw in these transport mechanisms.

18.4 Flows

Once a *landslide* has occurred, it is common for some further transport to take place by processes involving flow, especially where the material has a high proportion of clays (Fig. 18.6). This may involve much more distant transport than the initial landslip (Plate 18.3) and such flows may be very slow compared with water flow. Rates may be lower than 10 m yr^{-1} and although such movements may have the surface appearance of flows, there is evidence that sliding also takes place at the base of the mass movement. Such *slide flows*

Plate 18.2 Rock falls in the Grand Canyon of Yosemite National Park, USA, resulting in accumulation of extensive scree at base of slope.

may move much more frequently than the landslides feeding them, although the movement may still be discontinuous and display considerable seasonal variation. Explanation of the mobilisation of the clay following the initial slip remains unclear, though it is related to high water content and the resultant high pore-water pressures. These factors are encouraged by the tendency of landslides to discharge into valleys, where water tends naturally to concentrate.

In coarser materials, *debris flows* occur which tend to be faster than those mass movements just discussed and usually are only found on rather steeper slopes. Their significance (apart from their contribution to the debris cascade) relates to their hazard to human life caused by their speed. Recent work on debris flows is summarised by Innes (1983) who stresses the importance of remoulding of materials, incorporating water, the degree of inter-particle contact and the poorness of sorting.

Some clays may liquefy once disturbed and are known as *quick clays*. Failure is initiated by a slip, whose debris liquefies and moves away from the base; a further slip then occurs in the new face repeating the process. Although these clays result in a very active transport system, they occur in very few parts of the earth's surface.

18.5 Slow mass movements

In contrast to the mass movements described previously (which occur at discrete locations and usually intermittently), slow mass movements occur extremely widely. In section, they may be very obvious but their very slowness makes them a difficult phenomenon to study. Observations in the Pennines in northern England indicate that the downslope movement of new surface material was only 1 cm every 40 years (Young, 1974). A number of different processes contribute to slow mass movements and usually, more than one will contribute to the net transport of material.

In cohesive materials, continuous creep occurs due to the downslope stress exerted by gravity. Figure 17.3(d) distinguishes between materials which show a progressive strain rate with stress and plastic ones, which suddenly fail. As Fig. 18.7 shows, clays in this respect behave in a fashion intermediate between these two types of stress relationships. If stresses are sufficiently high, failure is likely to occur, but for lower stresses, where the stress–strain curve is more nearly parallel with the horizontal axis, strain will still occur but without the formation of slip planes. The underlying process is probably related to the breaking and reforming of electrochemical bonds between clay particle edges and water layers.

Nearer the surface, various heave processes occur. Explanations of how these heave processes cause downslope movement essentially rely on the upward movement being normal to the slope surface and the downward movement being more vertical, due to gravitation. Heaving occurs as a result of various systematic environmental climatic factors such as freeze-thaw, thermal expansion – contraction and wetting – drying, particularly where clay-sized particles are present. Although the net movements leading to soil creep may occur, random short-term movements are apparently large in comparison with downslope transport over time periods of the order of a year (Finlayson 1981). In part, these processes are

Plate 18.3 Flow of material from A to B following initial short-distance rotational slip near crest of slope, Basilicata, southern Italy.

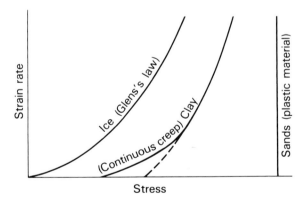

Fig. 18.7 Strain rate of clay as a function of applied stress. Note that ice has no minimum yield strength unlike both clay and sand.

caused by biotic factors such as the opening and collapse of animal burrows and plant rootways. Exposed rock particles on bare rock surfaces or in scree slopes also display slow movements down-slope. This *rock creep* has its origins in the pro-cesses which are important in slow movements within soil or regolith, such as thermal expansion–

contraction and freeze-thaw action. *Solifluc-tion* is a type of slow mass movement, discussion of which is postponed until Chapter 23.

18.6 Retrospect

The debris cascade on slopes, as a result of mass movements, involves a very wide variety of mech-anisms which result in transport at many different rates. The slowest mass movements are so slow as to be almost beyond the skill of the geomorphol-ogist's reliable measurement. Yet, such move-ments are by far the most widespread and, in many areas, may be the important component of the cascade. At the other end of the scale are rock slides and avalanches which are concentrated in mountain and coastal areas and tend to happen very infrequently. In rare cases, rock may fall in sufficient quantity and achieve a sufficient velocity so that air is trapped and acts as a cushion, on which the debris may then be supported and trans-ported very rapidly downslope.

Estimates of velocity for the Mount Huascaran landslip in Peru (1962) yielded maximum values of approximately 400 km hr^{-1}. For the first 15 km, relatively little deposition took place due to the air cushion. When deposition did occur, it led to the death of over 20 000 people and, furthermore, movement continued for a further 50 km. Fortunately, such landslides are extremely rare. Nevertheless, landslides in general are major hazards in many parts of the world. During earthquakes, landslides are one of the most important associated hazards and causes of death, with the shock waves leading to short-term increases in shear stresses and/or reductions in shear strength. Rapid mass movements can also be of considerable significance with the construction of new roads.

Key topics

1. The characteristics and mechanisms of mass movement.
2. The control of shear strength on mass movement.

3. The nature of rapid or catastrophic movement of surface materials.
4. The role of gradual mass movement in distributing surface materials.

Further reading

Chandler, R. J. (1977) The application of soil mechanics methods to the study of slopes, in Hails, J. R. (ed.) *Applied Geomorphology*. Pergamon; Oxford and New York, pp. 151–81.

Cooke, R. U. and Doornkamp J. C. (1974) *Geomorphology in Environmental Management*. Clarendon Press; Oxford (Ch. 6).

Gardner, J. S. (1982) Alpine mass-wasting in contemporary time: some examples from the Canadian Rocky Mountains, in Thorn, C. E. (ed.) *Space and Time in Geomorphology*. George Allen and Unwin; London and Boston, pp. 171–92.

Selby, M. J. (1982) *Hillslope Materials and Processes*. Oxford University Press; London and New York (Ch. 6).

Yatsu, E., Ward A. J. and Adams, F. (eds) (1975) *Mass Wasting*. Geo Abstracts, University of East Anglia, UK.

Chapter 19 Sediment transport on and within slopes by water

The previous three chapters have already demonstrated the key importance of water in the debris cascade, through its role in weathering mechanisms and its effect on the shear strength of materials. In this chapter, aspects of the slope debris cascade are examined in which water movement directly transports debris, and there are three quite distinct mechanisms involved (Fig. 19.1). Firstly, raindrops may transport materials by their impact and secondly, under high rainfall intensities, some slopes may suffer from

erosion as a result of water flowing across its surface. Finally, within slopes, movement of debris also takes place primarily in solution. Vertical movement of materials leads to the vertical differentiation of weathered products and the development of soils, as described in Part IV. Nevertheless, these processes acting both vertically and laterally (downslope) have importance as components of the debris cascade.

19.1 Rainsplash

Where bare ground is exposed, raindrops striking a slope have a variety of geomorphological effects (Fig. 19.2). On many surfaces, the first effect is the breakdown of soil aggregates, and Farres

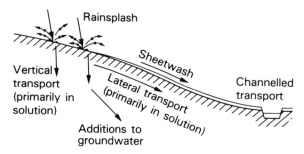

Fig. 19.1 Main mechanisms for transport of debris by water.

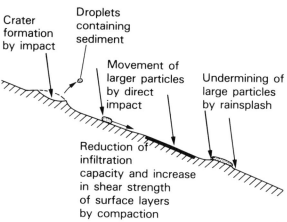

Fig. 19.2 Geomorphological effects of raindrop impact.

(1980) presents evidence to show that the most important process in this breakdown is slaking, which is caused by internal hydrostatic forces produced by water trapping air within the aggregates. The impact of raindrops tends to cause the formation of craters by materials sliding across the surface away from the impact and by splashes of sediment-laden water being thrown away from the impact. Only sand-sized or smaller particles can be moved by aerial transport and crater formation. Thus, transport is selective according to size and in some locations, a surface lag transport of coarser material may be left. Larger particles may be moved to a limited extent as a result of raindrops physically pushing particles downslope and by undermining due to rainsplash of the surrounding slope. Pebbles can exert a protective influence on the underlying soil leading to the formation of *earth pillars* and similar forms may be produced by some plants.

It should be apparent that all the mechanisms of movement by rainsplash can lead to upslope, across-slope as well as downslope movement. The resultant net sediment movement is controlled by the direction of raindrop impact and the slope angle, and the steeper the slope, the greater will be the net downslope transport (Table 19.1). For sediment within splashed droplets, this arises from their longer downslope trajectory compared with those moving upslope, In crater formation, the tendency for net downslope movement arises from the greater force required for upslope compared with downslope movement.

The transport caused by rainsplash fundamentally arises from the energy possessed by the raindrops. This kinetic energy (K_e) is defined as follows:

$$K_e = \tfrac{1}{2}\, mv^2 \qquad [19.1]$$

where m is the mass of the raindrops and v their velocity. Thus, the amount of material transported is a function of rainfall amount, the raindrop size and the velocity of the raindrops. Larger raindrops not only have greater mass but also higher velocities and hence will contribute most to rainsplash erosion. In tropical areas, rainfall intensities are usually much higher than in temperate areas so that the proportion of erosive rainfall is higher and the total kinetic energy is substantially greater for tropical areas (Table 19.2).

Such considerations are valid only for unvegetated surfaces since under most types of vegetation cover, the effects of rainsplash are reduced very significantly. Consequently, with increasing rainfall under natural conditions, the expected tendency is for increased rainsplash erosion to be counteracted by increasing vegetation cover. It has been estimated that peak transport under natural conditions lies at rainfall amounts of only 300–425 mm. However, for comparable areas which have been cleared of vegetation, then higher rainfall amounts will lead to greater transport rates. Under high canopy forest, there is evidence that drips from leaves (section 7.1) and direct rainfall can contribute significantly to rainsplash where the near-surface vegetation cover is incomplete. For example, Mosley (1982) found that under a beech forest canopy in New Zealand, splash detachment was three times that in open ground. Morgan (1982) has carried out experiments which suggest that for some crop types, there is an inverse relationship between soil detachment and the total kinetic energy of the rainfall striking the top of the plant canopy. This suggests that more complex relationships are operating than those which are described above.

As described in the following section, slopes

Table 19.1 Relation between slope angle and proportion of material transported downslope

Slope angle (°)	Proportion of material downslope (%)	Source
0	0	Mosley (1973)
6	75	Ellison (1944)
25	96	Mosley (1973)

Table 19.2 Sample values comparing the erosivity of rainfall of temperate with tropical areas. (Based on figures in Hudson 1971)

	Annual rainfall (mm)	Rainfall that is erosive (%)	Amount of erosive rainfall (mm)	Typical K_e of rainfall/mm (J m^{-2} mm^{-1})	Total K_e of rainfall (J m^{-2})
Representative temperate climate	750	5	37.5	24	900
Representative tropical climate	1500	40	600	28	14 400

may be covered by a thin layer of water after heavy rainfall called overland flow which, if sufficiently thick, can act in a protective manner against the impact of raindrops. A further important interaction between rainsplash and overland flow is that the impact of raindrops causes surface compaction, with a resultant decrease in infiltration capacity (section 6.1) and increase in shear strength of the surface layers. However, the effects of the former are generally believed to be much more important than the latter and thus the effect is primarily to increase the erosion rates. The formation of soil crusts is particularly prevalent on sandy and loamy soils.

19.2 Surface wash

Surface or sheetwash transportation originates in overland flow (section 7.1) which develops in two different ways, depending upon whether the surface layers are saturated or not. In the former case, entry of water into the soil is inhibited because the soil voids are occupied by water. Saturation results not only from inputs by rainfall, but also by water movement through the soil. Consequently, there is a tendency for saturated conditions to be found in sites where subsurface water flow concentrates. Such concentration tends to be in hollows and at the base of slopes and hence has a relatively limited areal extent within drainage basins. Knapp (1978) shows, moveover, that within a slope profile, the upper parts rapidly lose water after having been saturated. Conversely, at the slope base the moisture content drops relatively little and remains close to saturation for a considerable period afterwards. Consequently, saturation overland flow, following subsequent rainfall, is most likely to occur at the slope base. Saturation overland flow tends to be more typical of higher rainfall areas, with substantial vegetation cover. In large part because of this, sediment transport by saturation overland flow is usually limited.

The other type of overland flow is termed *Hortonian* after the American hydrologist R. E. Horton. He pointed out that overland flow can be initiated without saturation as long as the rainfall intensity exceeds the infiltration capacity (section 6.1), which is the maximum rate at which water can enter the soil at a particular moment. This qualification is important since the infiltration capacity of a soil is not a constant, but usually varies

substantially with time. Infiltration rates of initially dry soils typically fall at a rapid rate during a storm to a relatively constant value. Due to the high spatial variability of soil characteristics within even small areas, it follows that infiltration capacity, and hence the occurrence of overland flow, is highly variable. Overland flow will not immediately commence when rainfall intensity exceeds infiltration capacity because of surface microtopography and the surface depressions must be overtopped before downslope flow can occur.

It should be noted that the relationships between runoff rates and rainfall intensity are affected by antecedent conditions. For example, Emmett (1978) compared sediment loss for the same slopes which were initially dry and wet and found much higher loss for the same simulated rainfall intensity for wet slopes. From this, he concluded that the runoff rate is dependent on infiltration rates as controlled by the antecedent soil moisture and that the greatest erosion rates are related to the highest runoff rates. Since the rainfall intensity and hence rainsplash erosion were the same in both situations, it was argued that this shows the importance of overland flow for transport of sediment, compared with rainsplash.

Hortonian overland flow is more typical of semi-arid areas with high rainfall intensities, sparse vegetation cover and impermeable soils. In humid areas, infiltration capacity will often be too high ever to be exceeded by naturally occurring rainfall intensities. A preliminary understanding of the erosive forces acting on a slope as a result of sheetwash can be gained by referring back to Manning's equation (equation [17.26]). For a very wide channel or overland flow across a hillslope, the depth of flow (d) can be substituted for the hydraulic radius (R). Thus, equation [17.26] can be rewritten as follows:

$$\text{Velocity} = v = \frac{d^{2/3} s^{1/2}}{n} \qquad [19.2]$$

where s is the slope and n the coefficient describing the roughness of the surface across which flow is occurring.

Equation [19.2] can be rearranged to obtain an expression for the discharge flowing across a slope. Firstly it is apparent that for a unit width of slope (Fig. 19.3), the velocity (v) can be expressed as q/d, where q is the discharge per unit width. Secondly, the effects of slope angle (θ) can no longer be expressed by the slope (s) but must be represented by $\sin \theta$, since slope angles may be quite large on hillslopes (see equations [17.22] and

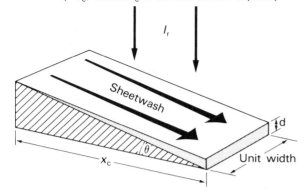

Rainfall intensity, I_r, excess intensity, $i = I_r - I_c$, where I_c is the infiltration capacity

Fig. 19.3 Main variables in determining width of belt of no sheetwash erosion.

[17.23] for explanation). Therefore, the following equation is obtained for the discharge/unit width (q):

$$q = \frac{d^{5/3} \sin^{0.5} \theta}{n} \qquad [19.3]$$

The shear stress needed to be overcome by the slope material's shear strength is given in equation [17.23]. As for equation [19.2], substitutions of d for R and $\sin \theta$ for s are made, giving the following equation for the shear stress:

$$\tau_o = \gamma_w d \sin \theta \qquad [19.4]$$

where γ_w is the unit weight of water. The discharge for a unit width of slope (q) can be expressed as a product of the rainfall intensity in excess of the infiltration capacity (i_{ex}) and the distance downslope from the divide (x), as follows:

$$q = i_{ex} x \qquad [19.5]$$

As with the depth of water, the shear stress will increase until the shear strength (τ_c) is exceeded and erosion will commence. The distance x_c, at which this depth occurs, defines the belt of no erosion (Horton 1945). Rearranging equations [19.3]–[19.5] gives this expression for x_c:

$$x_c = \left(\frac{\tau_c}{\gamma_w s_c} \right)^{5/3} \bigg/ n i_{ex} \qquad [19.6]$$

where $s_c = (\sin \theta / \tan^{0.3} \theta)$. It should be noted that this expresses the width (x_c) for slope material with a given shear strength. Furthermore, it will also vary according to the rainfall intensity (in excess of the infiltration capacity) and thus it is not a fixed slope parameter, even for given slope material.

The above theoretical considerations are now believed to provide a rather inexact prediction of how slopes are actually eroded in many circumstances. Thornes (1979) cites objections in terms of experimental results showing that discharge

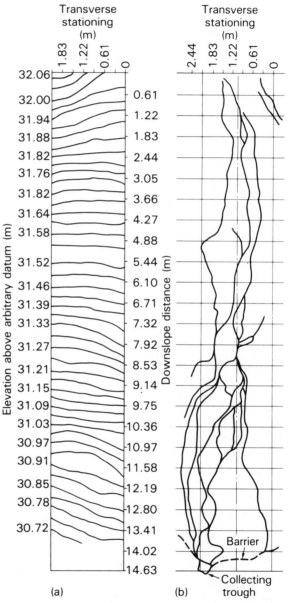

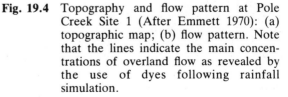

Fig. 19.4 Topography and flow pattern at Pole Creek Site 1 (After Emmett 1970): (a) topographic map; (b) flow pattern. Note that the lines indicate the main concentrations of overland flow as revealed by the use of dyes following rainfall simulation.

173

does not increase with distance from the divide, that the considerations fail to take account of the armouring effects of lag deposits on slopes and that flow on slopes may often be concentrated in rills where erosion rates are greater. Observations using artificial rainfall on natural slopes show that even when flow is not actually in channels, it is commonly concentrated into preferred lines (Fig. 19.4). Consequently, there will be considerable variations in depth of flow and degree of turbulence.

Despite these reservations, the discussion of the forces involved in sheetwash does indicate the variables which are important in controlling sheetwash rates. Both laboratory and field experiments (in which sediment is collected by sediment traps) have shown strong relationships with slope angle and slope length (Table 19.3). The reasons for the importance of the former presumably relate both to the influence of slope angle on the resistance of the material to motion and the shear stresses tending to move slope material. Slope length is related to sediment movement mainly because it acts as a surrogate for discharge (q) on slopes which are linear in plan form (i.e. without hollows or spurs). The difference in values of the coefficients in the equations of Table 19.3 should be noted. These are related to differences in experimental procedures and slope materials and where there are changes in infiltration capacity, then there may be very large changes in soil erosion rates with distance downslope (Townshend 1970).

A further variable of great significance in controlling sheetwash rates is the vegetation cover. In agricultural areas, where vegetation cover is limited, very considerable sheetwash rates may occur although under extensive natural vegetation, such erosion is unlikely to occur (Plate 19.1). Many

Table 19.3 Empirically determined relationship between sediment transport and slope characteristics. Adapted from (Carson and Kirkby 1972)

	Source
$s \propto x^{1.6} \tan^{1.4}\theta$	Zingg (1940)
$s \propto x^{1.35} \tan^{1.35}\theta$	Musgrave (1947)
$s \propto x^{1.73} \tan^{1.35}\theta$	Kirkby (1969)

s = sediment transport; x = distance from interfluve; θ = slope angle.

Plate 19.1 Sheetwash on a field of young maize (corn) in Indiana, USA.

Table 19.4 Average runoff and soil erosion from experimental erosion plots, Mpwapwa, Tanzania, 1946–54. (Based on Temple and Rapp, 1972 using data from Van Rensburg, 1955). Note the effects of different land types and conservation methods on % runoff and the much larger effect on soil loss

Land cover	% Runoff	Soil loss (m^3 ha^{-1})
1. Cultivated Plot with sorghum – no conservation methods	19.3	36.9
2. Cultivated plot with sorghum – with two 2 m wide grass belts across plot.	15.5	20.2
3. Upper $\frac{1}{2}$ under sorghum – as for (1) above; lower $\frac{1}{2}$ under grass as for 4	11.6	2.2
4. Grass covered (grass up to 1 m tall)	4.9	0.5
Mean annual rainfall	690 mm	

field experiments have shown how erosion rates are influenced by vegetation cover and the way in which land is cultivated and managed. These experiments are usually conducted by constructing barriers around several plots of the same size and then collecting the sediment and water in a sediment trap at the base of the plot. Different vegetative cover types are grown in each of the plots. Table 19.4 shows the results of one such set of experiments from semi-arid Tanzania and the consequences for erosion rates, of cultivating land with a reduction in soil productivity, can be clearly seen. It should be stressed at this stage that these measurements include both transport by sheetwash and by rainsplash, since separation of transport by the latter is technically difficult.

The dramatic influence of vegetation on surface erosion rates is exerted by several different mechanisms. Precipitation is intercepted by the foliage and a proportion never reaches the surface due to evapotranspiration. However, such effects tend to be less important for high precipitation rates which result in the occurrence of overland flow. Vegetation cover and the resultant litter check the impact of raindrops thus reducing the breakdown of soil aggregates by splash and hence hindering the formation of surface crusts with low infiltration capacities. As already pointed out, under some canopies the splash effects may remain significant. Vegetation also influences erosion rates through the accumulation of organic-rich surface soil materials with high infiltration capacities.

Root action is important in its effect both on soil structure, which raises the infiltration capacity, and soil cohesion which physically binds the soil together. However, the latter influence is very dependent on the density and type of rootlets present. Vegetation and litter also block and slow overland flow and consequently reduce erosion. As already discussed, however, individual plants can also physically concentrate flow locally, increasing erosion rates. The effects of rainfall on vegetation growth mean that as for splash erosion, sheetwash rates tend to be at a maximum in semi-arid areas where the erosive effects of increasing rainfall amounts begin to be counterbalanced by increasing vegetation cover.

If inappropriate land usage occurs, especially in semi-arid regions, erosion can rapidly remove the more fertile soil layers. This in itself will tend to hinder the regrowth of vegetation and leave more of the ground exposed, further encouraging more soil erosion. This positive feedback loop can lead to long-term reductions in land capability. The relative importance of sheet erosion and rainsplash is the subject of some controversy. Without rainsplash, rates of sheet erosion will be less because of the impact of rainsplash on the formation of a surface crust. However, with increasing depth of flow, the physical impact of rainsplash is much reduced by the presence of the water layer. This in turn will reduce the disturbance of slope materials for incorporation into the overland flow. In practice, as already mentioned, it is not easy to separate these two transport processes when making measurements.

19.3 Subsurface water erosion

Surface processes provide the most obvious evidence of slope erosion by water but below the surface, considerable transport is carried out by both physical and chemical processes. Within soils and rocks, the movement of water is usually very slow compared with surface rates and consequently its ability to transport material physically is correspondingly low. Chemical transport is usually a much more potent process in subsurface slope erosion.

(a) Physical transport of material

Physical transport of material takes place through holes and pores of various dimensions within the soils and these dimensions set thresholds to the size of material that can be transported. Usually, such holes and pores are quite small and thus the size and quantity of sediment that is transported is also small. In recent years, however, many locations have been found where pipes are common and contribute significantly both to the movement of water into channels and to sediment transport (Jones 1981; Bryan and Yair 1982). Such pipes may vary from a few millimetres in diameter to, in exceptional circumstances,a metre or more and they tend to develop where throughflow is concentrated by the presence of relatively impermeable layers beneath more permeable ones.

Thus, pipes are often found in peat-covered slopes in upland parts of the UK where the surface organic layers have very high permeability. On such slopes, a network of pipes may develop beneath the surface and its presence is indicated only where a pipe has collapsed leaving a surface depression. The development of pipes is poorly understood although it is apparent that some develop from steep stream banks, gully headwalls and the headwalls of landslides (Temple and Rapp 1972), where a junction between permeable and impermeable materials is exposed. Piping can also occur in clays where there is a large amount of exchangeable sodium which causes dispersion of clays and allows enlargement of existing pores and cracks. Other likely origins of pipes include holes made by higher animals like rodents, lower animals (such as worms) and the rotting of old roots.

If such pipes are absent, then only the finest materials will be transported, usually in relatively modest amounts. Moreover, there will be a tendency for pipes to be blocked by the movement of material itself. However, measurements by Pilgrim and Huff (1983) from a Californian field plot in the USA reveal that high concentrations (> 1000 mg l^{-1}) of suspended sediment can occur within the subsurface flow. The transported materials were mainly between 4 and 8 μm in size and moved within macropores caused by roots, worms and cracking and it was concluded that the materials originated not from within the soil itself, but from sediment detached originally by rainsplash erosion. This study therefore provides an interesting example of the interlinkage between the surface and subsurface components of the slope debris cascade.

(b) Chemical transport of material

Although a distinction between chemical and physical transport is commonly made, there is no sharp boundary between material which is in a true ionic solution and material in a colloidal state. Cations (Ch. 16) often attach themselves to colloids and thus will be included as part of the physically transported load. The chemical transport of materials is of considerable importance in the debris cascade and in many parts of the world, this type of transport is of greater significance than physical transport. Evidence for this can be obtained from measurements of stream load, since the immediate origin of the solution load is very largely from subsurface flow.

Even in Arctic areas, it has been found that solution can be an important component of debris transport (Table 19.5). The details of the processes involved have been discussed in part in Chapter 16 and in the chapter on soil-forming processes (Ch. 26). The latter processes are themselves of considerable geomorphological significance in ultimately contributing to the loss of material from slopes. In more arid areas, the infiltration may be inadequate to maintain a substantial flux of weathered materials from the soil profile and the near-surface horizons will become zones of accumulation of the products. This also means that transport of material in solution is usually of relatively minor significance in the debris cascade of such areas.

In humid tropical areas, the potential for chemical weathering is highest. However, the rate of chemical transport may be relatively small under rainforest conditions, because of the efficient nutrient recycling of the root system, and this results in only a small net loss of materials in solution to groundwater and thence to the stream

Table 19.5 Percentage of river load in solution. (Adapted from Statham 1977 and Derbyshire et al. 1977)

Region	%
Congo (Zaire)	76
Amazon tributaries	32
Xingu ⎫ Amazon tributaries	75
Jutai ⎭	11
Southeast Devon, UK (5 small catchments)	45–77
Southern Negev	0.6
Karkevagge Valley, Sweden	22
Mississippi	36

system. Low rates of chemical transport will also result if the constituent materials of the slopes have already suffered intense deep weathering. The most mobile cations (namely calcium, magnesium, sodium and potassium) tend to be moved preferentially compared with the relatively immobile components of iron, aluminium and silica. Within soils of extra-tropical areas, vertical movements of iron and aluminium can occur within soil profiles but in the humid tropics, even iron and aluminium can be transported laterally downslope.

The route taken by water in passing through the slope system affects the rate of loss of material in solution (Trudgill 1977 and Fig. 19.5). Overland flow moves rapidly and thus the water is not in contact with the materials for a sufficient time for saturation to occur and solute concentrations will be low. Movement by interflow (section 7.1) or throughflow will also be too rapid for much dissolution to be achieved. At the start of a storm, water which has been virtually stationary within the slope may be forced out by the addition of the new water input. Thus, throughflow and saturation overland flow, at the base of a slope or in

hollows, may initially possess higher than expected solute concentrations.

Groundwater moves much more slowly and hence saturation is much more likely to occur. Thus, concentrations of the dissolved load in groundwater will usually be much higher than for overland flow or throughflow. It should be apparent from the above discussion that the slower the movement of water, the greater will be the concentration of a given dissolved mineral until saturation level is reached. The rate at which such movement takes place is a function of the pore size and the frequency of occurrence of pipes and cracks. Local surface morphology and position on a slope profile also affect rates of solution. Detailed work on soils overlying quartzites in the Quantock Hills, Somerset, UK, by Crabtree and Burt (1981), indicate that denudation by solution is higher within a hillslope hollow than on the adjacent spurs due to higher rates of throughflow. The base of the hollow has a lower rate of solutional denudation due to the throughflow received being solute-rich as a result of solution upslope.

The effect of rock type on the importance of sub-surface solution is best shown in limestone areas, which often have little or no surface drainage outside of areas with frozen ground. The extent of sub-surface drainage is caused by the solubility of limestone's main constituent minerals (calcite and dolomite) to water in which carbon dioxide has dissolved. Thus numerous pores, joint planes and pipes are opened up and overall infiltration rates are very high. The solution of carbon dioxide in water occurs both in the atmosphere and within the soil: it depends directly on the carbon dioxide concentrations in the air, inversely on temperature and as a function of biotic activities in the soil. The hydrogen ions of the resultant weak carbonic acid cause the limestone to be dissolved, with production of calcium ions (Ca^{++}), bicarbonate ions (HCO_3^-) and if dolomite is present, magnesium ions (Mg^{++}) (section 16.2). If the amount of free carbon dioxide in solution falls below an equilibrium level, then the reaction reverses and calcium carbonate is reprecipitated. The rate of solution of limestone depends not only on the availability of hydrogen ions but also on the rate at which water passes through the regolith and rock.

In summary, transport of material by solution is a complex function of the materials present in terms of their solubility, the routeing of water that occurs and the rate of water movement. These in

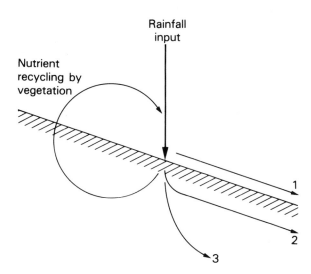

Type of flow	Speed of flow	Concentration of dissolved load
1 Overland flow	Rapid	Low
2 Interflow	Moderately rapid	Low
3 Groundwater flow	Slow	High

Fig. 19.5 Routeways of dissolved load from slopes. (Adapted from Trudgill 1977)

turn are controlled both by the climate and the actual vegetation which is present. It is certain that the dissolved load makes a significant contribution to the debris cascade and outside of the arid and semi-arid parts of the earth, it is often dominant.

19.4 Retrospect

The direct movement of debris by water includes not only the physical transport of solid debris but also transport in solution. The most important route taken by the dissolved load is within groundwater and this provides one of the most direct links between the debris and hydrological cascades. Compared with any other component of the debris cascade, transport by solution is by far the most continuous in time and space, and for humid areas at least, it never ceases. In contrast, sheetwash as a result of overland flow is a much rarer phenomenon and for Hortonian flow, its occurrence is strictly controlled by the threshold exerted by infiltration capacity and the extent to which the latter is exceeded by the rainfall intensity.

The relative importance of all the processes analysed within this chapter varies greatly across the globe. Climate exerts a major direct influence on this variability through the supply of water and through removal of water by evapotranspiration. The latter removes water from slope systems and hence prevents this water contributing any further to the debris cascade. The climatic influence is also exerted through its effect on vegetation. The second major influence on spatial variability of these processes relates to properties of slope materials, especially in terms of their shear strength and infiltration capacity.

Key topics

1. The significance of factors controlling rainsplash erosion and sheetwash in sediment transfer on and within slopes.
2. The role of saturation in overland flow.
3. The interrelationship between rainfall intensity and infiltration capacity.
4. The nature of subsurface physical and chemical water transport.

Further reading

Carson, M. A. and Kirkby, M. J. (1972) *Hillslope Form and Process*. Cambridge university Press; Cambridge (Chs 8 and 9).

Embleton, C. and Thornes, J. (1979) *Process in Geomorphology*. Edward Arnold; London (Chs 2, 6 and 7).

Hudson, N. (1971) *Soil Conservation*. B. T. Batsford; London (Chs 2–5).

Kirkby, M. A. (1979) *Hillslope Hydrology*. John Wiley and Sons; New York and London (especially Chs 1, 5 and 9).

Selby, M. J. (1982) *Hillslope Materials and Processes*. Oxford University Press; London and New York (Ch. 5).

Trudgill, S. T. (1977) *Soil and Vegetation Systems*. Oxford University Press; London and New York (Ch. 5).

Chapter 20 Sediment transport by water in channels

The action of mass movement (Ch. 18) and water on and below the ground surface (Ch. 19) usually results in sediment transfer across relatively short distances, before a stream or river channel is reached. The type of movement of sediment within a channel varies considerably according to its size. Furthermore, materials in solution or colloids will move at the velocity of the river and transport takes place more or less continuously, though at varying concentrations. The transport rates of very fine solid particles may also show relatively small variations, but for most particles, movement is highly intermittent and is most common during flood discharges. The continuity of movement of materials is further disturbed by temporary storage, especially in floodplains. Where such sediments are cut into and left as river terraces, 'temporary' storage may last for thousands of years. Although some channels are cut into bedrock, most river channels are alluvial and the size, shape and gradient of such channels interact with sediment transport in a highly complex way.

By way of introduction, it is important to define the *drainage basin* or catchment of a river (section 7.3) whose limits are set by the topographic divides which separate the drainage basin from slopes leading down to other river channels (Fig. 20.1). In terms of surface flow, the divides functionally limit the area from which the river system receives its inputs of water. However, and particularly for small drainage basins, there may well be significant additions and losses of groundwater from areas outside of the defined catchment area (section 6.3). Although topographic maps make the drainage network appear to be of fixed extent, in practice the length of active drainage lines usually varies considerably, even during indivdual storms, with the headward growth of the network. Normally, this growth will be into previously eroded channels but it may also extend above the existing channels. In headwater positions, the fluvial system is battling against the slope system which is continuously providing sediment to obliterate fingertip tributaries.

20.1 Solute transport

In the analysis of subsurface transport by water, the dominance of the dissolved or solute loads was stressed (section 19.3). Solution within slopes (compared with that in rivers) is important because of the slow movement of the water and lengthy contact with the slope materials. Once the water enters a river, movement is much more rapid and for most rock types, the majority of the dissolved load of rivers originates from subsurface solution, and little comes from the channel materials themselves. On the other hand, for soluble

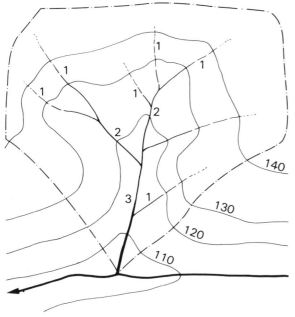

_____ Contours: height in metres

—·—·—·—·— Perimeter of catchment

_____ Channels containing flowing water

— — — — — Channels without flowing water

· · · · · · · · · · · Parts of valleys into which channels
may extend following high-intensity
infrequent rainfall

1 2 3   Orders of network according to the
Strahler system. Unbranched
elements are first order; two first
orders produce a second order;
order only then increases by one if
two streams or channels of the
same order join. Thus, when two
second-stream order streams join
they produce a third-order stream.
However, a first-order or
second-order stream joining a
third-order stream does not affect
its order

Fig. 20.1 Elements of a drainage basin and its
stream network.

rocks (such as limestones) the erosion of channels, either in surface or subsurface forms, by solution is undoubtedly important. Once the dissolved load enters a river, mixing rapidly occurs due to the turbulent nature of the river flow (section 17.5).

If samples of river water are taken during a storm and the dissolved load is determined, a decline in solute concentration will almost always be found as the discharge increases. This is due to dilution by the addition of rainfall during a storm, which has spent little time in contact with surface materials before running off to join the stream water. In some cases, an initial rise in concentration may occur (Oxley 1974) due to the flushing-out of solute-rich waters from bottom-slope positions and the removal of salts which have accumulated near the ground surface. For a given cross-section of a river, it is common therefore to find an inverse relationship between solute concentrations and discharge, although positive relationships, or even open V-shaped relationships, may be found depending on the extent and location of different rock types within the contributory drainage basin (Walling 1971). Even when there is a direct relationship between discharge and total dissolved load, it is likely that storm flows may be less important than flows nearer to the average, because of the much greater frequency of the latter. Conversely, Foster (1980) found that the concentration of potassium ions was approximately constant during baseflow, but quickflow detached these ions (K^+) from soil-clay colloids, leading to rapid increases in concentration.

The total dissolved load is a product of the concentration of solutes and water discharge. Consequently, even for rivers displaying the dilution effect and hence an inverse relation of concentration with discharge, there may well be a direct relationship between total dissolved load and discharge. In the USA, analysis of data from streams over a wide variety of climatic zones (Leopold *et al.* 1964) revealed a relationship between total dissolved load (in kg day^{-1}) and discharge (m^3 day^{-1}) with the following form:

$$TDL = aQ^{0.8} \qquad \textbf{[20.1]}$$

where TDL is the total dissolved load day^{-1}, Q is the discharge and a is a constant. For some individual rivers, the exponent is as low as 0.5, but it never exceeded 1.0 which indicates that dilution was present in all rivers.

On average, the most common ionic constituents of the solute load are the bicarbonate (HCO_3^-), sulphate (SO_4^{2-}), calcium (CA^{2+}) and silica (Si) ions. Of less, but still considerable importance, are the ions of magnesium (Mg^{2+}), sodium (Na^+), potassium (K^+), iron (Fe^{3+}) and nitrate (NO_3^-). In areas of intensive farming, the concentration of some ions (such as NO_3^- and K^+) may be very high seasonally, due to addition and subsequent washing-out of fertilisers. It is appropriate to note here that an estimation of the sediment loss by solution must be based not only on the measured solute concentration of river waters, but also on inputs from rainfall and artificial sources. In areas with rocks containing largely insoluble materials, dissolved materials in rainfall may contribute to a large proportion of the total dissolved load. Variation in rock type affects both the total amount of dissolved load and the relative proportions of different ions. In addition, similar rock types in different climatic environments will also result in considerable contrasts in the proportion of solutes.

20.2 Initiation of particle movements

Water flowing through a channel exerts a tractive stress (τ_0) on the materials which make up its bed and sides. However, movement of these materials will not occur until this tractive stress exceeds a threshold set by their resistance to movement. The largest particle size that can be moved by a given flow is called its *competence*. Once movement is initiated, then the stresses required to maintain movement are somewhat less because the coefficient of dynamic friction is less than the coefficient of static friction. Furthermore, once particles start to move, they encourage the movement of other particles by striking them.

In section 17.5, the following equation ([17.23]) was derived, which relates the tractive stress (τ_0) to the unit weight of water (γ_w) and the hydraulic radius (R), the latter being approximately the depth (d) for wide, open channels:

$$\tau_0 = \gamma_w RS \doteqdot \gamma_w dS \qquad [20.2]$$

At a distance (y) above the surface of the bed, the tractive stress reduces to:

$$\tau_0 = \gamma_w (d - y)S \qquad [20.3]$$

The critical tractive stress (τ_c) required to move a particle can be estimated by calculating the turning movements necessary to move a particle over the one in front, assuming that the particles are approximately spherical with diameter D (Richards 1982). This condition is shown in Fig. 20.2 and is expressed in terms of forces.

In terms of stresses (force per unit area), the expression for F becomes $\tau_c (D^2/n)$, where n is a packing coefficient, so that (D^2/n) represents the area of grain on average exposed at the surface. In determining the weight of the particle, its submerged weight must be used because of the buoyant effects of water. Thus, the unit weight of the

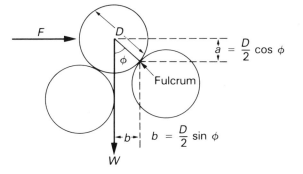

Angle ϕ tends on average to be the angle of internal friction. Condition for movement: Fa (moment due to force, F) $\geq$ Wb (moment due to weight, W)

i.e. when $\quad F\dfrac{D}{2} \cos \phi \geq \quad W \dfrac{D}{2} \sin \phi$

Fig. 20.2 Initiation of movement of a particle based on moments.

particle (γ_p) is reduced to $\gamma_p - \gamma_w$, where γ_w is the unit weight of water. Assuming the particle is spherical, its volume equals $D^3 \pi/6$ and its weight equals $(\gamma_p - \gamma_w)D^3\pi$.

Taking account of the expression in Fig. 20.4, and since τ_c equals τ_0 at the moment of movement, the following equation is obtained:

$$\left[\tau_c \frac{D^2}{n}\right]\left[\frac{D}{2} \cos \phi\right] = \left[D^3 \frac{\pi}{6}(\gamma_p - \gamma_w)\right]\left[\frac{D}{2} \sin \phi\right] \qquad [20.4]$$

Rearranging equation [20.4] gives an expression for the critical stress as follows:

$$\tau_c = \frac{n\pi}{6}(\gamma_p - \gamma_w) \tan \phi \, D \qquad [20.5]$$

For a given type of bed material, only the diameter (D) will vary on the right-hand side of the equation and hence experimental results would be expected to yield a simple linear relationship between the critical shear stress and particle size, with the following form:

$$\tau_c = kD \qquad [20.6]$$

where k is a constant.

Results from Baker and Ritter (1975) suggest that although the general form of the equation is correct, the mean stresses actually required to initiate movement are found to be significantly less than those estimated for particles which are sand-sized and coarser. This arises because, in turbulent flow, considerable variations in the instantaneous velocity occur. Moreover, the submerged weight will be counteracted by vertical forces which are associated with turbulence and pressure differences at the top and bottom of particles, due to the strong vertical velocity variations near the bed (Fig. 17.5).

A further complication is introduced by the fact that even in fully turbulent flow, there is a thin laminar layer close to the bed, with thickness δ. If the particle diameter (D) is small relative to the thickness of this layer, then the entrainment of the particles will be much less likely than if the particles protrude through this layer into the turbulent flow above.

A further complication is that finer particles may possess cohesion and thus require additional stresses before entrainment occurs. This is shown in the graph of velocity against particle size (Fig. 20.3), derived from experimental results by Hjulstöm (1935), where it is apparent that particles of approximately 0.2 mm in size are easiest to move. Conversely, particles which are larger than this size are more difficult to move because of their increasing submerged weight, and consolidated materials smaller than this (e.g. silt and clay-sized materials) tend to be more cohesive and hence more difficult to move. A comprehensive

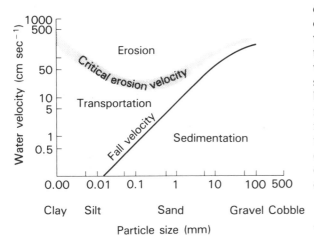

Fig. 20.3 Relationship between particle size and water velocity, indicating threshold velocities for initiation of movement (erosion) above the critical erosion velocity zone; the threshold for deposition (sedimentation) is limited by the fall velocity line and the area where transportation will continue to occur once movement has been initiated. (Adapted from Hjulström 1935)

understanding of the entrainment of materials will also need to take account of factors such as variations in particle shape, the progressive armouring of channel beds by coarse material (as finer materials are differentially removed) and whether a particle finds itself in a pool or on a river bar within a channel.

20.3 Bed material transport

Once shear stresses just exceed the critical level necessary for entrainment, movement will probably occur by sliding or rolling of particles across the bed. In such circumstances, the submerged weight of the particle is largely borne by the bed of the channel (hence the term *bed load* is used to describe it) and such material may intermittently lose contact with the bed by bouncing. In most environments, the bed material load is the least important of the three main components of stream load. However, in mountainous environments, where the supply of coarse materials from the slope system is high, it may be greater than the combined dissolved and suspended loads.

Derivation of equations to analyse the quantity of bed material moved in a channel has proved difficult, not only because of the inherent difficulty of mathematically describing the relationships involved, but also because of the practical difficulties involved in measuring the flux of bed material within real channels. For example, construction of satisfactory traps in the bed of a river channel is both difficult and expensive and moreover, it can readily disturb the natural movement. One of the most common approaches for estimating bed material movement has been to express the sediment discharge per unit width (q_s) in terms of the size of the tractive shear stress (τ_0), which is in excess of the critical tractive stress (τ_c), as follows:

$$q_s = C\tau_0(\tau_0-\tau_c)$$ [20.7]

where C is a constant.

This is the basic Du Boys formula and is most applicable to sand-sized and coarser materials, assuming that sufficient materials of an appropriate size are available for transport. However, the difficulties in deriving realistic values of τ_0 (which were discussed in the previous section) hinder the applicability of this approach. Moreover, the rate of transport is a function of bedform, i.e. whether dunes or a plane bed are present, with the latter leading to greater sediment discharge.

An alternative approach has been proposed by Einstein (1950), based on the probability of material being lifted from the bed (p), the size of the particles and a characteristic exchange time (t_e), which is the time necessary to replace a bed particle by a similar one. The number of particles eroded per unit area per unit time (N_E) can be expressed as follows (Statham 1977):

$$N_E = \frac{i}{A_D} \cdot \frac{p}{t_e} \qquad [20.8]$$

where i is the fraction of particles of the size range considered and A_D is the area of exposed cross-section. The transport rate is a function of the probability of movement and the size of the particles. Within stream channels, systematic changes in bed material size are usually observed in the progressive decline in particle size from headwaters to the mouth. These changes are due partly to the sorting of materials associated with the relationship between shear stresses and particle size and to the reduction in size due to the attrition of particles as they strike each other.

20.4 Suspended load transport

As particle movement across a channel bed increases in velocity, sliding and rolling are disturbed by vertical particle movements, followed by a more gradual descent. The resultant asymmetrical movements are known as *saltation* and initially will tend to occur close to the bed, because the vertical forces leading to these movements are greatest near the bed, as a result of velocity gradients being greatest here. Away from the bed, the vertical forces decline and the particles descend. As velocity and turbulence increase, particles spend a longer time away from the bed and then form part of the *suspended load*. Particles may again fall to the bed and may move by rolling and sliding for the same mean velocity. Thus, the distinction between suspended and bed material is not a sharp one.

The fall velocity (V_f) of particles, which has to be overcome by the upward fluid forces, is given at a first approximation by *Stokes' Law* as follows:

$$V_f = \frac{2}{9}(\gamma_p - \gamma_w)\left(\frac{D}{2}\right)^2 / v \qquad [20.9]$$

where v is the fluid viscosity.

For very small particles, their fall velocity may be so small that even at low discharges and veloc-ities, settling will not occur once such materials are introduced into the flow. This part of the load is termed the *wash load*. The other types of load that are transported in solid form are highly variable in space and time. Dependent upon channel dimensions, stream velocity, bedform and other conditions, particles may move as bed load or suspended load.

Reference to Fig. 20.3 shows that for a wide range of particle sizes and velocities, transportation in suspension will continue to occur even though velocities may be far below the critical velocity necessary for erosion. For coarser particles, the boundary between transportation and sedimentation approaches the zone of critical erosion velocity, showing that a relatively small drop in velocity below this critical level will result in sedimentation. It should be stressed that the Hjulström graph (Fig. 20.3) cannot be applied to natural streams in a rigid way. It is based on experimental work with uniform-sized materials, and hence will not take full account of the armouring of channel beds by coarser materials as discussed above. Moreover, materials composed largely of fines may not occur as individual grains but may be eroded as aggregates or even change into aggregates if solute concentrations increase in a process known as *flocculation* (section 24.3).

The measurement of suspended load, through the collection of samples of river water, is certainly easier than measurement of bed load. However, unlike the solute load, concentrations of suspended load are very variable within channels since they decline away from the channel bed due to vertical changes in velocity and turbulence. For coarser particles in particular, a high proportion of the suspended sediment load may be close to the bed and hence be difficult to sample accurately. For finer materials, the change in concentration vertically becomes less until for the wash load, there will be virtually uniform concentrations throughout the flow.

The amount of material transported as suspended load is largely controlled by the availability of material for movement. This can be appreciated by the simple observation that streams with clear water are common in most parts of the world although, from the Hjulström graph, it is apparent that fine material could be transported if available. During a storm, suspended sediment concentrations will increase due to the increase in critical entrainment velocities and also because of the addition of sediment to the stream waters from surface and subsurface flow. Mass movements, especially on slopes immediately adjacent to the channel, can also contribute materials to the suspended load and the erodibility of channel sides will also affect sediment supply. For all these reasons, estimation of the suspended sediment load simply on the basis of stream-flow characteristics, which are independent of sediment supply, makes little sense. Seasonal variations in suspended sediment are commonly found. They arise because of seasonal variations in the supply of available sediment and are also due to changes in temperature, since reductions in the latter lead to increases in viscosity, a decrease in fall velocities (equation [20.9]) and hence increases in sediment concentrations.

20.5 Channel response

The title of this section is, in a sense, misleading since it suggests that river channels change their form simply in response to the various processes described in the previous sections. In fact, as has already been indicated in several places, the channel characteristics themselves affect the way in which these processes operate. In a river channel system, we have an extraordinarily complex set of feedback mechanisms, operating through many different time-scales. The various interactions are so complex that the behaviour of river channels has been considered as indeterminate by some research workers (e.g. Maddock 1969). In this section, four types of alluvial channel response will be considered, namely the formation of different bedforms, channel cross-section, channel plan form and the river long profile.

(a) Bedforms

Within river channels, especially those with sand beds, a variety of bedforms are found as the intensity of flow increases. The idealised scheme (Fig. 20.4) is based on flume studies, and natural channels and variations in stream velocity, in different parts of the channel, mean that the same forms can coexist at the same time. Initially, virtually no movement takes place although, when the critical threshold for erosion is exceeded, small ripples develop (Fig. 20.4(a)). At greater flows, dunes develop with their steep face pointing downstream, since material is carried up the gentle *stoss slope*, over the crest (Fig. 20.4(b)) and then falls down the *lee slope*, giving rise to cross-

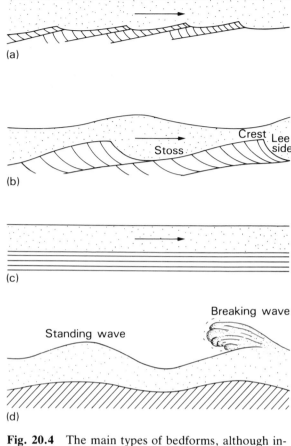

Fig. 20.4 The main types of bedforms, although intermediate stages do exist (adapted from Simmons and Richardson 1966; Allen 1970). (a) ripples; (b) dunes; (c) plane bed; (d) antidune – antidunes usually have indeterminate laminations with a tendency to move upstream.

bedding and forward movement of the dune. Both ripples and dunes contribute significantly to the channel's resistance to flow.

However, at higher stream energies, the dunes are washed out and a *planar channel bed* is formed, with even laminations (Fig. 20.4 (c)). This planar surface is replaced at higher flow intensity with antidunes, which commonly move progressively upstream. As they grow, they may be periodically washed out by the breaking of standing waves (Fig. 20.4(d)). The resistance to flow by antidunes, and especially in the planar form, is much less than for ripples or dunes (Allen 1970). For coarser sands, the first stage after sediment movement commences tends to result in a plane bed forming first, before the ripple stage.

The type of bedform is related closely to the *Froude number* (*Fr*) which is calculated as follows:

$$Fr = \frac{v}{\sqrt{gd}} \qquad [20.10]$$

where d is the depth of flow and v the velocity. In Fig. 20.4 bedforms a and b are typical of values of *Fr* which are less than 0.4 and bedforms c and d have typical values in excess of 0.7. In materials coarser than sands, dune bedforms are also found although their form becomes more irregular as particle size increases and sorting becomes weaker. A common adjustment to flow, especially of flat pebbles, is to adopt an *imbricated structure* in which they lie stacked and overlapping each other, at an angle to the stream, with their upper edges pointing downstream.

(b) Channel cross-section

Explanation of the form of channel cross-sections is related to two basic factors; firstly, the overall dimensions are adjusted to transfer the water and sediment discharges through the channel cross-section. Secondly, the shape of the channel is controlled by the resistance of the bank and bed materials to the distribution of erosive forces of the river and its sediment load. Although larger discharges are undoubtedly associated with larger channels, it is not easy to determine which particular type of discharge controls the channel's dimensions. One way to study this problem is to measure the discharge that results when the channel is at a *bankfull stage* and to attempt to determine the frequency of this discharge.

Although other discharges will be responsible for channel erosion, it is reasonable to assume that this bankfull discharge is important since, at greater overall discharges, overbank spillage will lead to proportionately smaller increases in depth and discharge within the channel itself. Evidence from many rivers around the world suggests that bankfull discharge occurs with a return period between one and two years and for many streams, it is close to one and a half years. The *return period* of a flood is the average interval of time within which the flow will be equalled or exceeded once (Gregory and Walling 1973). Although the recurrence interval of bankfull discharge varies somewhat even for different cross-sections of a single river, the evidence does suggest that the overall channel dimensions are generally adjusted not to particularly rare events or to very common ones, but to events of moderate magnitude which

occur once every one or two years (Wolman and Miller 1960).

This adjustment does not mean that only events with this particular recurrence interval are responsible for channel development, but it does imply that they are the dominant events. Also extremely rare events may lead to crucial thresholds being exceeded, which can result in alterations to channel dimensions that are apparent for very long periods of time. An important distinction should be made between channels with highly mobile beds and those with coarse bed material, in terms of their adjustment to increasing discharge at a cross-section during a flood. The former channels, which are more commonly found in semi-arid areas, will experience considerable scour during the rising stage to the extent that the depth of the channel may even be doubled. In more humid areas, the channel bed is often much more stable, with relatively little change in the absolute elevation of the bed until very high discharges are reached. This arises because of armouring of the channel bed by coarse bed material, due to the protective influence of vegetation, and because channels may be rock-floored or have only a thin veneer of mobile material above rock.

The shape of river channels is also a function of discharge and the examination of channel shape downstream will usually reveal that the width of the channel increases more rapidly than depth. In other words, as discharge increases downstream, the width–depth ratio increases. The materials forming the bank and bed of channels will also affect the resistance to erosion and hence the shape. Channels with banks composed of a high proportion of silt and clay are generally relatively narrow compared with their depth, whereas channels with banks composed predominantly of sands tend to be wide relative to their depth.

This difference relates to the greater cohesion of materials with a high silt–clay content. Richards (1982) derives the following empirically based equation relating the width–depth ratio (*F*) to the percentage of silt–clay materials in banks (*B*) and the mean annual flood (Q_m):

$$F = 800 \, Q_m^{0.15} \, B^{-1.20} \qquad [20.11]$$

The relatively small size of the exponent for Q_m, relative to that for *B*, indicates the much greater importance of particle size compared with discharge. The difference in signs of the exponents means that increases in discharge lead to somewhat wider channels, whereas increases in the amount of silt and clay lead to narrower channels as expected.

(c) Channel plan form

Three basic types of river channel plan form exist, namely straight, meandering and braided (Fig. 20.5(a)). In fact, the first type is unusual since nearly all channels are found to have sinuous courses. In braided channels, meanders are also found but their distinctive characteristic is that a single river course is replaced by two or more channels. Analysis of meandering and braided channels by Leopold and Wolman (1957) suggested that for a given bank-full discharge, braided channels tend to have steeper gradients than meandering channels (Fig. 20.5(b)). Work by Schumm and Kahn (1972) in flume experiments indicate that braided channels were characterised both by steeper slopes and by greater sediment loads compared with meandering channels.

The presence of less resistant banks has also been suggested as being important for the generation of braids, since rapid erosion of opposite banks in the same reach will naturally encourage a splitting of a single stream into two parts. From numerous field and laboratory observations, braided channels are characterised by high-energy environments with high sediment loads, high channel gradients for a given bankfull discharge and erodible banks associated with low width-depth ratios. Although several of these properties of braided channels are in accord with intuition, the relatively higher gradients of braided channels are not.

Numerous explanations of the cause of river meandering have been proposed, although none offer a complete answer as yet. Any explanation of river meandering must take account of the fact that meanders are found elsewhere in nature, varying in scale from the giant meanders of the Gulf Stream to the path of raindrops on a window pane. Langbein and Leopold (1966) found that meanders correspond closely to sine-generated curves, defined by the following equation:

$$\theta_s = W \sin \left(360 \frac{D_s}{D_R}\right) \qquad [20.12]$$

where θ_s is the orientation of a particular segment, D_s is the distance downstream, D_R is the total reach length and W is a variable describing the maximum deviation at the meander inflection points. As it increases, so the sinuosity (or 'bendiness') of channels increases.

The particular characteristic of note of this curve is that it involves the minimum variance of change of channel direction. One of the outstanding characteristics of river meanders is that they show little change in form with increasing size or, expressed more precisely, as channel width increases, the wavelength increases linearly (i.e. $\lambda \alpha W$). This proportionality is probably related to the fact that when the ratio of the radius of curvature (R_m) to channel width has a value near 2, the minimum resistance to flow due to curvature is found (Leopold et al. 1964). Not surprisingly, in view of a previous statement (section 20.5(b)) that discharges control channel width, it appears that discharges with return periods around one to two years are most important in the shaping of river channel meanders.

Explanations of the initiation of meandering are associated with the heterogeneities in the resistance of materials to erosion and to variations in stream erosion associated with secondary across-channel circulations, superimposed on the main streamflow. Once a deviation from a straight

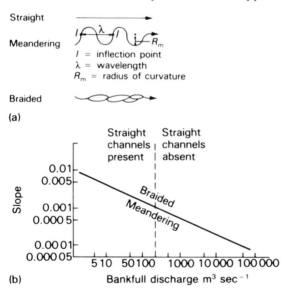

Fig. 20.5 Variations in channel plan form; (a) types of channel; (b) relationship between channel plan form, slope and bankfull discharge (Adapted from Leopold and Wolman 1957). Thus, for a given discharge, braided channels tend to have steeper slopes than meandering channels. Straight channels are found both sides of the line, separating braided from meandering channels, but apparently they are restricted to streams with lower discharges.

channel exists, then continued erosion and deposition of material will lead to an amplification of this deviation in a positive feedback loop. At the apex of meander bands, a rotational flow is found with the surface water moving to the outer bank and at the bed, water moving towards the inside of the bend. Maximum bed shear stress and erosive power are found in the apex of the bend and consequently, a sequence of *pools* and *riffles* is found with the latter being formed near the inflection point (Fig. 20.5). The wavelength of the riffle–pool sequence was found by Harvey (1975) to be most closely related to channel width and sub-bankfull discharges. The location of maximum shear stress is usually just below the apex of the bend, thus giving rise to the progressive migration of meanders downstream. On the inside of the river bends, active deposition takes place to produce the so-called *point-bar deposits*. These are composed of a series of *scroll bars* formed within the shallow water near the channel, and are associated with progressive lateral channel movement (Allen 1970).

(d) Channel long profile

Examination of field relationships, between stream gradient and various possible causative factors, has generally shown that the gradient of streams can be expressed as a function of discharge of bed material size. Using data from Appalachian streams in the USA, collected by Hack (1957), the resultant equation for stream gradient(s) is as follows:

$$S = k \ \frac{D^{0.78}}{Q_m^{0.53}} \qquad [20.13]$$

where D is mean particle size, Q_m is the mean discharge and k is a constant.

The implication of this relationship is that stream gradient is adjusted to the discharge and bed material sizes so that the river can transport the sediment entering a reach. Work in southeast England (Penning-Rowsell and Townshend 1978) has shown that the spatial scale at which stream gradient is measured should be considered in deriving an understanding of the factors affecting channel gradient. Channel reach was found to be most closely related to discharge (measured using drainage area as a surrogate) but for the local gradient over riffles, the strongest relationship was with particle size. In neither case did the shape of the channel significantly relate to slope, although

more open channels (with a higher width–depth ratio) would be expected to be associated with lower gradients, because of the lower resistance to flow of such channels. However, for a single channel reach without any variation in discharge, channel slope was found to be significantly related to the width–depth ratio. An important geomorphological consequence of the relationship between bed material size and gradient is that when irregularities due to changes in rock type are superimposed on the overall upward concavity of river profiles, they will not be eliminated with time. As long as material of different sizes is supplied to the channel by slope processes and channel bed erosion, the characteristic slopes on different rock types will be maintained.

20.6 Bedrock channels

The previous discussion in this chapter reflects the concentration of recent geomorphological research on alluvial channels, since relatively little work has been carried out on channels in bedrock. Lengthy sections of rivers in bedrock are not common, but it is possible to find short reaches without a mobile bed, especially in mountainous areas. In tropical areas, the presence of *rock bars* in major river channels is much more common than in temperate areas. For example, major rapids and falls are found close to the mouth of the Zaire River in West Africa, which have been attributed to the case-hardening of rock materials in tropical rivers (Tricart 1972).

Erosion of bedrock channels is dependent upon the physico-chemical properties of the rock present. Clearly, limestones are likely to be eroded primarily by solution whereas in other rock types, abrasion by material being transported through the channels is more important. The density and orientation of joint patterns control the local form of erosion, as is shown by the clear joint planes which are commonly revealed in bedrock channels, and differential weathering along such joints will often be a necessary precursor to erosion. Howard (1980) reports on bedrock channels in badland areas where rock resistance is low. He determined the relationship between erosion rates (E), drainage area (A) (which closely represents discharge), and gradient (S) as expressed in the following equation:

$$E = KA^{\phi} \ S^{\sigma} \qquad [20.14]$$

Table 20.1 Types of floodplain landforms and deposits (Based partly on Gregory and Walling, 1973)

Floodplain landforms and depositional environments	*Mode of deposition and characteristics of materials*
Active channels: Point bars Channel bars	In channels, active deposition occurs especially in the following locations: i) inside of river bends as point bars by lateral accretion ii) within channels as bars which may eventually lead to braided channels forming around islands at low water iii) at the junction of channels.
Deferred junctions	Deposition within the floodplain may result in the main river level being above the height of tributaries where they enter the floodplain. This can lead to the tributary flowing across the floodplain parallel to the main channel, until a junction is eventually reached.
Cutoff channels and abandoned channels	Low gradients and the meandering character of floodplain channels result in whole bends or parts of bends are cut-off producing oxbow lakes. Sometimes a long reach comprised of several bends may be abandoned. Through time, fine materials and organic matter will be deposited until no open water is left. The resultant deposits will therefore be fine in their upper part, but may well be much coarser at depth as the channel lag deposits are reached.
Meander scrolls	Over time a series of point bars are commonly deposited on the inside of river bends, each separated by intervening swales. Collectively these features are called a meander scroll.
Levees	At the margins of channels, the lower velocity leads to greater deposition, and can result in raised banks called levees. Progressive deposition of these finer materials and of channel bed deposits can ultimately lead to the water-level being raised above the floodplain. Many levees are partly or even wholly artificial due to dumping of dredged material from the channel.
Vertical accretion	Flood waters due to river banks being over-topped usually contain only suspended load so that the resultant deposits formed by vertical accretion are characteristically fine-grained. Once thought to be the main mode of floodplain formation, it is now recognized as relatively unimportant though more significant in some humid tropical floodplains.
Splays	Where a river breaks through its levee, the rapid flow of water may lead to a spread of channel deposits including coarser bed materials onto the adjacent floodplain.
Backswamps and lakes	Because of deposition of floodplain materials, damming of water flow frequently occurs forming backswamps and lakes, especially where tributaries are blocked.
Marginal features	The sediment transport systems of adjacent slopes bring materials onto the floodplain margins by mass movement wash processes etc. These colluvial deposits usually are more poorly sorted than alluvial floodplain deposits. Where a heavily sediment-laden tributary channel reaches a floodplain and its gradient is markedly reduced, alluvial fans may be deposited.

where K is a factor including the effects of inherent bed erodibility and the magnitude and frequency characteristic of flow: ϕ and σ were found empirically to be 0.44 and 0.68. If erosion rates are nearly constant throughout the network, then the following relationship will result:

$$S \propto A^{-\phi/\sigma} \qquad [20.15]$$

20.7 Floodplains

The understanding of sediment transfer processes in river channels is important not only so that erosional processes can be understood, but also so that the way in which materials are deposited can also be comprehended. Floodplains viewed from

Plate 20.1 Landsat view of the lower Mississippi floodplain, Arkansas, USA.

above (Plate 20.1) reveal much of the complex recent history of fluvial erosion and deposition. Recently abandoned meanders can clearly be seen, but other much fainter scars of considerably older channels can also be detected and the scroll pattern of point bars deposited on the inside of channels is also apparent. Each of these types of landform is composed of a distinctive suite of materials which is illustrated in Table 20.1. The deposition of materials in floodplains is a consequence of a whole complex of processes. It does not arise, as was once thought, because of a decline in river velocity, since numerous observations have shown that river velocity changes (on average) very little from stream source to the limits of tidal influences. If any overriding cause of floodplain formation is to be sought, it is more likely to be found in terms of the increasing demand for transport of material for each reach downstream. This results from the need not only to transport its own materials but also those of the whole drainage basin above. This leads, not surprisingly, to material being temporarily stored (as if it were in queues) in the lower part of a river's course.

20.8 Retrospect

The word 'complexity' has been used several times in this chapter but it needs to be mentioned once more in introducing the concept of complex response, proposed by Schumm (1979). Largely on the basis of laboratory experiments, it was found that an initial single change of input to a channel can result in a complex response involving multiple changes. Thus, a lowering of the outfall of the channel (following a change in base-level) led to the generation of not a single terrace but two sets of terraces. If this sort of complex response com-

monly occurs for river channels, it makes the task of understanding the behaviour of river channels even more difficult than previously thought. Moreover, complete derivation of environmental changes which give rise to a river basin's development begins to appear well-nigh intractable.

Key topics

1. The origins and forms of solute load.
2. The factors affecting particle movement in channels.
3. The characteristics and processes of movement of materials in channels.
4. The influence of geology, discharge and load characteristics on channel morphology.

Further reading

Coates, D. K. and Vitek, J. D. (1980) *Thresholds in Geomorphology*. George Allen and Unwin; London and Boston (Chs 4–7).

Derbyshire, E., Gregory, K. J. and Hails, J. R. (1977) *Geomorphological Processes*. Dawson, Folkestone and Westview Press, Boulder, pp. 44–106.

Embleton, C. and Thornes, J. (1979) *Process in Geomorphology*. Edward Arnold; London (Chs 3 and 7).

Goudie, A. (1981) *Geomorphological Techniques*. George Allen and Unwin; London and Boston, pp. 47–61, 196–211.

Morisawa, M. (1973) *Fluvial Geomorphology*. George Allen and Unwin; London and Boston.

Schumm, S. A. (1977) *The Fluvial System*. Wiley-Interscience; New York and London.

Chapter 21 Aeolian transport systems

The driving force behind all of the processes analysed so far in sediment transport systems is gravity. However, in the case of aeolian transport systems, the force provided by the wind is of at least equal significance. Although katabatic winds (sections 5.2 and 14.2) moving under the force of gravity may be of considerable local geomorphic significance, aeolian erosion and deposition arise primarily because of movements of the atmosphere (Ch. 11) in response to unequal receipts of energy in different parts of the world (section 2.2). In this chapter, the prime objective is to analyse the effect of wind on loose particulate surfaces, which results in distinctive depositional landforms. Additionally, evidence for the erosional action of winds will be presented even though our knowledge of such processes remains imperfect. Furthermore, although few landforms are created by the wind away from coasts in humid temperate areas, entrainment of particles by wind can be a major cause of soil loss in agricultural areas.

21.1 Principles of aeolian transport

Many of the principles discussed in Chapters 17 and 20 with reference to water flow are also applicable to air movement since air, as a transporting agent, acts as a fluid in many respects. However, the main difference (compared with water) is that air is much less dense than water: in fact, it is approximately 1000 times less dense (Statham 1977). This means that air is less able to entrain and transfer particles and, moreover, its action never becomes as spatially concentrated as water action in channels. In vertical profile, air moves in a similar way to that of turbulent water (see Fig. 17.5). A thin layer of air, moving in laminar fashion, gives way upwards to air in which considerable vertical and horizontal motions are always superimposed upon the overall forward movement.

The velocity profile of the air above the ground can be derived from the following equation (Allen 1970):

$$v = K U^* \log \left(\frac{h}{h_0}\right) \qquad [21.1]$$

where v is the velocity for height (h) above the ground; h_0 is the height above the ground of a layer of air with zero velocity, which has a thickness of only approximately one-thirtieth of the diameter of the surface particles; K is a constant and U^* is the so-called drag or shear velocity, which has the units of velocity ($m\ s^{-1}$). This velocity is defined by the following equation which involves the shear stress (τ_t) at the base of the air, caused by the turbulent flow across the surface, and the density of the air ρ_a:

$$U^* = \sqrt{\frac{\tau_t}{\rho_a}} \qquad [21.2]$$

The physical meaning of U^* can be better understood by reference to Fig. 21.1, which shows the graph of the logarithm of height plotted against velocity. The steepness of this line is inversely proportional to U^* which becomes a measure of the velocity gradient of the air. Direct observation of U^* is difficult, but it can be obtained from the following expression (Bagnold, 1941):

$$U^* = \alpha\,(v_z - v_t) \qquad [21.3]$$

where v_z is the velocity at height z, v_t is the threshold velocity at height k^1 and α is a constant equal to $(0.174/\log(z/k^1))^3$. As the velocity of air moving across a surface increases, a *critical velocity* is reached at which particle entrainment begins.

In terms of the shear velocity, U^* can be defined as follows for particles greater than 0.1 mm:

$$U^*_{\text{crit}} = A\sqrt{\frac{\rho_s - \rho_a}{\rho_a} \cdot Dg} \qquad [21.4]$$

where ρ_s and ρ_a are the densities of the particles and the air respectively. In the previous chapter, it was convenient to use unit weights (γ) rather than density (ρ), but it can be recalled from section 17.5 that the two variables are simply related by γ, i.e. where g is the force of gravity. It should be noted that the large difference between the densities of water and air means that very large shear velocities are needed to move larger particles and hence wind tends to move particles no larger than sand-sized.

In Fig. 21.2, the relationship in equation [21.4] is plotted graphically. However, for particles with diameters less than approximately 0.1 mm, an inverse relationship is found between particle size and shear velocity U^*. This arises primarily because of the cohesive forces present between smaller particles, especially as a result of moisture effects, which result in suction by capillary action. Supplementary factors are the inter-particle cohesion, which is due to the action of weak chemical bonds and a lower surface roughness. Hence, the grains are more likely to remain within the less erosive laminar layer. A similar effect to the latter was described for water flow. Thus, for smaller particles below 0.1 mm, the critical shear velocity increases, which is a similar relationship to that found for particle movement in water, as depicted in the Hjulström diagram (see Fig. 20.3). Consequently, not only is there a restriction on the upper limit to the size of particle that air can move but there is also a lower limit on the fineness of particles that can be moved. Thus, wind transport is restricted to a relatively narrow range of particles usually varying only from silt to sand sizes.

Beneath the curve in Fig. 21.2 is another, lower, line called the *impact threshold curve*, in which lower values of U^* are found for a given particle size D. This curve is for the situation where particle movement into the air has already been initiated and is more appropriate therefore for most aeolian transport systems. The bombardment of the surface by these particles reduces the shear velocity at which movement is initiated. The

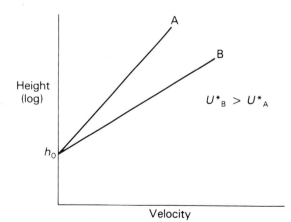

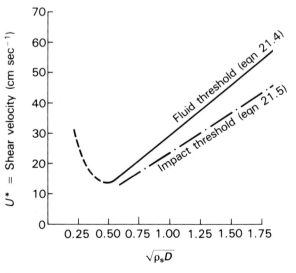

Fig. 21.1   Interpretation of shear velocity in relation to height above the ground surface and wind velocity. Air with more rapid changes with height has higher values of U^*.

Fig. 21.2   Relationship between threshold shear velocity (U^*) and particle size (D = diameter, ρ_s = density) for aeolian transport. (Adapted from Allen 1970)

trajectory of these particles has already been described in Chapter 20 and the movement is known as saltation (Fig. 21.3). The impact threshold ($U^*_{\text{I crit}}$) is estimated from a complex function of particle diameter (D), as follows:

$$U^*_{\text{I crit}} = 680 \sqrt{D} \log (30/D) \qquad [21.5]$$

Movement of sand-sized particles is usually by saltation rather than as genuine suspended load, in which the weight of the particle is supported by upward turbulent forces for some time. Once entrainment of very fine particles occurs, then materials may be taken up to great heights, since their fall velocity is low, and very long distance intercontinental transport may then result. For example, even the southern UK receives occasional deposits of Saharan dust which are most noticeable if precipitated from a mid-tropospheric southerly Tc air flow (section 12.4) by dust-laden raindrops. Because sand-sized particles tend to be moved by saltation close to the surface, rather than by suspension, sand usually moves close to the gound.

Saltation is the most effective process of aeolian transport since, in comparison, suspension only affects a relatively small proportion of the transported materials. However, movement of material at the surface is relatively slow and although movement occurs simply as a direct result of wind shear, surface movement (called *creep*) is primarily due to the impact of saltating grains. The quantity of sediment (q_s) moved in g cm^{-1} width hr^{-1} can be estimated as a function of the shear velocity. For materials with a diameter of 0.25 mm, the relationship (Bagnold 1941) is as follows:

$$q_s = 1.1 \frac{\rho_a U^{*3}}{g} \qquad [21.6]$$

For other sizes of materials in the size range above 0.1 mm, the following weighted modification of equation [21.6] can be used:

$$q_s = C \sqrt{\frac{D}{D_{0.25}}} \frac{\rho_a}{g} U^{*3} \qquad [21.7]$$

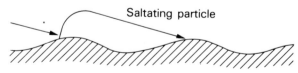

Fig. 21.3 Aeolian ripples. Mean wavelength corresponds approximately with the mean length of the sand-grain trajectory. .

where C is a constant dependent upon sorting and $D_{0.25}$ is the diameter of standard (0.25 mm) sand. As the sorting improves (i.e. as the variation in particle size around the mean declines), the value of C declines, presumably because smaller particles stimulate the movement of larger ones as the action of saltation declines.

Vegetation cover has a dramatic effect on aeolian transport systems and a complete cover eliminates all wind erosion. Even partial vegetation cover has a marked influence because of the increased friction and reduced wind velocity at the ground surface (section 8.4). Vegetation also has an ameliorating effect since it maintains the soil moisture at a high level which reduces wind erosion because of the increase of shear strength due to capillary action. Methods for the prevention of wind erosion include planting of vegetation together with other methods of surface stabilisation such as gravel, chemical sprays, water and even oil (Cooke *et al.* 1982). Windbreaks that are semi-permeable to reduce eddy effects (section 8.6) can be produced by planting appropriate vegetation (especially poplar trees) or by the construction of fences to divert the flow or impound the particulate movement.

21.2 Sources of wind-blown materials

The erosive action of wind (in contra-distinction to entrainment of particles from loose surfaces) will be discussed in section 21.4. However, it is generally accepted that the origins of materials transported by winds can often be a result of differential movement of sediments produced by other sediment transport systems. These include river and glacial deposits, the latter including both glacial *till plains* and *outwash plains* derived from fluvio-glacial action (Ch. 22). Other sources include marine beaches and dry lake beds, such as the *playas* of the western USA and *Chotts* of North Africa. Removal of vegetation in humid areas for agriculture can result in considerable wind erosion where the soils are vulnerable. These soils are of two main types, namely sandy soils, where the absence of fines means that cohesion is low, and organic soils where, once they dry out, the low density of the materials encourages erosion. The absence of windbreaks and deep-draining, leading to desiccation, wind erosion and oxidation, has led to a major depletion of soil fertility in the Fenlands of eastern England in recent years.

Since the absence of vegetation is a necessary condition for the initiation of wind erosion, the changing distribution of arid areas in the Pleistocene means that large areas of the world have been affected by aeolian sediment transport systems. Fossil dune systems are found, for example, in areas now covered by savanna vegetation, as far north as 16° S in Angola and Zambia, and some dune fields are even found within the Zaïre rainforest (Grove 1969). Loess is composed of predominantly silt-sized materials and is found over very wide areas. Its origin lies in the effects of wind erosion on glacial materials in the Pleistocene and in mid-latitude deserts during periods of the Pleistocene when they were much more extensive than at present (Derbyshire 1983). Increasing evidence of the significance of wind abrasion (section 21.4) suggests that a large proportion of materials transported by wind also owe their initial detachment to aeolian action.

21.3 Accumulation features

The landforms resulting from aeolian action vary in size from just over a centimetre to several kilometres and forms of any intermediate size can be found. However, for particles of a given size, three distinct scales of features are usually found which are given the names *ripples*, *dunes* and *draa*. Thus, for any particular particle size, landforms with three distinct wavelengths are found with landforms of intermediate size being absent.

Ripples are small-scale turbulence features, produced by the impact of saltating grains, and their wavelength is a function of the flight-path length of the grains (Fig. 21.3). Theoretically, the mean length (L_t) of the trajectory is given as follows:

$$L_t = k \; \frac{(U^* + U^*_{crit})^2}{g} \qquad [21.8]$$

where k is a constant, which is insensitive to grain size (Allen 1970). Thus, the spacing of ripples is related primarily to wind speed.

Dunes display a very wide variety of forms and this diversity relates to the fact that although air motion is usually very irregular, secondary currents can be superimposed upon the forward motion of the wind. Several origins have been suggested for the development of these circulations and they include the superimposition of free and forced convection cells (section 3.1) on horizontal motion and the formation of lee waves (section 14.6) behind ridges. The creation of accumulation features leads to these secondary cur-

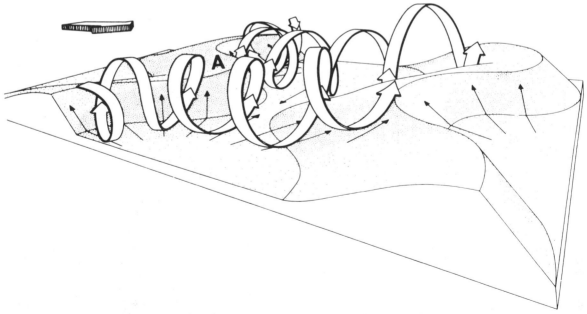

Fig. 21.4 A sinuous dune ridge itself propagates vortices which distort a subsequent ridge. (After Warren 1979)

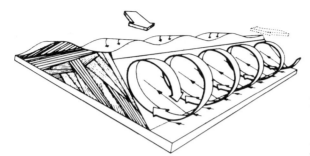

Fig. 21.5 The formation of a seif dune by winds blowing from two principal directions, either at different seasons or at different times of the day. The slip face and shallower beds dip in two directions. (After Warren 1979)

rents being reinforced and thus the dune forms continue to grow (Fig. 21.4) due to this positive feedback loop. Furthermore, some dunes owe their origin to the presence of winds from two or more predominant directions (Fig. 21.5). The accumulation of sand on dunes is encouraged by the fact that sand particles are far more likely to bounce off rock surfaces than off a sand-covered surface, where the momentum at impact is absorbed by the movement of other particles. Draa are the largest and least understood accumulation features but their origin is presumably related to long vortices which form part of the secondary circulation of the wind, since the landforms are themselves regularly spaced.

21.4 Aeolian abrasion

Erosion of bare surfaces by abrasion, as distinct from transport from loose particulate surfaces, has long been the subject of controversy. Although originally some authorities ascribed much potency to the erosion by wind, the absence of convincing evidence for the erosivity of wind-borne particles has hindered general acceptance of the concept. In recent years (Goudie 1983), observations from air photography and space images have shown the widespread occurrence of landforms orientated parallel to the wind. They are discordant both to structure and drainage pattern and the resultant aerodynamically shaped mounds are known as *yardangs*.

Near the Tibesti Plateau in North Africa, quasi-parallel grooves are found up to 1 km wide and tens of kilometres long, which have an aeolian origin. On a much smaller scale, wind-faceted stones may be formed by wind erosion. These features are called *ventifacts* and, as an example, they are most conspicuous in the ice-free 'dry' valleys

Plate 21.1 Ventifacts near Lake Vanda in Victoria Land, Antarctica.

of Victoria Land, Antarctica (Plate 21.1). The mechanisms by which wind erosion occurs relate to the abrasive impact of particles although because of the small size of wind-borne particles, their kinetic energy is low (equation [19.1]). Materials which have been subjected to weathering, especially salt weathering in arid areas, may have a low cohesion with the solid rock so that detachment is facilitated. Nevertheless, wind-tunnel experiments demonstrate the efficiency of wind abrasion even on unweathered rocks.

21.5 Retrospect

Aeolian transport systems are most usually found in subtropical arid areas but they also operate extensively in polar regions and, on a more localized scale, in coastal dunes. However, it is important to stress though that in all arid areas many other transport systems are found, and fluvial erosion, although infrequent, is often of considerable importance. Slope processes, especially sheetwash, also play a major role in eroding the landscape. As for the other transport systems that have been described, the use of basic physical concepts gives considerable assistance in understanding the operation of aeolian transport systems. Modelling of the entrainment and movement of individual grains is, however, much better founded than the formation of accumulation forms such as dunes, where current knowledge remains semi-quantitative at best. Although until recently wind was regarded largely as an agent of transference of material from one place to another, increasing evidence suggests that it may often be a very potent agent of abrasion as well.

Key topics

1. The mechanisms of entrainment of particulate matter into the atmosphere.
2. Modification of wind erosion by soil moisture and vegetation cover.
3. The characteristics of substrates liable to transportation by wind.
4. The effects of grain size on the magnitude of accumulation features.

Further reading

Cooke, R. U. and Warren, A. (1973) *Geomorphology in Deserts*. B. T. Batsford; London, pp. 229–327.

Derbyshire, E., Gregory, K. J. and Hails, J. R. (1979) *Geomorphological Processes*. W. Dawson and Sons Ltd; Folkestone and Westview Press, Boulder (Ch. 4).

Mabbutt, J. A. (1977) *Desert Landforms*. Australian National University Press; Canberra, pp. 215–51.

Petrov. M. P. (1977) *Deserts of the World*. John Wiley and Sons; New York and London (Ch. 11).

Chapter 22 Sediment transport by glaciers

On a global scale, contemporary sediment transport by glaciers is areally relatively unimportant since only 10% of the land is covered by ice, with almost all of this area being in Antarctica and Greenland. Its significance lies both in its potency and importance relative to other transport systems in these locations, where contemporary glaciers do occur, and also in its much greater areal extent during the Pleistocene when ice sheets covered more than three times their current area. The rate of movement of glaciers and their potency as agents of sediment transport are a direct function of their energy and mass balances, which are influenced by the atmospheric fluxes discussed in Chapter 2.

In summary, the discharge of ice is a function of the input of snow, especially in the glacier's upper *accumulation zone*, and the loss of mass from the glacier by melting and vaporisation, particularly in the lower *ablation zone* which is closely related to local energy balance (Fig. 22.1). The conversion of freshly fallen snow to ice in a moving glacier is an extremely complex process involving a profound metamorphosis, passing through intermediate forms such as *névé* and *firn*. This leads to changes in density of up to 100 times and even larger changes in strength of between four and five orders of magnitude. In the first part of this chapter, the basic ideas of ice flow (which were introduced in Ch. 17) are developed and are followed by an analysis of glacial erosion and glacial deposition.

22.1 Deformation of ice

When laboratory experiments are performed on ice to assess how it deforms under an applied stress, it is found that however small is the applied stress, some strain will result (section 17.4). In this respect ice acts like a fluid, albeit a very viscous and slow-moving one: for example, the outlet glaciers of the Greenland ice sheet are among the world's fastest-flowing yet they move only a few kilometres per year. Like a fluid, when ice is moving down a valley its velocity tends to decrease with depth and towards the sides but, because of its high viscosity, its Reynolds number is very low and its motion is essentially laminar (section 17.5). However, the analogy with a fluid must not be taken too far since, as Fig. 17.4 (d) shows, a fluid should show a linear relationship between stress (τ) and the strain rate ($d\epsilon/dt$), which clearly ice does not.

It is apparent that the behaviour of ice can be modelled by a simple power relationship between these two variables, as described by Glen's Law (equation [17.17]), namely $d\epsilon/dt = A\tau^n$. The value of the exponent (n) varies considerably from values below 2 to over 4, largely as a function of the

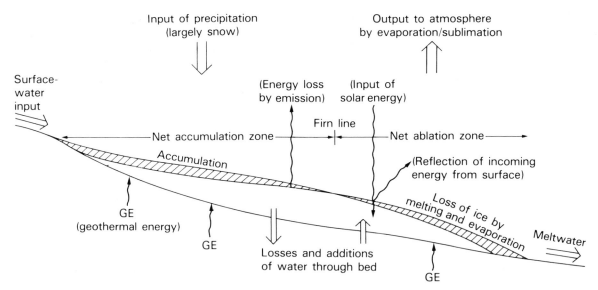

Fig. 22.1 Main inputs and outputs of mass and energy for a glacier. Depending on local environmental conditions, the magnitude of the inputs and outputs changes enormously.

amount of entrained debris within the ice, and A is a constant dependent upon temperature. Thus, as stress increases, the strain rate increases very rapidly and as is shown in Fig. 17.4(d), the line describing the relationship between τ and $d\epsilon/dt$ becomes more nearly parallel to that of an ideal or rigid plastic. It will be recalled from section 17.3 that in the latter, no strain occurs until a critical threshold stress is reached. This change in behaviour of ice is related to recrystallisation at higher stresses, which results in a changed orientation offering less resistance to strain. Given the changes in n with the amount of entrained debris, it follows that glaciers containing large amounts of debris behave more like plastics at lower applied stresses than does clean ice.

The simple power relationship of Glen's Law is a useful approximation but it has been replaced by more complex relationships, such as that of Colbeck and Evans (1973) as follows:

$$d\epsilon/dt = 0.21\,\tau + 0.14\,\tau^3 + 0.055\tau^5 \qquad [22.1]$$

Colbeck and Evans show that this relationship is best near to the *pressure melting-point* (section 22.2). However, it should be emphasised that a considerable scatter of points was found around the best fit line. The strain rate ($d\epsilon/dt$) has also been shown to be dependent on the length of time that the stress is applied. In Fig. 22.2, three main types of strain change are identified which are termed transient or primary creep, secondary creep and tertiary creep. In the first, the strain

rate declines with time whereas in secondary creep, there is a more or less constant increase in strain with time. If high stresses are applied, then the strain rate can increase overall with time, probably in association with recrystallisation, although there may be rapid fluctuations as well.

The shear stress exerted by a glacier on its bed can be derived from the following equation:

$$\tau_0 = \rho_i gR \sin\theta \qquad [22.2]$$

where ρ_i is the density of ice and R is the hydraulic radius.

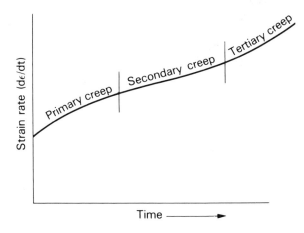

Fig. 22.2 Types of creep showing the relationship between variations in strain rate and time. Note that similar relations exist for rocks.

Futhermore, combining equation [22.2] and Glen's Law and assuming that the ice sheet is very wide (so that depth, d, can be substituted for the hydraulic radius, R), the following expressions are obtained for the change of shear stress (τ_o) and velocity (v) with depth at a distance (y) from the surface (Allen 1970):

$$\tau_0 = \rho g d \frac{y}{d} \sin \theta \qquad [22.3]$$

$$\text{and } v = \frac{kd\,(\rho g d \sin \theta)^n}{n+1} \left[1-(y/d)^{n+1}\right] \qquad [22.4]$$

Thus, the shear stress varies linearly with distance from the bed while the velocity increases rapidly at first and then increases more gradually. These equations are based on the assumption that the basal velocity is zero, which is strictly only true for cold polar glaciers in which no melting occurs at the base (see next section). Compared with a river, the shear strength of a glacier is very high, but its low velocity means that its overall power (rate of doing work) is similar to that of a river (Allen 1970).

22.2 Glacier movement

The previous section on ice deformation is based largely on laboratory experiments on small specimens of ice, although an understanding of glacier movement has also to take account of the movement of ice in relation to the rock floor and sides. *Basal sliding* of glaciers (i.e. the sliding movement over the rock floor due to the weight of ice and slope gradient) contributes very variable amounts to the total movement, and the main control of its effectiveness is whether or not the base of a glacier is frozen to the underlying rock. This in turn is affected by whether the pressure melting-point is exceeded or not. The pressure, and hence depth of ice, is important since the temperature at which melting takes place is lowered by increases in pressure.

The term pressure melting-point is somewhat misleading since the pressure at which melting takes place is not a fixed constant, but depends on the presence of impurities such as salts, CO_2 and air content (Paterson 1981). If the ice is frozen to the rock, the glacier is termed *cold or polar* and if not frozen at the base, it is called *warm or temperate*. In the former case, very little movement results from *basal slip* whereas for temperate glaciers, it can produce up to 90% of the total move-

ment with the remainder being derived from *ice creep*. The vertical temperature profiles of glaciers vary considerably but overall, they generally increase with depth as a function of geothermal heating at the base and the heat generated by internal friction within the ice mass.

In temperate areas in particular, the pressure melting-point is often reached at the glacier's base whereas at higher latitudes, cold glaciers are more common. However, because of the influence of pressure on the melting-point of ice, basal melting is found under great thicknesses of cold ice and, for example, evidence from radio echo-sounding suggests that in Marie Byrd Land, Antarctica, a few metres of water are present at the base of the ice sheet (Drewry 1983). Glaciers are also found where the upper and lower parts have contrasting thermal regimes, which are illustrated in Fig. 22.3. In this conceptual model of Boulton (1972), the glacier is regarded as being composed of four different zones and the relationships between them are best seen in type (d) in Fig. 22.3.

In zone A, there is net basal melting with water passing directly into the next zone downslope and in part being transmitted through the unfrozen bedrock. In zone B, there is an approximate balance between melting and freezing and the existence of frozen ground downslope means that if the discharge of meltwater is sufficiently high, then water accumulates underneath the glacier, which freezes to the basal layers (zone C). This superimposed ice raises the temperature at the glacier's base and basal slipping can occur whereas in zone D, the subglacial materials are frozen to the glacier.

The presence of water reduces the shear stress necessary for movements by friction and by the upward, buoyant hydrostatic force exerted by the water. Although the precise mechanism is not understood, the accumulation of water is a necessary precursor to the very rapid surges which characterise some glaciers (Paterson 1981). Because of the dependence of melting temperature on pressure, ice movement is facilitated around obstacles on the glacier's floor. For example, on the upstream face, the increased pressure near an obstruction will lead to melting and a flow of water around the obstacle or through the rock itself, if it is sufficiently permeable. Conversely, on the downstream side, the pressure reduction will lead to refreezing and the release of latent heat of fusion. For small obstructions (i.e. less than 10 cm long), the flow of released latent heat from the downstream to upstream sides is important in

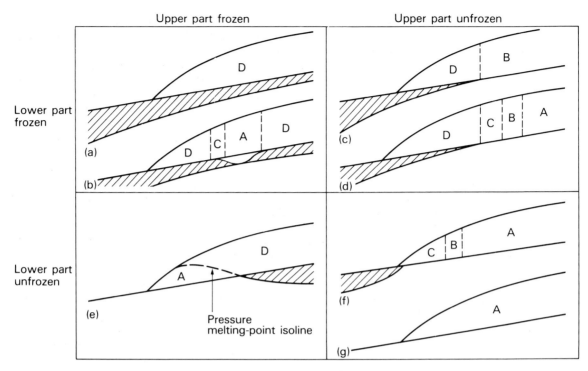

Fig. 22.3 Types of thermal regime under glaciers. (Adapted from Boulton 1972; Embleton 1979)

Zone	Thermal regime of ice	Main erosional processes	Entrainment/transport processes	Depositional processes
A	Net basal melting	Crushing and abrasion by particles embedded in ice. Subglacial fluvial action. Plucking relatively unimportant	Regelation behind obstacles. Physical forcing upward of particles by bed irregularities. Both lead to entrainment in a thin layer of basal ice only	Subglacial lodgement till. Initially against obstructions to form drumlinoid features and then in low areas of floor
B	Balance between melting and freezing	Similar to A, but crushing and abrasion less if whole of glacier is in regime B	Similar to A	Without introduction of meltwater, rate of subglacial lodgement higher than in A, but with meltwater similar rate
C	Water freezes to moving glacial sole	Crushing and abrasion plus plucking important	Water freezes at low points of bed, underneath debris-charged ice and thus debris lifted	Supraglacial deposition due to surface melting releasing englacial debris. If lodgement till deposited, then lifted into glacier
D	Subglacial materials frozen to ice	Crushing and abrasion unimportant. Possibility of plucking of very large erratics from floor	Incorporation by large-scale plucking only	Lodgement till unlikely. Ablation and flow till more important

maintaining this process which is called *regelation*.

Movement around obstacles is also aided by the fact that the increased stress results in a more plastic deformation (see Fig. 17.3) (d)) and the velocity of the ice consequently increases. Lliboutry (1979) has stressed the importance of cavities filled with water, mud and stagnant regelation ice,

such that the moving ice is physically separated from the bedrock. Without such separation, it is argued that frictional forces would be too high to permit typical observed velocities of glaciers. The presence of entrained rock material within the moving ice will greatly reduce the sliding velocity (Paterson 1981). Consequently, movement by basal sliding is commonly not a continuous process but occurs as a series of jerks, separated by periods of much slower movement or even no movement at all.

22.3 Sediment erosion

The direct erosion of intact rock by pure ice is generally believed not to be possible, because the shear strength of ice is much less than that of the rock. The main processes of erosion are thought to be *abrasion* and *crushing* by debris within the ice and the detachment of materials by the *plucking* action of ice. Abrasion includes scratching, grooving and polishing by rock debris and some materials may also be reduced by crushing. Where basal slip is absent, then abrasion will be much less important, although movement of larger particles, due to glacier creep in higher layers, may lead to limited abrasion.

The frictional drag or shear exerted by a particle within the ice on the rock floor is a function of the normal pressure (i.e. the one acting at right angles to the floor). This pressure equals $\gamma_i h - u$, where γ_i is the unit weight of the ice and u is the basal upward counteracting pressure of the water. The frictional drag or shear (τ) is given by the following expression:

$$\tau = A\mu(\gamma_i h - u) \qquad [22.5]$$

where A is the area of contact and μ is the coefficient of friction. Thus, for increasing thicknesses of ice, there will be greater pressure exerted by the embedded particles on the rock floor which results in more abrasion and crushing.

However, as the pressure continues to increase, a condition will be reached where τ exceeds the cohesion between the ice and the particle and Boulton (1972, 1979) suggests that the ice will flow around the particle and abrasion will cease. Thus, as the depth and basal pressure increase, this hypothesis suggests that abrasion rates will rise to a peak and then decline, with consequent deposition. Larger particles being carried within the ice may also erode by physically removing smaller protrusions on the rock floor.

The plucking of blocks from the floor and sides of the glacial trough will clearly be a function of the jointing that is present, although some jointing may itself arise from the weight of the ice. The process essentially depends on the refreezing of water around a block, for example in the lee of an obstruction following regelation, and its physical removal by continued movement. Boulton (1972) has suggested that in cold glaciers, where the depth of freezing coincides with joints or other major planes of weakness, plucking of very large pieces of the bed may occur. At the base of temperate glaciers, freeze-thaw action is important in loosening materials and making them available for subsequent transport. Under temperate glaciers moving across non-resistant rocks, sediment transport also takes place as a result of the bedrock being progressively deformed by movement of the glacier (Fig. 22.4).

Where water is present at the base of glaciers, fluvial erosion can occur by the action of subglacial meltwater streams. The existence of the latter depends on the melting of their cavities being greater than the rate at which the ice moves by plastic deformation to close them. Erosion on the channel floor occurs by the processes discussed in Chapter 20, although abrasion by coarse bed material is probably most important. Additionally, *cavitation* can occur which is the creation of bubbles in rapidly moving water, when the local pressure becomes smaller than the vapour pressure. The collapse of these bubbles generates extremely erosive shock waves which are capable of eroding solid rock. These different erosional processes vary significantly according to the thermal regime of the glacier (Fig. 22.3). One particularly interesting possibility occurs when a hollow or trench is found in the profile, which leads to a release of pressure melting and more active erosion, tending to enlarge the trench. Thus, a positive feedback loop is created leading to greater deepening of the trench (Fig. 22.3(b)).

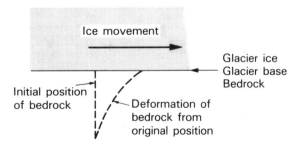

Fig. 22.4 Deformation of bedrock in non-resistant rocks at the glacier base. (Adapted from Boulton 1979)

22.4 Transport of material within ice

The debris carried by ice consists of much more material than is eroded by the actual glacier since all the different transport systems discussed in this part of the book deliver material to the glacier. The environment adjacent to glaciers is normally very harsh since there is usually very little vegetation to hinder erosion. This material is delivered to the upper surface of the glacier and becomes incorporated into the glacier when it is covered by the addition of snow (and eventually ice in the upper accumulation zone) and when material falls down crevasses.

The mechanism for incorporating debris into glaciers from their base must take account of the essentially laminar flow of moving ice since turbulence, which would add material to the ice mass, is apparently absent. Regelation of ice around obstacles will supply materials from the lee side, but only to the depth set by the upper limits of regelation. Where ice is added to the base of the glacier by freezing of water derived from below (as in zone C in Fig. 22.3), then debris can be raised into the glacier and where the movement of a glacier is confined by a narrowing of its trough, then both lateral and vertical movement of debris will occur. Melting of the glacier at the base will have the opposite effect, since it concentrates debris in the lowest layers, and the movement of ice around large obstacles by ice creep will lead to debris being concentrated into streams (Boulton 1975).

Another process which aids the incorporation of debris is the shearing of glacier ice which overrides the ice in front, bringing up debris frozen to the base of the glacier. The process of plucking large erratics by cold glaciers also leads to the incorporation of debris and the pressure of surrounding ice also squeezes debris up into cavities at the base of the glacier. It is apparent that the entrainment and transfer of debris in glaciers are very different from those in water or air, although they all act fundamentally as fluids. The high viscosity of ice means that once materials are entrained, they will move downwards only extremely slowly and will require no continued upward turbulent forces to maintain their position. Indeed, considerable amounts of material are transported on the surface of glaciers and since entrainment is independent of hydraulic processes, there is little tendency for any sorting of the resultant sediments. The type of entrainment associated with each of the glacier zones, defined according to their thermal regime, is summarised in Fig. 22.3.

22.5 Deposition of glacial materials

Debris in glaciers can be deposited in several different ways giving rise to a great variety of sedimentary materials (Embleton 1979a), which are generally called *till*. Perhaps, the most intuitively obvious process of deposition is that occurring at the glacier's base due to melting, which usually occurs in stagnant ice. Deposition at the base of an active glacier produces *lodgement till*, which usually consists of thick clay-rich materials, although the precise mechanism by which this till occurs remains unclear. However, high basal pressures are probably important since they lead to the movement of ice around particles rather than their transportation by the mechanism discussed in section 22.4. It has also been suggested that clay materials will tend to adhere better to previously deposited material than to the ice and low glacier velocities will also favour deposition of this till. The presence of cavities or crevasses may mean that materials are squeezed preferentially into them, or are squeezed out at the margins (Boulton 1979). The movement of the ice over these materials results in streamlined forms known as drumlins although they may be formed by other processes.

In the ablation zone of the glacier, materials will be exposed at the surface by the melting and evaporation of ice and these till deposits may then insulate the underlying ice and delay its melting (Plate 22.1). Subsequent mass movements of this *supraglacial till* are very common and form *flow tills* which, despite their name, usually have slip-planes at their base. The advance or readvance of ice produces features known as *terminal moraines*. It is now believed that much of the material found within linear ridges at right angles to the glacial advance in fact owe their origin to the deposition of supraglacial materials rather than to the push action of glaciers (Paul 1983).

At the sides of glaciers, sediment concentrations are high due mostly to the transfer of materials by subaerial processes (such as rock fall) from the adjacent slopes (Fig. 22.4) and also to the detachment and entrainment of materials along the margins of the ice. Melting of the glacier leaves the materials as upstanding features or *lat-*

Plate 22.1 Supraglacial till on the Tasman Glacier, New Zealand (formed by melt-out) almost completely masking the glacier ice beneath, which is only exposed at moulins (meltwater pot-holes).

eral moraines, and where glaciers coalesce, then *medial moraines* can form. Meltwater flowing at the margins of glaciers deposit materials which, following the final ablation of the ice, form a *kame terrace* and the deposition of materials from a subglacial stream results in long sinuous ridges of sorted materials known as *eskers*. Reference to Fig. 22.3 shows the types of till associated with different thermal conditions within glaciers.

based on scanty evidence and inadequate theory. Nevertheless, it is clear that the processes acting within glaciers are exceedingly complex and undoubtedly vary both according to the nature of ice flow and the physical characteristics of materials which are eroded by the glaciers. It is worth repeating that glaciers transport much more material than they erode and that a large proportion of the material finally deposited as till has its origin in other parts of the debris cascade.

22.6 Retrospect

Compared with many of the other processes analysed in earlier chapters, those associated with glacial erosion are relatively poorly understood. Indeed, many of the processes are also very poorly observed, simply because of the great practical difficulties involved in directly observing subglacial and englacial environments. Consequently, much of our knowledge of glaciers is unfortunately

Key topics

1. The dynamics of glacier movement.
2. Contrasts between polar and temperate glaciers.
3. The role of glacial erosion in moulding the landscape.
4. The factors determining the deposition of glacially transported materials.

Further reading

a

c

b

Plate V   (a) Solonchak soil (Salorthid) near Baghdad, Iraq. Note saline crumb under crusty surface and efflorescence of salts in profile. Long dimension of blackboard *c.* 20 cm. (b) Rendzina soil (Rendoll) on chalk, Wye, Kent, UK (scale provided by lens cap). (c) Low oblique aerial photograph showing a soil catena/toposequence along the Pangani River, Tanzania. Foreground – Ferruginous soils of the savanna with nomad settlements; middle distance – salinised soil with grassland; along the river – flooded alluvium with gleyed clay soils; beyond the river – same sequence in reverse order.

a

b

c

d

Plate IV   (a) Chernozem (Mollisol, suborder Ustoll, great group Argiustoll) with a textural B horizon (23–75 cm) and a carbonate horizon (75 cm), Hastings, Nebraska, USA (rainfall *c.* 620 mm). Marks on staff are 10 cm apart. This is a southern chernozem soil somewhat lower in organic matter than would be typical further north. (b) Brown chestnut soil (Mollisol, suborder Xeroll) with a prismatic textural B horizon (17–50 cm), Akron, Colorado, USA (rainfall *c.* 370 mm). Marks on right hand staff are inches; on left 10 cm intervals. (c) Desert sand (Entisol, suborder Psamment) Erg Ubari, Libya. Long dimension of blackboard is *c.* 20 cm. (d) Humic gley soil (Histosol) with shallow water table, Silchester, Berkshire, UK.

Plate III   (a) Podsol (Spodosol) with thin ironpan, Exmoor, Devon, UK. (b) Brown earth (Alfisol) containing flints overlying chalk, Hungerford, Berkshire, UK. (c) Brown Mediterranean soil (Alfisol, suborder Xeralf) developed on stratified terrace deposits, Basilicata, Italy. Scale is provided by coin (white spot) and light meter (black oblong). (d) Ferrallitic soil (Oxisol), Sedantry, Ife University, Nigeria.

Plate I   The active layer with ice wedges and the permafrost table at Frobisher Bay, NWT, Canada.

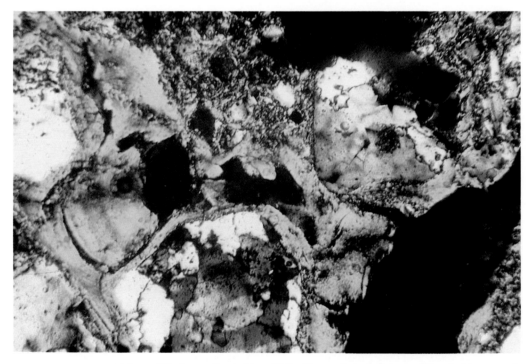

Plate II   Micrograph of a soil from Kintbury, Berkshire, UK, illustrating skeleton grains (white areas), voids (black) and argillans (yellow). The large grains are approximately 0.8 mm across.

Chapter 23 Ground ice and nival transport systems

Over 20% of the land area of the northern hemisphere is underlain by permanently frozen ground, known as *permafrost*, and beyond this zone there is a considerable area frozen for a substantial proportion of the year. These cold conditions have profound effects on the sediment transport systems which operate. The general name given to this type of environment is *periglacial*, although it is in a sense misleading since very cold conditions can be found in localities far distant from glaciers. All of the sediment transport systems (except by definition the glacial system) operate in periglacial environments, according to the basic principles already described. Additionally, a number of distinctive processes occur associated with ice below the surface and with snow. Consideration of the sediment movement caused by frozen ground and snow is the subject of this chapter.

23.1 Frozen ground and ground ice

Permafrost refers to ground whose temperature remains below 0 °C for at least two summers. If pore water is present, then the surface materials become cemented together, adding cohesion and hence strength to them. Given the above definition for permafrost, it is clear that the limits of permafrost can vary over a number of years. In warmer areas, permafrost does not form a continuous layer (Fig. 23.1) but is discontinuous with islands of permafrost surrounded by unfrozen ground. Although some permafrost may be comparatively young, much is believed to have originated during the Pleistocene. Above the permafrost layer is a zone where annual freezing and thawing take place and this is known as the *active layer* (Plate I) which, during the summer, melts and becomes very mobile.

As shown in Fig. 23.1, there may sometimes be an unfrozen layer between the seasonally frozen active zone and the permanently frozen ground. This unfrozen layer, and indeed any unfrozen material within or below the permafrost, is known as *talik*. Due to geothermal heating, unfrozen ground is always found below permafrost, although it may be several hundred metres beneath the surface. So far as the hydrologic system (section 2.1) is concerned, the presence of permafrost (with an active layer above) means that vertical movement of water is hindered and movement takes place preferentially in the surface layers. Even where the ground is only temporarily frozen below, the surface layers can become saturated and their strength then lowers due to reductions in interparticle friction.

Mass movement of materials which are saturated is known as *solifluction* and it occurs most commonly in periglacial conditions. Where it is associated with permafrost or seasonally frozen ground, the name *gelifluction* is used and despite

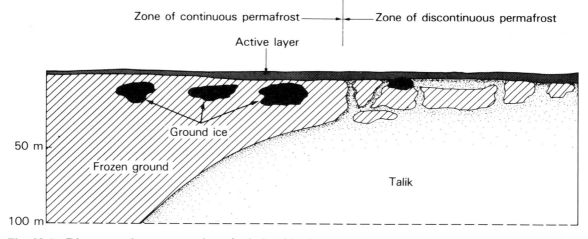

Fig. 23.1 Diagrammatic representation of relationships between frozen ground, ground ice and talik. Note that the change in depth of frozen ground to discontinuous permafrost takes place over several hundred kilometres.

its widespread occurrence, it remains unclear whether this movement is predominantly by flow or by failure along shear planes. It appears likely that the relative importance of the two types of movement varies according to environmental conditions and the type of materials that are being transported (section 17.4), although Harris (1981) suggests that the term solifluction should be restricted to flow movements alone. Closely associated with the process of gelifluction is that of *frost creep*. This is the movement of materials caused during a freeze-thaw cycle, due to displacement of materials normal to the slope during freezing and movement in a more vertical direction during thawing (section 18.5).

Frost heaving operates through the pressures caused by ice crystal growth as a result of pore-water migration. As a consequence of this process, stones are thrust upwards from fines and may form an almost continuous surface layer with almost stone-free materials below. In hot, arid areas moistening and drying cycles can produce morphologically similar stone pavements and such cycles may also be important in periglacial areas. The movement of materials by frost action leads to them being sorted according to size. Thus on slopes, *stripes* or alternately coarser and finer lines of stones may be found, whereas on more level ground, circular and *polygonal forms* may be produced with coarser materials forming the boundaries of the feature. In the upper part of the permafrost layer, the ground may not only be frozen but it may consist almost entirely of ice. The

origins of this *ground ice* are not simply a result of water freezing *in situ*, but arise from the movement and concentration of ice into preferred localities. Thus, if the ground ice melts, there will be more water than can be held in the soil pores and major geomorphic effects can result (section 23.2). Although ground ice can occasionally result from the burial of glacier ice and other types of surface ice, its origin is primarily associated with processes of ice segregation beneath the surface.

The segregation of ice and water results from the movement of liquid water through pores to the *freezing plane* which separates the liquid and solid phases of water. Although there is disagreement about the ultimate cause of this suction-like effect that ice has on water, many laboratory experiments have confirmed its existence. The growth of ice lenses results in a displacement of soil material and the expansion of the ice–water interface can only occur during freezing if the radius of the interface is sufficiently small. This radius is controlled by the size of the continuous capillary pore openings which are connected with other parts of the soil or surface materials which supply the water to be frozen. The conditions for ice segregation can be expressed mathematically as a function for the difference in capillary pressure exerted by ice (P_i) and water (P_w), as follows:

$$P_i - P_w > \frac{\sigma_{st}}{r_c} \qquad [23.1]$$

where σ_{st} is the surface tension between ice and water and r_c is the maximum radius of the continuous pore openings. Segregation tends to produce ice with layers parallel to the freezing plane and hence are usually horizontal in form. Vertical *ice wedges* are also found whose origin lies in the penetration of surface water into cracks, which are usually formed by thermal contraction. Freezing of the water that has entered the crack is followed by successive cycles of thermal contraction, producing cracks, and the addition of surface water can eventually produce large ice wedges (Plate I). Small features of this kind are known as *ice veins*, although cracks may also fill with sediment and thus *sand* or *soil wedges* may result. Furthermore, extensive development of ice wedges can result in a polygonal network being formed over large areas.

A third process by which ground ice may form is by the intrusion of water under pressure (French 1975). Although horizontal layers may be formed by this process, the formation of discrete lens-shaped masses of ice is best known which gives rise to the surface form known as a *pingo* (Fig. 23.2), which can rise to as high as 70 m. The origin of some of these forms is believed to be in the movement of water in liquid and gaseous phases under hydrostatic pressure beneath permafrost. However, where it emerges near the surface, it freezes as intrusion ice and if the upward pressure is too great, a pingo can rupture. A pingo thus formed through growth from below, under upward artesian pressure is called an *open-system pingo* (Fig. 23.2(a)). This is distinct from so-called *closed-system pingos* (Fig. 23.2(b)) which are formed by the freezing of entrapped water within a lake that has had sufficient sedimentation to raise the lake floor into the zone where annual freezing occurs. Small surface mounds within peat are commonly found in periglacial areas and are similar in form to small pingos. They are known as *palsas* and have a somewhat different origin from pingos, since they have separate ice lenses rather than a core of clear ice. Although the mechanism for their formation is unclear, they are presumably caused by frost heaving.

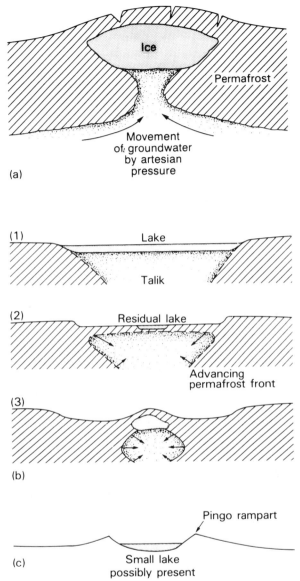

Fig. 23.2 Formation of a pingo (type (b) adapted from MacKay 1972): (a) open-system pingo; (b) closed-system pingo; (c) form after melting of ice core.

23.2 Thermokarst

When ground ice melts, the pores in the surface materials are unable to contain all the water and consequently rapid changes in surface form result. This leads to substantial short-distance sediment transport and a characteristic suite of subsidence landforms known as *thermokarst*. The formation of thermokarst is of considerable economic importance since the construction of buildings, roads, runways, etc. can cause ground ice to melt and in turn can lead to the destruction of the constructions themselves. Natural formation of thermokarst can arise from erosion of the ground surface in the active layer, which heats lower layers and then melts the ground ice. Climatic warming can obviously disturb the thermal equilibrium of the ground ice on a wide spatial scale. Furthermore, changes in vegetation cover control the albedo (section 2.2) and latent heat flux (section 2.4) at the earth's surface, which also affect the thermal status of the ground ice. If such changes are a result of burning, the heat of combustion can also result in the melting of ground ice and the formation of thermokarst. The presence of standing water during the summer in lower locations, such as along the thermal contraction cracks of ice wedges, encourages more extensive thawing which can then lead to the development of thermokarst.

Whatever the cause, the thawing of an area extensively dissected by ice wedging will usually lead to the formation of linear and polygonal troughs, and if the formation of these troughs is sufficiently pronounced, upstanding mounds will be formed which are known as *baydjarakhs*. Continued degradation of the land can lead to the integration of the surface depressions to form *alases*, which are large steep-sided, flat-bottomed depressions up to several kilometres across. Lakes may form in the alas which can encourage further thawing and sedimentation of such lakes can subsequently be associated with pingo development and renewed ice wedging. Pingos, whose core ice melts, will leave rounded depressions often with a raised rim or rampart surrounding a small lake (Fig. 23.2(c)).

Headward scarp retreat can also result from thermokarstic action. An initial exposure and thawing of ground ice, by whatever process, leads to continued backward retreat by slumping as materials are moved away from the base of the scarp by gelifluction, flow and wash processes. The resultant scarps are usually less than 2 m high but, despite their low relief, they are important because of the rapidity of the resultant erosion (French 1975). *Thaw lakes* owe their formation (or at least their enlargement) to the thawing of frozen ground. They are very shallow, rounded and usually somewhat elongated depressions which are often orientated with the long axis parallel to the dominant wind direction. The latter presumably causes preferential erosion due to wave and current orientation. Compared with most lakes, they are very dynamic in size and location. Drainage into adjacent depressions can be very rapid or alternatively, progressive sedimentation can take place along with colonisation by vegetation.

23.3 Nival processes

These processes include all those associated with snow and hence include both slow movements, associated with snow patches, and extremely rapid movements during avalanching. *Nivation* is the term used to describe the various processes operating under and close to snow banks and includes frost action, chemical weathering, gelifluction and slope wash. Although freeze-thaw processes (leading to physical weathering) are often regarded as being the most important processes involved in the breakdown of slope materials, Thorn and Hall (1980) analysed both Arctic and Alpine nivation and concluded that chemical weathering is of considerable importance. In particular, they stressed the importance of chemical weathering during exposure of materials to meltwater and the relatively high local temperatures that can be found at that time.

Freeze-thaw processes are likely to be more important under shallow rather than deep snow (Harris 1974) because of the insulating effect of the latter. By the same argument, it is difficult to see how freeze-thaw action can be greater beneath snow cover than elsewhere and it may well be that freeze-thaw action is lower in such locations than elsewhere. As discussed in Chapter 18, the physical shattering of rocks may in fact result not so much from freeze-thaw activity as from hydration, which causes expansion of crystals when water is added. The most common landform resulting from nivation is a shallow hollow, aligned approximately parallel to the contour, which is enlarged to form a large, semicircular shaped *nivation cirque*. On steeper slopes, sliding of snow occurs which is apparently capable of causing erosion by the movement of entrained debris across bedrock (Embleton 1979b), even though the movement is slow. Where snow has accumulated on steeper slopes, debris falling on to the snow will show a tendency to slide across the surface and accumulate at the base of the snow-covered surface. The resultant ridge is known as a *pro-talus rampart*.

Failure of snow bodies results in avalanching which is now increasingly recognised as an important agent of erosion (Embleton 1979b). The instability of the snow mass arises from many factors which include the addition of snow, internal structural weakness and changes associated with percolating snow melt. Some snow avalanches have virtually no geomorphic significance whereas others may pick up considerable volumes of rock debris, especially if the surface upon which they impact and move over is bare of protective snow. The resultant mixed snow and debris avalanche can then cause further erosion downslope with the uprooting of large trees and even the destruction of buildings. Whereas some snow avalanches fall through the air like powder, it is more common for a turbulent flow to result. When the avalanche is very wet, it will move in a manner akin to a debris flow (Ch. 18). The resultant deposits of debris will tend to be very badly sorted and may form a ridge-shaped mound, not dissimilar in form from glacial moraines with which they may be confused.

23.4 Retrospect

The sediment transfer processes analysed in this chapter have received increasing attention with the more intensive development of the Arctic areas for their natural resources. The ease with which the thermal equilibrium of frozen ground can be disturbed, and the harmful consequences of such disturbance, have encouraged a much more scientific investigation of the processes that are operating. Nevertheless, our understanding of many of these processes remains weak and is often based on inadequate, qualitative observations and limited quantitative measurements. Furthermore, physically based theory has unfortunately played relatively little part in the investigation of these phenomena as yet. However, a more rigorous approach to the analysis of these sediment transport systems will doubtless come in the future, despite the practical problems of working in these harsh environments.

Key topics

1. The factors determining the distribution of permafrost.
2. The control of ground ice on surface textures.
3. The role of thermokarst in modifying periglacial landscapes.
4. The consequences of gradual and catastrophic nival processes in periglacial environments.

Further reading

Carson, M. A. and Kirkby, M. J. (1972) *Hillslope Form and Process*. Cambridge University Press; Cambridge (Ch. 12).

Embleton, C. and King, C. A. M. (1975) *Periglacial Geomorphology*. Edward Arnold; London.

Embleton, C. and Thornes, J. (1979) *Processes in Geomorphology*. Edward Arnold; London (Chs 6 and 9).

Washburn, A. L. (1979) *Geocryology*. Edward Arnold; London.

Part IV
Soil systems
Chapters 24–27

The following four chapters introduce some of the basic concepts of soil science with particular reference to their role in earth surface processes. The physical properties of soil materials are considered along with the organic and inorganic processes which determine soil profiles. The approach is geographical and emphasis is placed on the interactions of geological, climatic and vegetal factors on the development of soil profiles. Particular attention is given to the processes which form some of the world's main zonal and intrazonal soil types. The concluding chapter analyses the soil factors which influence the growth of plants, with special reference to the agricultural crops and forest trees of importance for human use.

Satellite view of desert and irrigated soils in the Nile Delta and Sinai.

Chapter 24 Soil physical properties and morphology

Soils have been studied for thousands of years but systematic work really began during the nineteenth century with attempts to relate soil chemistry to plant yields. The foundation for the modern view of soils was laid by Dokuchaev (1886) with his recognition that they were independent natural bodies with morphologies dominantly reflecting the effects of climate. This concept still underlies soil science and has led logically to the consideration of soils as an active system whose basic energy is derived from the sun.

The soil system is thus part of the environmental 'super-system' with its own internal processes and laws and resembles an organism as much as an organisation. The basis of study is the soil profile consisting of a vertical sequence of horizons which vary in their degree of separation from surrounding influences. The physical, chemical and biological processes are the interlocking subsystems which determine the 'state' of the whole. This is expressed in terms of its properties such as temperature, chemical composition, micro-organic life and profile morphology. The soil system is complex and can only be fully specified with reference to many properties and variables. Some of these are more important than others and are known as *key variables*, but these are often not those which are easiest to observe and measure.

Key variables can be both external or internal to the soil system. These can be subdivided into *external variables* which can be considered as *independent* or *control variables* and *internal variables*, which are mainly dependent on the external but can be causative of others. Examples of the former are climate and lithology and of the latter are soil structure and soil texture. Climate is independent of soil processes but although soil texture controls the way in which plant nutrients are moved, it is itself determined by parent materials. Colour, on the other hand, is an example of an ordinary dependent internal variable which has relatively little effect on others. Because of the difficulty of measuring processes while they are active, however, it is usually necessary to analyse and classify those features of soil systems which can be readily observed and measured, emphasising those which are indicative of other soil properties and the processes which cause them. For example, the study of the biochemical processes which form organic matter can conveniently be approached through observation of its composition and morphology and the changes in soil water from the colours in affected horizons.

The soil can be defined as the collection of natural bodies on the earth's surface containing living matter and supporting (or capable of supporting) plants. Its upper limit is characterised by air or water and at its lateral margin it grades into deep water or barren areas of rock, ice, salt or shifting sand dunes. However, its lower limit is the hardest to define since it includes all horizons differing from the underlying rock material as a result of interactions between climate, living organisms, parent materials and relief. This limit is generally, therefore, the lower limit of the common rooting of the native perennial plants, a diffuse boundary that is shallow in deserts and tundra and deep in the humid tropics.

Soils are essentially the living medium resulting from the action of climate and vegetation on landforms and consist of a vertical profile containing the components shown on Figs 16.1 and 24.4. The formative processes are pedological and although there is some overlap, these differ from geomorphological processes in two major ways. First, they mainly concern the vertical movement of water and air with their dissolved or suspended constituents within the soil profile, while geomorphological processes are dominated by the lateral movements of these over the earth's surface. Secondly, pedological processes tend to be mainly chemical and biological in character, while geomorphic processes are dominantly physical and inorganic. Soil development thus acts as a link between the broader environmental controls of climate and landform on the one hand and plant and animal life on the other.

The soil system is in a state of perpetual change and although the observer sees it at only one point in time, observations of key variables governing its present state provide data for the interpretation of its formation and probable future evolution. Soil is composed of mineral components derived originally from the weathering and transportation systems described in Part III; its organic constituents and processes are considered in Chapter 25 and non-organic processes in Chapter 26. It is necessary, however, to begin by specifying the basic physical properties of soil mineral matter which forms the substance of this chapter. It considers the nature of soil composition, its physical properties with particular reference to the movement of fluids (which has been considered in Ch. 6), the combination of primary particles into structural aggregates at both large and small scales and the subdivision of *soil profiles* into *soil horizons*.

24.1 Composition of mineral soils

The soil material consists of four components: mineral matter, organic matter, water and air. As a rough approximation, about half the volume of a normal soil consists of solid matter, about one-quarter of water and one-quarter of air. The solid matter consists of mineral and organic components and, for example, mineral surface soils may contain anything from a trace to 15 or 20% of organic matter, although in humid temperate regions values generally lie between 1 and 6%. On the other hand, organic soils (such as peats) often contain over 90% of organic matter.

The mineral matter consists of a mixture of particles of various sizes, which can be grouped into size ranges known as *separates* and widely used examples of such groupings are shown in Table 24.1. These classes have relevance to soil processes in that each stone or gravel fragment may contain a mixture of minerals. Furthermore, sand and silt particles usually contain only a single mineral in comminuted form while clays consist of

Table 24.1 Soil textural grades (conventionally, textural description relates only to the fine earth fraction, i.e. less than 2 mm in diameter)

Particle size range (mm)	US Department of Agriculture (1975)	International as used by Soil Survey of England and Wales (Fenwick and Knapp 1982)
Over 75	Stones	Stones
2–75	Gravel	Gravel
1–2	Very coarse sand	Coarse sand
0.5–1	Coarse sand	
0.25–0.5	Medium sand	
0.2–0.25		
0.1–0.2	Fine sand	Fine sand
0.05–0.1	Very fine sand	
0.002–0.05	Silt	Silt
Under 0.002	Clay	Clay

particles formed of the clay minerals, which were described in Chapter 16. They are largely the weathering products of silicate and aluminosilicate rocks.

Since the absorption of water, nutrients and gases (and the attraction of soil particles for each other) are all surface phenomena, the size and arrangement of such particles have an important influence on soil behaviour. Stones and gravel provide the skeleton which makes the soil more permeable but they are relatively inert. Sands are loose, non-plastic and permeable, with a low water-holding capacity, and likewise do not enter much into soil reactions. The smaller size, larger total surface area and the more irregular (often platy) shape of silt particles gives them greater cohesion and water-holding capacity, but makes them less permeable. Clays show these characteristics to a greater degree and when dry, they form hard aggregates whereas when wet they are plastic and sticky. These differences are due mainly to the surface area of the particles. For example, a grain of silt has about 50 times, and a grain of clay perhaps 10 000 times, the surface area of that of the same weight of medium-sized sand.

Soil mineral matter is a mixture of sand, silt and clay whose relative proportions define the *soil texture*. Where any one of these separates is dominant, the soil will have its characteristics as outlined above but when they are relatively evenly mixed, they are called *loams*. Each of these major categories, however, contains considerable internal variations so that twelve textual classes are now recognised. These are shown in Fig. 24.1 in relation to the percentages of sand, silt and clay that they contain.

Soil textures may change as a result of sedimentary additions, differential weathering, chemical translocation or the activities of soil fauna, particularly earthworms and moles. The activities of ice, wind and running water add such characteristics as till, loess and alluvium respectively but they may also remove textures selectively. In arid climates or where vegetation is sparse, for instance, the wind will remove sand and silt but concentrates both stones and clay, which for different reasons are less mobile. *In situ* weathering generally tends to reduce the size of particles, but often preserves textural patterns which reflect the original internal arrangement of the parent materials, as for instance in the persistence of layering in soils formed on sediments. In humid climates, the downward translocation of *colloids* will cause internal textural differentiation in the profile.

24.2 Density and pore spaces of mineral soils

Soil processes result from the movements of both air and water through pore spaces. The overall volume and the sizes of individual pores thus control the amount and rate of such movements. The volume of pore spaces can be calculated from the difference between the *particle density* and the *bulk density* of the soil. The particle density is the specific gravity of soil solids and since most of these are formed of quartz, feldspars and mineral silicates, densities range between 2.60 and 2.75 g cm^{-3}. Unless there is an unusually high proportion of heavier minerals, a mean value of 2.65 g cm^{-3} can normally be assumed.

The bulk density, however, is the specific gravity of the soil material as a whole which includes not only the solids but also the intervening *pore space*, which is expressed as a percentage of the total volume not occupied by solid matter as follows:

$$\% \text{ pore space} = 100 - \% \text{ solid space} \quad [24.1]$$

This is difficult to measure and is normally estimated by comparing the weight of a given volume

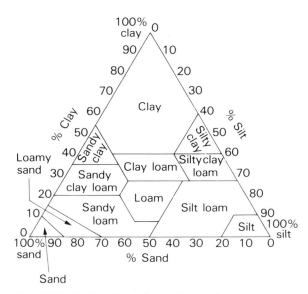

Fig. 24.1 Soil texture classes in relation to percentages of clay (below 0.002 mm), silt (0.002–0.05 mm) and sand (0.05–2.0 mm). (After USDA 1975)

215

of soil with an assumed value for particle density (i.e. 2.65 g cm⁻³). Since:

$$\% \text{ solid space} = \frac{\text{bulk density}}{\text{particle density}} \times 100 \quad [24.2]$$

the following formula can be written for pore space:

$$\% \text{ pore space} = 100 - \frac{\text{bulk density}}{\text{particle density}} \times 100 \quad [24.3]$$

In most soils, pore space occupies about half the total volume, so that a bulk density of about half the particle density is normal. Substituting in equation [24.3], a soil having a bulk density of 1.2 g cm⁻³ and a particle density of 2.65 g cm⁻³ will thus have a pore space of 55%. Pore spaces vary in size. Water and young roots penetrate freely only through pores wider than about 0.3 mm whereas water drains by gravity out of a soil only if it can move through pores larger than 30–60 μm. Root hairs and large soil micro-organisms (such as protozoa and fungi) need pores larger than about 10 μm to grow into or move in, and smaller micro-organisms need pores greater than 1 μm for movement. The very finest pores (<2 μm) are important in giving good tilth, especially in clays. Thus, very roughly, pores can to divided into coarse (>200 μm), medium (20–200 μm), fine (2–20 μm) and very fine (<2 μm). Therefore, although clays have more total pore space than coarser textures, the large size of pores in the latter allows much greater permeability and aeration.

24.3 Soil structure

The influence of texture on soil processes depends on the way in which discrete particles of sand, silt and clay aggregate or bond together. Such aggregates are known as *peds* and their shape, size and disposition define the soil structure. This greatly affects porosity and thus governs soil properties such as water-holding capacity, permeability, aeration, heat transfer and the growth of roots and soil organisms. Although a whole profile may have a uniform structure, more often the horizons will differ since soil structure is determined by three variables: strength, shape and size. A soil horizon will normally contain a mixture of peds of different sizes, with some loose material. Where the

proportion of the material which is held in the peds is large, the structure is strong. However, where it is small, the structure is weak with intergrades being definable between the two. Two types of soil are structureless: those which are solid and without internal differentiation are called *massive soils* whereas those which collapse into a heap of loose grains are termed *single grain*.

Where aggregation does occur, four major structural forms (Fig. 24.2) are recognised: *platy*, *prism-like*, *block-like* and *spheroidal*. Platy structures are formed of relatively thin laminae or lenses, which are almost invariably horizontal. They are sometimes inherited from previous depositional layers and sometimes result from crusting by raindrop impact or from the evaporation of soluble salts in arid areas. Prism-like structures are termed *columnar* where the tops are rounded, or *prismatic* where they are level. Columnar structures are especially characteristic of certain subsoils in arid areas where the downward translocation of soluble salts has led to the *deflocculation* of clays (i.e. destruction of fine aggregates or *flocs* in which neighbouring clay particles normally cohere in soil) in the presence of high levels of exchangeable sodium. Block-like structures are termed *angular blocky* if the edges of the peds are sharp and *sub-angular blocky* if some are rounded. Spheroidal structures are known as *crumb* if highly water absorbent and *granular* if only moderately so. They are most characteristic of surface soils where organic matter is present, where cal-

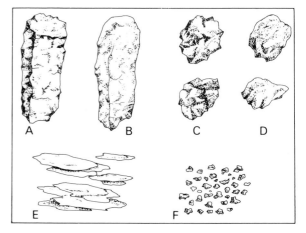

Fig. 24.2 Drawings illustrating some of the types of soil structure: (A) prismatic; (B) columnar; (C) angular blocky; (D) sub-angular blocky; (E) platy; and (F) granular. (After USDA 1975)

cium is the dominant cation and where there are marked alternations of temperature and moisture.

Peds display considerable variation in size, even within the same horizon and there are usually different orders of size with larger peds containing smaller aggregates. However, the most stable size is generally regarded as definitive of the whole. Spheroidal structures, and the short dimension of platy structures, vary from less than 1 mm to over 10 mm and are smaller than the other types which may be as large as 50 mm for blocky and 100 mm for prism-like types. Aggregates tend to be most stable where they are small, where organic activity is at a maximum and where clays (especially the hydrous micas and kaolinite) are present.

Some structures are inherited from the parent rock but the main formative processes are organic and colloidal. Plant roots and fauna (such as earthworms) create voids in the soil mass which separate the surrounding material into peds, which are then cemented and stabilised by the secretions of micro-organisms. The alternations of wetting – drying and freezing – thawing in clays cause the expansion, contraction and cracking which separate them into peds. The process is affected by the way in which the cations in the soil solution cause the clay particles to flocculate. This occurs when cations nullify the negative charge on clay particles or link them by giving a positive charge to one, which can attract a negative charge on another. This tendency is increased by the role of *polyvalent cations* in forming bridges between or-

ganic and mineral colloids (Fig. 24.3). In general, *trivalent cations* and *divalent cations* will permit this aggregation, but *monovalent cations* (those with a single positive charge, and particularly Na^+) are held loosely and will blanket positive charges on the exposed edges of the particles, causing the flocs to disperse. This causes impermeability and explains why sodium-dominated clays tend to become 'puddled', impermeable and agriculturally poor.

The aggregation of larger peds (including both coarse mineral particles and organic matter) is due not only to clay bonding but also to cementation by materials, which are moved in the soil solution and deposited around particles of all sizes. The main cementing agents appear to be certain organic compounds, ferric oxide and calcium carbonate. The most important are probably the large molecules formed by *polysaccharide gums* (complex cohesive sugar compounds produced by the micro-organisms which decompose roots), which may act as bonds between clay particles. Sesquioxides are increasingly mobilised as the pH falls below 4 or 5 and form *chelates*. These are organic compounds which combine with and protect the metallic cations from further reactions, while keeping them in solution at much higher pH values than their ionic forms. Chelates contribute to structure when adsorbed on to clay or other particles. This can apparently be caused by drying, rises in the cation–anion ratio or the oxidation of contained cations, notably iron.

24.4 Soil micromorphology

If the focus is narrowed to a microscopic examination of the arrangement of mineral particles in small samples of soil, it is possible to observe phenomena which aid the understanding of larger-scale soil processes. Clay colloids are not mixed evenly with larger particles but form patterns of arrangement around silt and sand grains which occur in the size ranges needing microscopes with magnifications of $c. \times 50 – \times 500$ for their observation. Such patterns constitute soil micromorphology.

The soil material consists of three components: the *S-matrix*, *plasma* and *voids*. The S-matrix is composed of *skeleton grains* which are the relatively inert sand and silt particles. The plasma is the surrounding, more mobile, colloidal mass in which active processes can be seen to occur, and

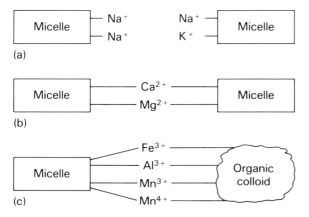

Fig. 24.3 Simplified diagram of the effects of cations on the flocculation and deflocculation of soil colloids: (a) monovalent cations separate micelles; (b) divalent cations link micelles; (c) polyvalent cations link micelles and organic colloids.

the voids are the empty spaces which provide av-
enues for the movement of fluids (Plate II). The
plasma may be arranged into *plasma fabrics* and
when these are oriented, they are called *sepic* and
can indicate the direction followed by the water
from which they were deposited. However, where
the fabric arrangement is not oriented, they are
known as *asepic*. An example of sepic plasma fabric is the *cutan*, which is also known as the *clay
skin* or *argillan*. It is an accumulation of clay particles oriented in sub-parallel fashion about a ped
so that when viewed under a hand lens (or even
with the naked eye), it appears shiny. It is sometimes possible to interpret the processes of formation in soils and *paleosols* (buried or preserved
fossil soils) by studying the form and orientation
of cutans.

24.5 Soil profile morphology and nomenclature

When a soil begins to form, it has the inherited
characteristics of the local lithology, is unadapted
to its environment and is therefore considered to
be young. As it develops, *pedogenetic* (soil-forming)
factors gain ascendancy, horizons are formed
and it becomes mature. This persists until further
changes (such as the removal of the surface horizons by erosion) cause it to become degraded.

Soils not only grade vertically into horizons,
but they merge laterally into other soils often
making it difficult to determine their mutual
boundaries. This is recognised in the concept of
the *pedon* which is the smallest volume that can
represent the soil at a given site (US Department
of Agriculture 1975). It is normally defined as a
block of soil of 1 m^2 at the surface extending to
the bottom of any horizons present. However, if
the soil shows any cyclical changes which recur
with a linear interval of 2–7 m, the pedon includes
one-half of the cycle. *Polypedons* are contiguous
groups of pedons of like character which can form
the basis of soil-mapping units.

The nature and distribution of soil horizons
provide a key to understanding the processes of
soil formation. Their classification and description
are given in Table 24.2, although explanations of
their processes of formation are considered in subsequent chapters. The *master soil horizons* are
designated by capital letters and their subdivisions
by suffixing lower case letters and sometimes arabic numerals. However, not more than a rela-

Table 24.2 Soil horizon nomenclature. (Adapted from Hodgson 1976)

Litter layers and organic horizons

L	Fresh litter	
F	Decomposing litter (fermentation layer)	
H	Well-decomposed litter	
O	Peaty horizon	
	Of	fibrous peat
	Om	semi-fibrous peat
	Oh	amorphous peat
	Op	ploughed peat

Omh or Ohh – Om or Oh horizons containing translocated colloids

Mineral horizons – datum for depth measurement

A	Mineral horizon incorporating humified organic matter or disturbed by cultivation, or both	
	Ah	uncultivated
	Ap	ploughed
	Ahg or Apg – where gleyed or mottled	
AB A, B	] Transitional	
AC	A transitional horizon that lacks the characteristics of E or B	
E	Eluvial horizon lighter in colour and with less organic matter than A	
	Ea	colour mainly determined by uncoated grains, i.e. bleached
	Eb	brownish from iron oxide
	Eg	gleyed
	Ebg	combining features of Eb and Eg
EB E, B	] Transitional	
B	has one or both (1) alluvial material, (2) alteration by solution, clay synthesis, or structural peds	
	Bf	thin iron pan or placic horizon
	Bg	gleyed
	Bgf	gleyed with abundant secondary Fe
	Bh	humic accumulation
	Box	residual accumulation of sesquioxides
	Bs	sesquioxide accumulation
	Bt	clay accumulation
	Btg	combining features of Bt and Bg
	Bw	altered but not qualifying as one of the above
	Bx	fragipan: a dense but brittle layer caused by compaction
BC B, C	] Transitional	
C	Unconsolidated or weakly consolidated parent material	
	Cu	no evidence of gleying, salt accumulation or fragipan
	Cr	too dense for root penetration

Cg	gleyed
Cgf	gleyed with abundant secondary Fe
CG	intensely gleyed
R Bedrock	

Additional suffixes

m	continuously cemented, otherwise than by Fe
x	fragipan
ca	horizon enriched in carbonates
cs	horizon enriched in gypsum
sa	horizon enriched in salts more soluble than gypsum

Lithological discontinuities are shown by using Roman numerals I, II, III, etc. as prefixes to the other horizon designations, numbering from the top down. Where two or more horizons in the same profile have the same nomenclature, they are distinguished by suffixing arabic numerals, i.e. Bh_1, Bh_2, etc.

tively small number of horizons are normally present in any profile. Lithological discontinuities, caused by stratification in the profile and not pedogenesis, are recognised by the use of roman numerals (Table 24.2). A typical soil profile is represented in Fig. 24.4. The *L, F* and *H horizons* overlie one another in this order since the decomposition of litter proceeds from below. *O horizons* occur on waterlogged and treeless sites and the subdivisions are based mainly on the internal structure of the peat and normally reflect the nature of the vegetation from which it is formed.

The *A horizon* is the uppermost mineral layer. In a typical soil in the humid temperate zone, this is characterised by organic processes which break down the leaf litter and other organic residues into humic material and acid and mix them with the mineral soil. At the same time, downward-percolating rainwater mobilises and moves bases and sesquioxides, clay colloids and humus out of the surface layers by the process known as *leaching*. Where the leached layer is clearly distinct from the organic topsoil, it is designated by the letter E (for eluvial). *A horizons* are mainly subdivided according to whether or not they are ploughed or gleyed (e.g. showing grey or bluish colours as a result of the reduction of iron and other elements). *E horizons* are generally classified on the basis of the degree of leaching or gleying, as reflected in visible soil colours.

The underlying *B horizon* is dominated by the reception and concentration of materials leached from above, but may also owe its distinctness to chemical alteration or the residual concentration of certain components. Where the character is due to the first, the letter I (for illuvial) is sometimes used to designate the horizon, but this is less common than the use of E. The substances washed in from above accumulate either together or as separate layers in the B horizon and, if they are separate, they are designated as sub-types of the horizon. Special symbols are also used for changes induced by gleying or other chemical change. The subsoil horizons of red tropical soils, for example, are considered to be B horizons although their distinction is due to a residual concentration of sesquioxides left behind by the removal of other constituents, and not to material deposited from above.

The *C horizons* consist of unconsolidated parent materials which do not show the properties of the other master horizons although there may have been some weathering, cementation, gleying or enrichment by soluble materials. The designation CG is used for intensely gleyed horizons in situations where there has been continuous subsoil waterlogging for a long period, i.e. in some Holocene sediments which have not dried significantly since they were deposited. The *R horizon*, sometimes called the *D horizon*, is the unaltered underlying bedrock.

24.6 Retrospect

Soils undergo continuous change but contain components which vary widely in the rate and extent of such change. Certain physical properties of mineral soil components are relatively enduring and provide a stable framework within which processes occur. This chapter has considered some of the most important of such properties for the understanding of the organic, inorganic and nutritional activities of soils discussed in the following three chapters.

The differences of particle size between sand, silt and clay are responsible for their widely different physical and chemical properties, so that their relative proportions, constituting soil texture, determine many of its most fundamental characteristics and its reaction to the influence of external variables. With the input of energy and materials, the soil acquires a more or less stable structure which then itself becomes a controlling variable for subsequent processes. The result at a

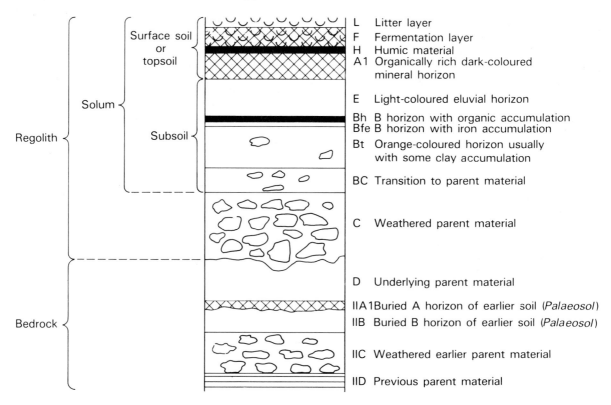

L	Litter layer
F	Fermentation layer
H	Humic material
A1	Organically rich dark-coloured mineral horizon
E	Light-coloured eluvial horizon
Bh	B horizon with organic accumulation
Bfe	B horizon with iron accumulation
Bt	Orange-coloured horizon usually with some clay accumulation
BC	Transition to parent material
C	Weathered parent material
D	Underlying parent material
IIA1	Buried A horizon of earlier soil (*Palaeosol*)
IIB	Buried B horizon of earlier soil (*Palaeosol*)
IIC	Weathered earlier parent material
IID	Previous parent material

Surface soil or topsoil — Solum — Subsoil — Regolith — Bedrock

Key to symbols used in soil profile diagrams Fig. 24.4 and Figs. 25.3, 26.2, 26.3, 26.4 and 26.5

	Undecomposed plant litter
	Fermented litter
	Humic material
	Humus-rich horizon with strong organic/mineral bonding, crumb structure
	Humus-rich horizon without strong organic/mineral bonding
	Ashy or bleached horizon
	Gravel
	Mottled
	Gleyed
	Kaolinitic clays
	Montmorillonitic clays
	Accumulation of ferric hydrate
	Accumulation of dehydrated ferric iron
	Secondary accumulations of calcium carbonate or gypsum
	Silt and loess
	Calcareous bedrock
	Non-calcareous bedrock

Fig. 24.4 A typical soil profile showing horizon names and including a key to the symbols used in all soil profile diagrams.

particular site is the formation of the pedon, grading laterally into other pedons and containing a vertical sequence of soil horizons. The definition of such horizons is usually the starting-point for the recognition and interpretation of pedological processes.

Key Topics

1. The relationship between the soil system/soil processes and earth surface systems/processes.
2. The physical and chemical characteristics of soils.
3. The significance of soil micromorphology in pedogenic processes.
4. Soil profiles as the key to pedogenesis.

Further reading

Brady, N. C. (1990) *The Nature and Properties of Soils*. Macmillan; London and New York (basic text with an edaphological rather than a pedological or geographical emphasis).

Buol, S. W., Hole, F. D. and McCracken, R. J. (1980) *Soil Genesis and Classification*. The Iowa State University Press; Ames.

Duchaufour, P. (1982) *Pedology*. George Allen and Unwin; London and Boston (translated from the French by T. R. Paton).

Wild, A. (ed.) (1988) *Russell's Soil Conditions and Plant Growth*. Longman; London and New York.

Chapter 25 Soil organisms and organic matter

The organic components of soils consist of living plants and animals, undecomposed remains and *humus*, which is a term reserved for organic material at an advanced state of decomposition. The processes which link these components form the pedological parts of the biogeochemical cycles, which will be considered in Chapters 30 and 31. They can be simply expressed in the form of a flow diagram (Fig. 25.1) which indicates the order in which they are considered in this chapter. It should be noted, however, that organic processes in soils are particularly difficult to observe and study. This is because of the complexity of the biochemical reactions, the microscopic scale of most organisms and the difficulty of reproducing field conditions in the laboratory. Knowledge about them is therefore still very limited.

25.1 The decay of organic residues

The decay of higher plants provides most of the organic residues processed by soil organisms, whose bodies are built from essentially simple compounds of carbon, oxygen, hydrogen, nitrogen and a number of other elements. Energy is mainly provided by the combustion of carbon compounds, whose availability critically controls the numbers and activity of soil organisms so that these can be measured in terms of the volume of carbon dioxide emitted from the soil in a given period. This tends to be at a maximum in areas with abundant natural vegetation.

Organic residues are mainly derived from plants although the bodies and waste of animals also contribute to a lesser extent. Plant materials added to the soil are highly variable in form and amount. For example, grasses and other annual plants add considerable residues to the soil from both roots and leaves. Trees and perennial plants, on the other hand, only yield part of their tissues annually and their woody parts are returned to the soil much more slowly. Leaf fall is important in forest areas and it is estimated that it can amount annually to 2–4 t ha^{-1} in temperate areas, 5 t ha^{-1} in subhumid tropical areas, and 10 t ha^{-1} in equatorial rainforests. There are also considerable differences between the types of litter deriving from different trees. For example, that from deciduous trees in the temperate zone is generally both greater in amount and relatively richer in metallic ions than that from needle-leaf trees (see Fig. 27.4). Roots left behind by harvested crops may also contribute several tonnes of dry matter per hectare. Plant residues generally contain between 60 and 90% of water, although the average value is about 75%, and some approximate figures for the composition of their dry matter are given in Table 25.1.

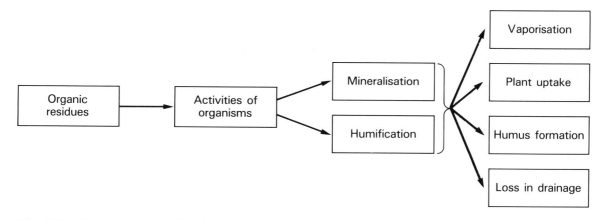

Fig. 25.1 Organic processes in soils.

Table 25.1 Typical composition of dry matter in some plant residues

Compounds	%
Carbohydrates	
Simple sugars and pectins	1–6
Hemicelluloses and polysaccharides	12–30
Cellulose	20–50
Fats, oils, waxes, resins	1–7
Lignins	10–30
Nitrogen compounds: proteins, amino acids	2–18
Mineral matter (consisting of C 44%, O 40%, H 8%, others 8%)	1–5

25.2 Soil organisms

Soil organisms consist of both plants and animals and some micro-organisms which have some of the characteristics of both. The total weight of living micro-organisms in soil is about 2% of the total weight of organic matter so that a soil with 5% of organic matter will contain about 2.5 t ha⁻¹. Soil animals can be subdivided into those of macro and micro size. Macro-animals include the *herbivores* (such as slugs and the larvae of some insects) which eat living plant tissues, *detritivores* (such as mites, beetles, millipedes, woodlice and springtails) which live mainly on decayed or decaying plant tissues and *predators* (such as moles, spiders, many insects, nematodes and protozoa) which live mainly on other animals.

Soil flora are more numerous and provide a greater *biomass* than soil fauna and are themselves dominated in both respects by the microflora, in spite of the small size of individuals. Table 25.2 gives approximate ranges for the numbers and biomass of some of the more important groups of microflora. The numbers of both soil flora and fauna are at a maximum where there is a dense vegetation cover, where they are not restricted by extreme temperatures or pH levels and where water fills most of the soil pores, but drainage is adequate to prevent waterlogging. Forest soils have a more diverse fauna than grassland soils but grassland fauna are metabolically more active and their total weight per hectare is greater. Cultivated

Table 25.2 Approximate estimates of relative numbers and biomass of microflora commonly found to the depth of a 'furrow-slice' (*c.* 15 cm) in temperate region soils, assuming 2.25 million kg dry earth per hectare of soil. (*Sources*: Brady 1974; Clark 1967; Russell, E. J., 1961; Russell, E. W., 1974)

Name	Approximate range of numbers (m^{-2} soil)	Biomass (kg ha⁻¹)
Protozoa	10^9–10^{10}	50–200
Nematodes	Few–$2 \times 10^{6*}$	10–200
Fungi	10^5–10^6	550–5500
Actinomycetes	10^7–10^8	440–4400
Bacteria	10^8–$10^{9\dagger}$	550–4400
Algae	10^{10}–10^{12}	A few ten-thousandths of soil volume

* Highest in grasslands.
† Occasionally as many as 10^{14}.

fields are generally lower both in the numbers and weight of soil organisms than are uncultivated areas.

Although all macro-animals that live on or in the soil play some part in its processes, only a few significantly affect the development of soil profiles. *Earthworms* are the most important of these in temperate regions and can number anything from 30 to 100 m^{-2} to plough depth, giving a biomass of 110–1100 kg ha^{-1}. They constitute between 50 and 75% of the total mass of animals present and are the main agent in mixing plant debris with the soil. They thrive best in soils where organic matter is abundant (with a pH above 4.5) but are intolerant of both drought and frost and so tend to be most active in spring and autumn.

Earthworms may pass through their bodies as much as 40 t ha^{-1} of soil in a year, subjecting the ingested organic and mineral matter to oxidation and grinding action. The oxidation is *enzymic* since it is catalysed by the presence of complex molecules secreted by micro-organisms, known as *enzymes*. The earthworms leave casts enriched in both organic matter and nutrients of as much as 20 t ha^{-1} which improve soil structure by leaving holes which increase aeration, bring subsoil materials to the surface and mix them with organic components. *Termites* should also be considered here. They occur only in the tropics but, in places, can be the dominant fauna and have effects which have sometimes caused them to be considered as the tropical analogue of the earthworm. In Africa, for instance, certain species construct nests which may reach 3 m in height and contain 2–3 t of fine-textured and base-rich material burrowed from the top metre of soil. When abandoned, the nests degrade gradually to ground level but may leave behind a low mound up to 10 m or so in diameter.

The most abundant soil macro-animals are *nematodes* and *protozoa*. Nematodes (sometimes called threadworms or eelworms) are important in reducing microflora to organic matter and feed on bacteria, the cells of other members of the soil population and plant roots. All consume liquid components only and they mainly inhabit soil water films. Furthermore, because of their relatively large size (0.5–1.5 mm long, 10–30 μm thick), many of them require moderately large pores. They are particularly active in coarse sandy soils in wet weather and may attack the roots of growing crops. Protozoa are single-celled organisms that live in the water films between soil particles. Most depend on bacteria for food and although they thus arrest some soil reactions, there is also evidence that they may hasten the turnover of readily available nutrients.

The most abundant soil flora are the roots of higher plants. They attract soil water and its dissolved nutrients through *osmotic suction* (section 4.3) and contribute to soil structure by penetrating into (and enlarging) cracks and voids. While the roots of many field crops extend only within the top 1–2 m of soil (where growth is vigorous and the material is penetrable), some roots may go as deep as 5–7 m and lucerne, for instance, may reach over 10 m. Trees root considerably deeper (tens of metres being common) and, for example, over 50 m depth has been recorded for *Tamarix* spp. in desert areas. Other soil flora are mainly of microscopic size and include *algae, fungi* and *actinomycetes*. Most algae resemble higher plants in being capable of performing photosynthesis and are therefore mainly found in surface layers into which light can penetrate. They are exceptionally tolerant of waterlogging, even in the presence of high salinity, and can fix atmospheric nitrogen under such conditions. This gives them special importance in areas of swamp rice.

Most soil fungi are moulds characterised by long filaments which may have a total length of as much as 100 m in 1 g of soil. They therefore form a major proportion of the total volume of soil micro-organisms, although they are unable to photosynthesise and must be parasitic on other forms of organic life. Furthermore, they are *aerobic* in that they will grow only in the presence of molecular oxygen. They are more tolerant of acidity than most other micro-organisms and are important in soils both because of the wide range of organic residues they can attack and the number of stages of decomposition they can enter. They are the main agents in destroying lignin, the woody tissue of plants, and so play a key role in forest biogeochemical cycles.

Actinomycetes resemble bacteria in being unicellular and fungi in being filamentous and, like both, they are unable to photosynthesise. Table 25.2 indicates that they are almost as numerous as bacteria and can exceed them in hot, dry conditions. They are important in the decomposition of organic matter, in the liberation of nutrients and in reducing stable compounds (such as cellulose and chitin) to simpler forms. They are, however, highly sensitive to acidity with growth being impossible below a pH of about 5.

Bacteria are single-celled organisms of great importance in soils. They participate in all organic reactions and are the sole agency for nitrogen fixation, nitrification and sulphur oxidation. They are very small, ranging from 5 μm to less than 2 μm in length, and have a capacity for very rapid multiplication. Most bacteria are heterotrophic, deriving their energy and body materials from other organic sources, and some are even predatory on other bacteria. Others are autotrophic and obtain their energy from the oxidation of mineral constituents (such as ammonium, sulphur and iron oxides) and most of their carbon from carbon dioxide. Some are known as *facultative autotrophs*, which are able to use both organic and inorganic materials for energy and growth. In general, they thrive best in the presence of abundant organic matter, moisture, both gaseous and chemically combined oxygen, at temperatures of 20–40 °C and at pH values of 6–8, which are approximately the same conditions as those required by most higher plants.

(such as beetles, mites and earthworms) which expose them to more rapid attack by other organisms. The primary consumers of the decayed products are the microflora, which are even active within the digestive tracts of some of the animals. Together with earthworms, they play an essential part in converting these products into humus and account for 60–80% of soil metabolism. These primary consumers are themselves food for predatory secondary consumers. Centipedes, for instance, eat small insects and moles eat earthworms whereas some mites, springtails, termites and protozoa eat the microflora. These predators are themselves prey for the tertiary consumers since ants eat centipedes and spiders prey on both primary and secondary consumers. These carnivores are in their turn subject to attack by the microflora and so tend ultimately to be oxidised to the organic matter and nutrients of the next biogeochemical cycle.

25.3 The trophic sequence

There is a close interaction between plants and animals since each is to some extent nourished by the products of the other. They form a sequence of *trophic levels* a simple example of which, in the humid temperate zone, is shown on Fig. 25.2. The early stages of decomposition involve the tearing and chewing of plant residues by macro-organisms

25.4 Soil organic matter decomposition

The term *soil organic matter* includes both recognisable plant and animal residues and their decay products (known as humus) which have lost the character of the original materials. It influences the physical and chemical properties of soils far

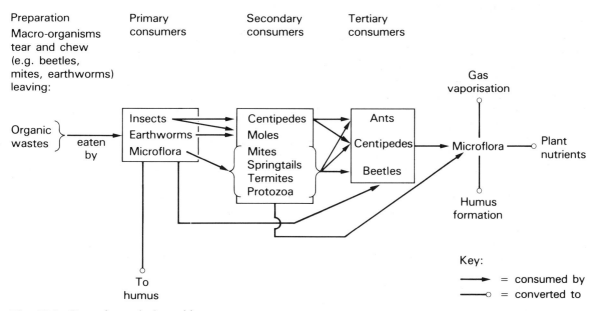

Fig. 25.2 Part of a typical trophic sequence.

out of proportion to the amounts present, which range from a trace up to 15 or 20% of the solum. As well as supplying energy and body-building materials, it often accounts for more than half the cation exchange capacity of soils, increases water absorption and is important in maintaining the stability of soil aggregates.

The decay of residues to humus can be considered to have two stages, namely *mineralisation* and *humification*.

(a) Mineralisation

Mineralisation is essentially a process of oxidation or 'burning' by soil micro-organisms, which leads to the formation of soluble or gaseous compounds (such as carbon dioxide, ammonia, nitrate, sulphate and phosphate) and simple organic molecules. Plant and animal residues vary widely in the rate at which they are broken down in the soil, in the approximate order shown in Table 25.3. The main chemical reactions can be expressed approximately as:

$$C,4H + 2O_2 \xrightarrow[\text{oxidation}]{\text{enzymic}} CO_2 +$$

carbon and oxygen-containing compounds

$$2H_2O + \text{energy} \qquad [25.1]$$

Furthermore, proteins contain nitrogen and so yield ammonium compounds which can be ultimately oxidised to nitrates, as follows:

$$R\text{-}NH_2 + H_2O \xrightarrow[\text{hydrolysis}]{\text{enzymic}} R\text{-}OH + NH_3 + \text{energy}$$

amino combination with organic materials (R)

ammonia

$$[25.2]$$

$$2NH_3 + H_2CO_3 \rightarrow 2NH_4^+ + CO_3^{2-}$$

ammonia   carbonic acid   ammonium carbonate ion

$$[25.3]$$

$$2NH_4^+ + 4O_2 \rightarrow 2H_2O + 2NO_3^- + 4H^+ + \text{energy}$$

ammonium nitrification ion

$$[25.4]$$

Table 25.3 Organic materials listed in decreasing order of their usual rate of decomposition in soils

1. Sugars, starches and simple proteins
2. Crude proteins
3. Hemicelluloses
4. Cellulose
5. Lignins, fats and waxes

Also produced are relatively simple humic molecules of what Kononova (1966) has called *'strictly humic substances'* which have lost the character of the original materials from which they were derived. They are of two main types. First, there are *aromatic* ring structures based on the hexagonal benzene (C_6H_6) ring. These are usually in the form either of *polyphenols*, in which hydroxyl (OH) groups have replaced some of the hydrogen atoms, or of *quinones* in which oxygen (O) atoms have replaced some of them. Secondly, there are nitrogen compounds which are mainly *proteins, amino acids* and *saccharides* (sugars). These two types of compound form the basis of three organic acids, with a key role in humification and other soil processes, and should be thought of as members of a continuous series. At one extreme are the *fulvic acids*, which are formed dominantly of nitrogen compounds and have only a weak tendency to *polymerisation* (the aggregation of smaller molecules of the same basic elemental composition into larger complexes). At the other extreme are the *grey humic acids*, which are dominated by aromatic ring structures and are strongly polymerised. *Brown humic acids* are intermediate between these two extremes.

In well-drained soils which are neutral to alkaline in reaction and have abundant organisms, these reactions proceed relatively rapidly. The carbohydrates and lignin are converted into a wide range of simple soluble and gaseous products suitable for subsequent humification. The primary protein is successively transformed into amino acids and ammonia and then nitrified by bacteria into nitrites and nitrates. In addition, microorganisms secrete polysaccharide gums which are the most active component in stabilising soil structure. These reactions are, however, seriously inhibited in highly acid or anaerobic conditions, i.e. waterlogged soils without free oxygen. For example, soils with a pH below about 5.5, often associated with coniferous or ericaceous vegetation and their lignin-rich litter, have a reduced bacterial population. In this case, most of the mineralisation is performed by the fungi, yielding decay products which tend to combine into stable complexes with protein and do not form humic acids. This seriously inhibits the conversion of protein to ammonia and its oxidation to nitrates.

Similarly, anaerobic conditions inhibit mineralisation and give rise to characteristic humic products. As in acid soils, the micro-organic population is reduced but, unlike them, fungi are almost absent. Mineralisation is then dominated by the reduction of oxygen-containing compounds by anaerobic bacteria and actinomycetes, which release gaseous products and leave a residue rich in lignin and lignified tissues. Although the ammonification of protein is strong, the low population of nitrifying bacteria makes nitrification minimal. Where plants continue to grow and yield litter under such conditions, organic material can accumulate to considerable depths (Fenwick and Knapp 1982).

(b) Humification

Mineralisation produces relatively simple compounds which are mainly soluble whereas humification builds these into complex colloidal molecules of low solubility and distributes them within the soil profile. These are generally brown to dark brown in colour and behave somewhat as clay particles. In biologically active conditions, where the pH is near neutrality, micro-organisms produce humic substances at a rapid rate. In addition, fauna (especially earthworms) can thrive and mix the soil with mineral matter to form a moderately deep layer of humified soil, known as *mull* (Fig. 25.3(a)). Organic and mineral colloids are generally bonded into stable aggregates and there is a rapid digestion and incorporation of the annual leaf litter.

Under acid conditions, micro-organic activity is repressed and there are few earthworms. The mainly coniferous and ericaceous vegetation yields litter with a large proportion of lignin. From the surface downwards, one usually finds the sequence of litter, fermentation and humic horizons overlying a narrow, humus-rich mineral horizon which is sharply separated from the underlying mineral soil. Acid accumulation with this sequence forms a soil known as *mor* (Fig. 25.3(c)), whereas *moder* (Fig. 25.3(b)) is an intergrade between mull and mor. Although some of the organic fraction is decomposed and incorporated into the mineral soil, the soil structure is weakly developed and there is some accumulation of undecomposed litter at the surface.

Under anaerobic conditions, the suppression of micro-organisms (especially fungi, which are the only ones capable of decomposing lignin) and the absence of nitrifying bacteria slow the humification process. The lack of aeration retards polymerisation so that only fulvic and sometimes brown humic acids are formed and the intermediate products of mineralisation, especially the polyphenols, accumulate in the soil. The bulk of

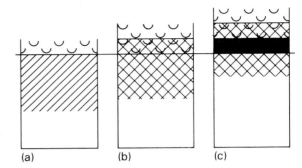

Fig. 25.3 (a) Mull; (b) moder; and (c) mor horizons. Key in Fig. 24.4. (After Fenwick and Knapp 1982)

organic residues, however, remain intact. The resulting carbonaceous humus form is known as *anmoor* if it is temporary and contains some mineral matter, or peat if it is permanent and purely organic. Peat is therefore widespread on sites with impeded drainage in humid climates.

25.5 The carbon–nitrogen ratio

The carbon–nitrogen (C/N) ratio (representing the relationship between total carbon and total nitrogen in a soil) is a valuable index of the nature and stage of its organic processes and in normal humid temperate soils, it ranges between 8 and 13 with a mean of about 11. A mull humus will usually have a value of about 10, whereas a mor will have a value between 20 and 30 with the value for moder intermediate between them. Plant and animal residues have high values, ranging from 20 to 30 in legumes and farm manure to 100 in strawy residues. The values for litter from different trees and shrubs vary widely. For example, *Robinia* sp. (a legume) has a value of 16; elm and beech have values between 30 and 45 and heather and *Pinus sylvestris* have values exceeding 65 (Duchaufour 1982). These values help to explain the relationship which is often discernible between vegetation and humus type. Vegetation types giving relatively low values tend to be associated with mull humus, oak and beech with either mull or moder and heather and conifers with mor.

The seasonal variations of the C/N ratio at a site can indicate the progress of humification. If the annual cycle can be considered to begin at a time of year when both microbial numbers and activity are low (i.e. early spring), then the introduction of fresh decomposable tissue immediately widens the C/N ratio. This allows a multiplication of micro-organisms which use the carbon for energy but cause nitrates to be rapidly converted to living tissue. As carbon dioxide is evolved and nitrogen retained, the C/N ratio narrows until it is again controlled by the lack of easily oxidisable carbon. Microbial activity sinks again towards quiescence, associated with the formation of humus and the release of simple products such as carbonates, nitrates and sulphates. As the rate of loss of carbon dioxide and nitrate stabilises, the C/N ratio

returns to the level it had at the beginning, although the total amount of organic matter has increased.

25.6 Organic matter and climate

When viewed globally, the distribution of organic matter in world soils can be regarded as a reflection of the control exercised on vegetation by climate. In low-latitude deserts, there is too little

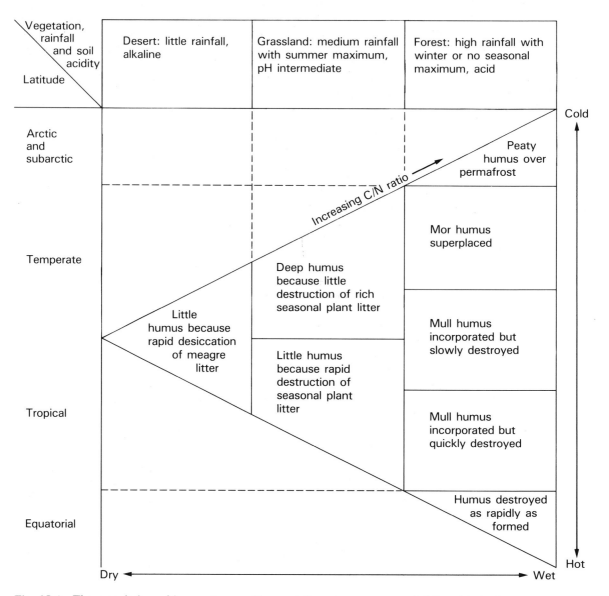

Fig. 25.4 The association of humus types with vegetation, temperature, rainfall and latitude.

vegetation for substantial organic matter to be produced. However, transects towards regions of higher rainfall show an increase in organic content (Fig. 25.4) but this increase is generally greater in temperate than in tropical latitudes. As rainfall increases in the former zone (as from west to east in North America or from southeast to northwest in Europe), there is a steady increase in organic matter attaining its maximum (which can range from 5 to 20%) in the zone of black humus-rich soils of the grasslands in southern Russia and central North America. These are known as *chernozems* or *Mollisols*, according to the nomenclature of the US Department of Agriculture or USDA (1975). The high content of organic matter (extending to a normal depth of over 1 m) is provided by the presence of an extensive mat of shallow grass roots under conditions which permit their conversion to humus, without allowing much leaching or destructive oxidation.

As rainfall increases, from the grassland to the forests in temperate latitudes, both the content and the normal depth of organic matter decrease to less than 0.5 m of mull in deciduous woodlands, to considerably less in areas with moder or mor humus. The tundra soils of high latitudes generally have a relatively high organic content because the underlying permafrost causes seasonally anaerobic conditions which favour the accumulation of peat. Values are highly variable and can range from a trace to perhaps 20% of organic matter in the soils of these regions. In tropical climates, the increase is less marked than in temperate climates and appears to rise continuously from values of zero in deserts to up to 10% in some soils under tropical rainforest, although values of under 5% are usual. The lower rate of increase is due mainly to two causes. Firstly, the abundance of both flora and fauna is greater under high than low temperatures, so that a larger proportion of the available nutrients are held in the living biota themselves. Secondly, the organic residues are more rapidly destroyed and reused and so have a shorter 'residence time' in the soil.

25.7 Retrospect

Soil organisms are a key agency governing soil processes and they vary in size and complexity from single-celled organisms to large plants and animals, through a series of trophic levels. They interact in highly complex ways, but are mutually complementary within that part of the biogeochemical cycle which occurs in the soil. The first stage is the mineralisation of residues which, initially, is accomplished by larger fauna and then by the microflora. These have special importance in converting complex tissues into simpler organic molecules, nutrients assimilable by other organisms and soluble and gaseous products (such as carbonic acid) which can assist in rock weathering.

The second stage is humification which forms large, stable organic molecules which act as bonding agents in soil structure. They also form the basis of soil humus when distributed through the profile by the churning and digesting action of macrofauna, such as earthworms. The geographical variations of soil organic matter reflect the broad climatic zones of the earth through their control on the distribution of vegetation. Within these zones, its local character is conditioned by the complex of factors which govern the temperature, moisture, texture and acidity of the soil.

Key topics

1. The nature of organic material in soils.
2. The role of organisms in pedogenic processes.
3. The decay of residues to humus in soils.
4. The relationship between climate and organic matter in zonal soils.

Further reading

Bridges, E. M. (1979) *World Soils*. Cambridge University Press; Cambridge.

Buol, S. W., Hole, F. D. and McCracken, R. J. (1980) *Soil Genesis and Classification*. Iowa State University Press; Ames.

Gerasimov, I. P. and Glazovskaya, M. A. (1965) *Fundamentals of Soil Science and Soil Geography*. Israel Programme for Scientific Translations; Jerusalem.

Trudgill, S. T. (1979) *Soil and Vegetation Systems*. Clarendon Press; Oxford.

Young, A. (1976) *Tropical Soils and Soil Survey*. Cambridge University Press; Cambridge.

Chapter 26 Soil-forming processes

An understanding of soil processes can best be introduced by reference to the concept of *zonality* introduced by Sibirtsev, Dokuchaev's closest collaborator (Afanasiev 1927) which recognises the dominant effect of climate in soil formation. In *zonal soils*, differences in rock formation or geological origin have been largely masked or rendered subordinate by the overriding effects of climate. As the name zonal implies, they tend to be recurrent and regional in extent. Where annual precipitation exceeds potential evapotranspiration, salts and colloids will tend to move downwards to the soil profile but, where evapotranspiration is dominant, the movement will be upwards. Because downward movement tends to lead to the residual concentration of iron and aluminium compounds in the soil profile and upward movement to the accumulation of calcium compounds, soils with the former characteristic have been called *pedalfers* whereas those with the latter are known as *pedocals*.

Where a local factor such as a marshy situation or the occurrence of certain rock types dominates the development of the soil profile, it is then termed *intrazonal*. *Azonal soils*, by contrast, have no horizon differentiation because they develop on bare rock or recently deposited materials such as alluvium or sand dunes. In this chapter, soil processes are considered within the context of the world's main zonal and intrazonal soils. Each can be regarded as a part of a soil–vegetation system with a distinctive cyclical movement of energy and materials known as a *biogeochemical cycle* (Chs 30 and 31). The energy is provided directly by the sun and by gravity and indirectly by the combustion of organic compounds. Furthermore, materials are provided by the atmosphere and the underlying rocks. In zonal soils, the system dominantly reflects the regional climate and vegetation, whereas in intrazonal soils it reflects the overriding influence of a local factor in soil formation. There are two processes, however, which it is convenient to consider in introducing these types, namely *cation exchange* in zonal soils and *gleying* (the anaerobic reduction of iron and other compounds) in intrazonal soils respectively.

26.1 Cation exchange

Previous chapters have described how clay micelles and humus particles carry negative chemical charges on their surfaces, which cause them to act as large complex anions. The soil solution contains a number of cations, some of which are adsorbed on to these surfaces and can exchange with others when their relative concentrations are altered. The character of soils is strongly influenced by the nature of the particle surfaces on to which the cations are adsorbed (known as the *exchange complex*) and by the overall capacity of this complex to hold cations (known as its *cation exchange capacity*). It is also influenced by the proportions of the different cations held which have been used in defining major soil categories in many classification schemes, including those of FAO/Unesco (1974) and USDA (1975).

In most soils in the humid temperate zone, the dominant cations are usually hydrogen (H^+), calcium (Ca^{2+}), magnesium (Mg^{2+}), potassium (K^+) and sodium (Na^+). They differ in the order given in the energy with which they are adsorbed on to the clay complex so that, if they are present in equal concentrations, the proportion of hydrogen so adsorbed will be the highest and that of sodium the lowest. Since H^+, Na^+ and K^+ ions all have single positive charges, then equal numbers of one will replace those of another. However, two of each of these monovalent ions would be needed to replace each divalent cation, such as Mg^{2+} or Ca^{2+}, and these reactions can be expressed as follows:

$$H^+ + Na : clay = H : clay + Na^+ \qquad [26.1]$$
$$2H^+ + Ca : clay = 2H : clay + Ca^{2+} \qquad [26.2]$$

Furthermore, because the atomic weights of hydrogen, sodium and calcium are different, it will require only 1 mg of hydrogen to replace 23 mg of sodium (atomic weight 23) or 40/2, which equals 20 mg of calcium (atomic weight 40).

In order to simplify calculations, therefore, exchange reactions and cation exchange capacity are normally quoted in terms of milliequivalents of particular or of total cations per 100 g of soil. Thus, while 1 milliequivalent of H^+ ions is 1 mg 100 g^{-1} soil, 1 milliequivalent of Na^+ ions is 23 mg 100 g^{-1} soil, 1 milliequivalent of Ca^{2+} ions is 20 mg 100 g^{-1} soil and so on. The *percentage base saturation* is the percentage of the exchange complex occupied by cations other than H^+. It is, for instance, said to be 80 when the cation exchange capacity of a soil is 50 milliequivalents per 100 g and the exchange complex is occupied by 40 milliequivalents of metallic cations and 10 milliequivalents of hydrogen.

Soil materials differ widely in their cation exchange capacities and some representative figures are given in Table 26.1. Zonal soils reflect these differences and in both tropical and mid-latitude forests, there is considerable leaching. Clays are largely kaolinitic and there is relatively little organic matter. The exchange capacity is therefore low, but the percentage base saturation is higher in soils under broad-leaved than under coniferous

Table 26.1 Typical cation exchange capacities of soil materials

Material	Range (milliequivalents per 100 g soil)
Fine sand	1–5
Kaolinite clay	3–10
Illite clay	15–50
Montmorillonite clay	70–100
Organic matter	150–400

woodlands. Where evapotranspiration is higher in relation to rainfall (as in the semi-arid tropics), the soils tend to be montmorillonitic with a relatively high cation exchange capacity and percentage base saturation, but may have a relatively large proportion of sodium ions. Grassland soils generally have both a relatively high cation exchange capacity and percentage base saturation, with calcium as the dominant cation. The following section considers in more detail the soil processes characteristic of the main climatic zones.

26.2 Soil processes characteristic of main climatic zones

Zonal pedalfers can be broadly divided into latitudinal zones reflecting the way in which temperature differences, acting through the vegetation cover, differentiate the soils. For example, *tundra soils* (USDA* *Inceptisols*) are found in the subarctic zone, *podsols* (USDA *Spodosols*) in the coniferous forest or *taiga* zone, *brown forest soils* and *brown earths* (USDA *Inceptisols* and *Alfisols*) in the zones of temperate and subtropical deciduous woodland and *latosols* or *ferrallitic soils* (USDA *Oxisols*) in the equatorial rainforest zone. All develop soil horizons which result from the chemical effects associated with downward-moving soil water.

(a) Processes in the subarctic zone

The subarctic tundra is intermediate both in temperatures and precipitation between the frozen and relatively arid Arctic and the zone of cool temperate deciduous woodland which covers wide areas, especially in northern Eurasia and North

---

* When USDA equivalents are given, they are generally the *soil orders* which include those after which they are bracketed.

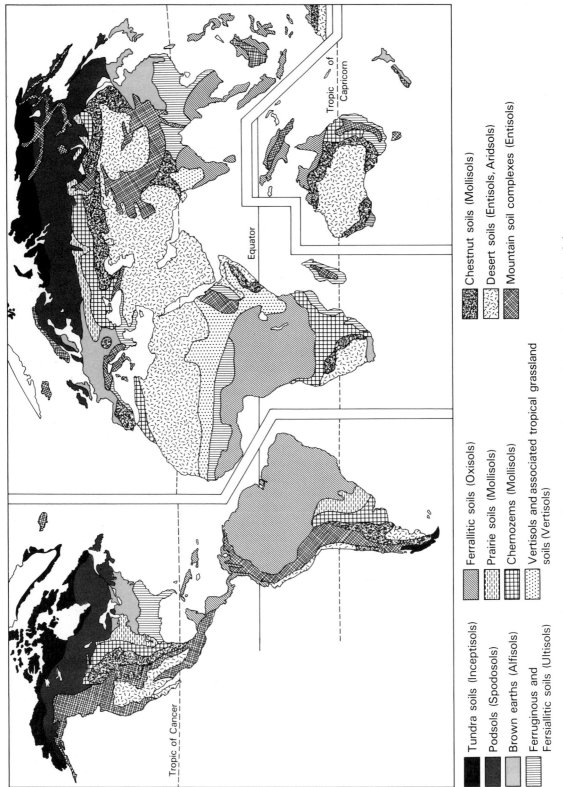

Fig. 26.1 Main soil regions and approximate equivalents of USDA nomenclature, 1975. (Adapted from Finch and Trewartha 1942 and Strahler 1969)

Tundra soils (Inceptisols)

Podsols (Spodosols)

Brown earths (Alfisols)

Ferruginous and Fersiallitic soils (Ultisols)

Ferrallitic soils (Oxisols)

Prairie soils (Mollisols)

Chernozems (Mollisols)

Vertisols and associated tropical grassland soils (Vertisols)

Chestnut soils (Mollisols)

Desert soils (Entisols, Aridsols)

Mountain soil complexes (Entisols)

America (Fig. 26.1). The brevity of the growing season prohibits annual plants but woody perennials grow and leave peaty carbonaceous residues, whose destruction and incorporation into the mineral soil are severely restricted by the relative inactivity of soil organisms in conditions of low temperatures and poor aeration. The poor aeration is due to the low evaporation and the superfluity of water caused by summer melting and the restricted drainage imposed by the underlying permafrost, which leads to gleying. Where the summer thaw proceeds to somewhat greater depths, such as under water bodies or large trees, the surface of the permafrost layer lies deeper. This makes its horizontal profile uneven, containing subsidence features called thermokarst (section 23.2).

The normal annual freezing comes from above and is uneven because it forms ice lenses, wedges and needles which cause localised compression and heaving. This is most intense in fine-textured materials, because of their high coefficient of water retention, and results in differential expansion which causes large objects (such as stones) to be sorted upwards and outwards. Thawing leads to subsidence so that the soil experiences a yearly churning known as *cryoturbation* and a mixed and indefinite soil horizon develops, with sometimes polygonal patterns forming in the surface materials.

(b) Processes in the humid temperate zone

In the humid temperate zone, the dominant features of the soil profile are associated with the downward movement of soil water. This removes organic and mineral colloids, sesquioxides and soluble salts from the upper horizons and deposits them in lower horizons. The chemical and physical processes involved are complex and a number of terms have been used to describe the different aspects. Leaching, for example, is the general term for the downward translocation of soil materials and the term *lixiviation* is sometimes used for that part of the process which removes soluble salts, especially those containing the main metallic cations. This leads to their replacement by hydrogen (H^+) and aluminium (Al^{3+}) ions on the exchange complex and when this happens, the profile becomes both *desaturated* and *acidified*. The leaching of calcium carbonate is known as *decalcification*, that of chelates as *cheluviation* and that of clay as *lessivage* or *pervection*. Clay thus translocated may

be redeposited lower in the profile as cutans (section 24.4).

The relative importance of different types of leaching varies according to the climate, as reflected in the vegetation. The high-latitude taiga is characterised by coniferous and ericaceous plants which give rise to a mor humus over an acidified profile, especially on coarse-textured and permeable parent materials. The resulting soils are termed podsols, from the Russian for 'ash soils' (Fig. 26.2(a) and Plate III (a)) where the A horizon is thin and poorly developed and the E horizon is bleached by the removal of all soluble salts and carbonates. Furthermore, some sesquioxides, humus and clay enrich the B horizon, usually giving it a reddish brown colour. The term 'podsol' is often qualified by a word which indicates the dominant character of the B horizon. For example, where it is most strongly influenced by accumulations of humus, iron or a combination of the two, it is known as a *humus podsol*, an *iron podsol* or a *humus–iron podsol* respectively. Likewise, hydromorphic conditions can give rise to *gley podsols* and *peaty podsols*.

Under somewhat warmer conditions in the humid temperate zone, especially where the materials are base-rich, the characteristic vegetation is deciduous woodland. The soils have a tendency towards *brunification* forming brown earths or brown forest soils, approximately equivalent to the *grey-brown podsolic soils* of earlier US classifications and including both Alfisols and Inceptisols (USDA 1975). These are characterised by a mull humus directly overlying the B horizon without the bleached layer of the podsol (Fig. 26.2(b) and Plate III (b)). These soils cover wide areas of western Europe, northeastern USA, northern China and Japan and thus contribute to the environment of a substantial proportion of the world's population (Fig. 26.1).

In brown earths, the relatively base-rich leaf litter leads to a higher pH and percentage base saturation than in the podsol and this allows nitrification to proceed and earthworms to thrive. Two sub-types are recognisable. The first is sometimes called *sol brun acide, Braunerde, burozem* or Inceptisol and is more base-rich. Iron, freed by weathering in hydrated oxide form, insolubilises humic compounds and forms bonds between humus and clay particles, leading to the wide dissemination of mull humus with a stable crumb or blocky structure. The second sub-type is called *brown earth with Bt, leached brown soil,* or *sol brun lessivé*. This soil tends to occur in cooler or

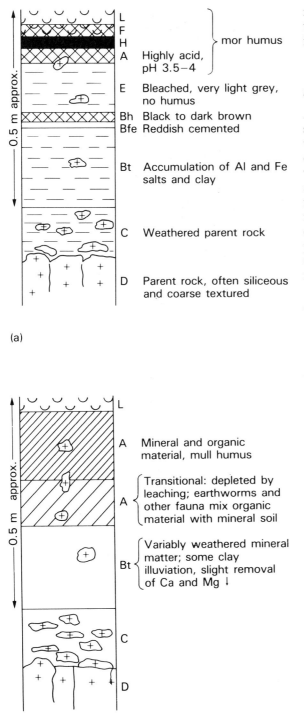

L
F
H mor humus
A Highly acid, pH 3.5–4

E Bleached, very light grey, no humus

Bh Black to dark brown
Bfe Reddish cemented

Bt Accumulation of Al and Fe salts and clay

C Weathered parent rock

D Parent rock, often siliceous and coarse textured

(a)

L

A Mineral and organic material, mull humus

A Transitional: depleted by leaching; earthworms and other fauna mix organic material with mineral soil

Bt Variably weathered mineral matter; some clay illuviation, slight removal of Ca and Mg ↓

C

D

(b)

Fig. 26.2 (a) Podsol (USDA Spodosol); and (b) brown earth soils (USDA Alfisol). Key in Fig. 24.4.

more humid climates and is somewhat transitional to the podsol. The structure is less stable and there is some lessivage, forming B_t and E horizons.

(c) Processes in humid subtropical, tropical and equatorial climates

As mean annual temperatures increase, the destruction of plant litter becomes more rapid and the weathering intensifies and extends to greater depth. The organic layer is thus shallower and weathering goes increasingly beyond the immediate influence of organic acids produced at the surface. It is thus mainly geochemical and has the form of a neutral or slightly acid hydrolysis. This causes a higher concentration of free oxides, especially of iron and aluminium, and a leaching of silica, especially where the parent materials are not acid. The ratio of silica (SiO_2) to sesquioxides (Fe_2O_3 and Al_2O_3) is known as the *silica-sesquioxide ratio* and it narrows usually to values of less than 2 in humid equatorial climates. The abundance of residual iron gives the soils their characteristic red colour to considerable depths.

The dominance of geochemical processes leads to variable rates of pedogenesis more closely related to parent materials and topography and less to surface organic processes than in temperate climates. Consequently, the variety of soils in a given region is often greater and climatic zonations are less easily discernible. Nevertheless, the three stages in the process of iron concentration (which can be designated in order as *fersiallitisation, ferrugination* and *ferrallitisation*) lead to the formation of three soil types which can broadly be associated with climatic zones representing progressive increments in temperature and rainfall.

Fersiallitic soils (Fig. 26.3(a)) are characteristic of the Mediterranean climates where there is a strong seasonal contrast between a mild wet winter and a hot dry summer, and include those called *red Mediterranean* and *brown Mediterranean soils* (Plate III (c), USDA *Xeralfs*). Weathering and the leaching of carbonates in winter liberate iron which attaches to clay particles and, combined with some lessivage, rubifies (gives red and ochreous colours to) the B horizon. These soils remain less acid than the temperate brown earths due to the inhibition of the leaching of bases during the dry season, which permits calcium to remain on the exchange complex and sometimes to form a carbonate horizon in the subsoil.

Ferruginous soils (Fig. 26.3(b)) develop under warm temperate climates without a dry season

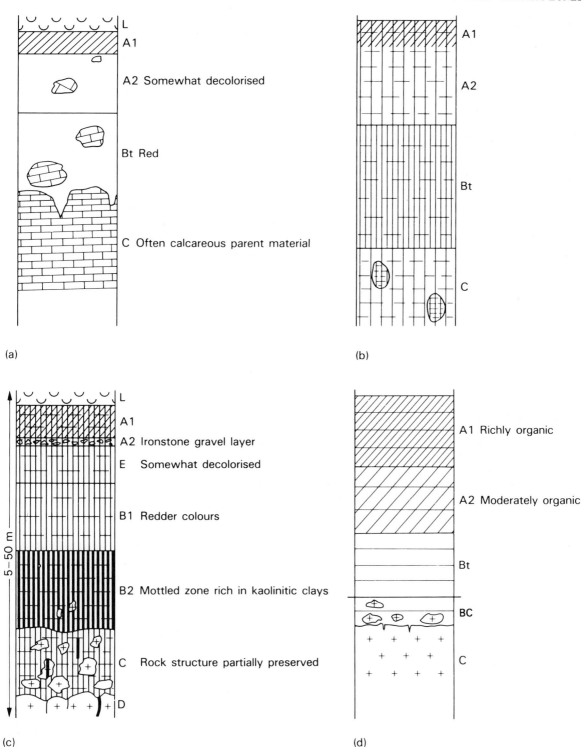

Fig. 26.3 (a) Fersiallitic (USDA Inceptisol); (b) ferruginous (USDA Alfisol); (c) ferrallitic (USDA Oxisol); (d) prairie or brunizem (USDA Mollisol, suborder Udoll) soils. Key in Fig. 24.4.

(USDA *Ultisols*) or in the tropical savanna or bushland zones with a dry season (USDA Alfisols), apparently because both climates allow a growth of broad-leaved woodland under conditions of adequate moisture and relatively high temperatures. These conditions, combined with relatively high pH levels, also contribute to a greater mobilisation of silica than in fersiallitic soils and more is lost in drainage, a process known as *desilicification*. Most of the clay fraction is newly synthesised from silica and alumina and is dominantly of kaolinitic type, but there is also some lessivage.

The third stage is the development of ferrallitic soils (USDA Oxisols; Fig. 26.3(c) and Plate III (d)) which are characteristic of the equatorial rainforest, where the destruction of organic residues is especially rapid so that little humus accumulates. Mature profiles may be tens of metres thick and almost all the primary minerals have been hydrolysed to sesquioxides, silica and bases. Under these conditions of high rainfall, high temperature and pH levels around neutrality, most of the bases and some of the silica are leached away so that the profiles largely consist of four components: ferric oxides, kaolinite, quartz and gibbsite. The B_t horizon is usually absent because of the strong resistance of the clays to dispersion and the resulting mixture of sesquioxide-rich materials is now often called *plinthite*. When exposed at the surface, either in groundwater seepage zones or from the erosion and exposure of subsoil horizons, plinthites are sometimes indurated into hard layers. These were formerly called *laterite* but are now more usually referred to as *ferricrete* (associated with the formation of the mineral *goethite*) which forms a surface carapace or shield covering a considerable area.

(d) Processes in subhumid and semiarid climates

As an observer proceeds from humid to drier climates, there is a change towards greater seasonal temperature variations and an increasingly marked concentration of rainfall during the summer months. In temperate climates, the vegetation changes first to open woodland with *prairie* (*brunizem*) *soils* (USDA Mollisol, suborder Udoll; Fig. 26.3(d)), which should still be regarded as pedalfers, then into pedocals with *chernozem soils* (USDA Mollisol, suborder Boroll; Fig 26.4(a) and Plate IV (a)) under hygrophyllic grassland. This

is followed by xerophytic steppe with *chestnut* (*Kastanozem*) *soils* (USDA Mollisol, suborder Xeroll; Fig. 26.4(b) and Plate IV (b)) and ends in unvegetated *desert soils* (USDA *Entisol/Aridisol*; Fig. 26.4(c)) or *desert sand/erg* (Plate IV (c)). The first three (Mollisol) soils are known as *isohumic* because they have a relatively even distribution of humus down their profiles.

The amount and depth of organic matter tends to be greatest in chernozem soils. This is because grasses produce a large amount of organic matter both by excreting water-soluble compounds from roots and by their decomposition *in situ*, aided by the intense activity of earthworms. The organic residues are thus rapidly mineralised and humified. Furthermore, the alternation of wet and dry seasons not only increases the speed of decomposition of their most unstable components but also stabilises certain other humic compounds against degradation (Duchaufour 1982). It also immobilises clay and iron oxides within aggregates and prevents the formation of Bt horizons. The high level of carbon dioxide rapidly decalcifies the underlying parent material but the lack of leaching permits the calcium and magnesium ions to remain in the soil, sometimes forming subsoil carbonate horizons. In climates more arid than those of the chernozem zone, vegetation becomes sparser and more xerophytic. The root mat becomes less dense so that both the quantity and depth of organic matter decrease. The soils become chestnut in colour, the carbonate horizon becomes shallower and a gypseous one may develop.

In tropical climates, the vegetation decreases along the wet–dry transect through a series of zones from a dense forest cover through various types of woodland, savanna, Sahel and desert. The soils differ from those with similar rainfall regimes along a comparable middle-latitude transect as a result of the higher temperatures. This gives them a lower content of organic matter because of the more rapid decomposition of organic wastes and the reduced leaching in the presence of higher evapotranspiration. It also allows more soluble salts to be retained which can have the effect of raising the pH and rendering silica more mobile which, in turn, can lead to the preservation and synthesis of montmorillonitic clays. This process is especially marked on intermontane plains and basins where seasonal alternations of wetting and desiccation occur under alkaline conditions. The resulting dark-coloured cracking clays are now known as *Vertisols* (Fig. 26.4(d)).

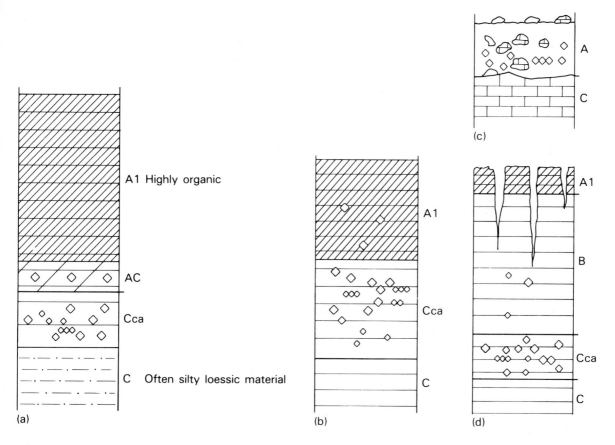

Fig. 26.4 (a) Chernozem (USDA Mollisol, suborder Boroll); (b) chestnut (USDA Mollisol, suborder Xeroll); (c) desert (USDA Entisol, Aridisol); and (d) Vertisol soils. Key in Fig. 24.4.

(e) Processes in arid climates

In desert regions, the climate is constantly too dry and the vegetation too sparse for much chemical weathering or appreciable humus formation, and most desert soils (USDA Entisols and Aridisols; Plate IV (c)) thus show little horizon development. The presence of iron in the oxidised but unhydrated *haematite* state leads to the prevalence of reddish colours. Leaching occurs only occasionally after the sporadic rains and the movement of moisture is soon reversed by evaporation. This allows the soils to retain alkali cations, such as Mg^{2+} and more particularly Na^+ (which in humid climates are more easily lost than calcium), and leads to the formation of shallow *gypsic* and *calcic* horizons which often harden into crusts known as *gypcrete* or *calcrete* respectively. In cooler climates, mineral matter is less weathered and more

organic matter may be present so that they are grey in colour and are called *sierozems*.

26.3 Intrazonal soil processes

Intrazonal soil processes are those in which a site factor overrides the effects of the local climate and vegetation and develops a distinctive soil profile. One such factor is the occurrence of waterlogging in basin sites. In humid regions, these tend to be permanently marshy (i.e. *hydromorphic*) and in arid regions they are *salinised* or *halomorphic*.

(a) Processes under hydromorphic conditions

Wherever soils suffer waterlogging, air is excluded and the supply of oxygen is reduced. This inhibits

237

aerobic activity (section 16.2) and encourages the multiplication of anaerobic micro-organisms, which extract their oxygen by reducing chemical compounds. The most conspicuous manifestation in mineral soil profiles is gleying which is caused by the reduction of the ferric (Fe^{3+}) ion to the ferrous (Fe^{2+}) and the accumulation of the latter, and gives a conspicuous greenish or bluish-grey colour to the soil. The exact nature and extent of the reducing action depend on both the pH and on the redox potential (section 16.2).This is because the tendency to reduction is increased by acidifying so that, for instance, while at pH 7 iron will be in the ferric (Fe^{3+}) state at an *Eh* of 0, at pH 3.5 it will be in the ferrous (Fe^{2+}) state even at an *Eh* of 600 mV (Duchaufour 1982).

Gleying may result from either subsoil or surface waterlogging and when the waterlogging is due to a high groundwater table (Plate IV(d)), there is little oxygen and so the *Eh* is low. Iron can thus be reduced in neutral or slightly alkaline conditions and it forms ferrous carbonate or insoluble complexes which are unstable, unless a fall in pH leads to further reduction of the compounds. Where the waterlogging is due to surface water, it usually contains dissolved oxygen and so has a relatively high *Eh* and gleying will only occur in strongly acid conditions. Since the waterlogging is only temporary, the iron will be reoxidised to the ferric state during its periodic disappearances, forming mottled patches or concretions along capillaries and voids (*pseudogley*) (Fig. 26.5(a)). In cold climates, however, surface water is very acid and can hold little dissolved oxygen, so that reducing conditions similar to those in the subsoil are created (*stagnogley*) (Fig. 26.5(b)).

(b) Salinisation and alkalisation under halomorphic conditions

In arid regions, basin sites tend to give intrazonal soils rich in sodium, which can determine their dominant characteristics, and soils of this type were called *salsodic* by Servant in 1975 (USDA Aridisols). There are two distinct types: firstly, *saline soils* which are characterised by a high concentration of soluble salts, sometimes called *solonchaks* (USDA *Salorthids*) (Fig. 26.5(c) and Plate V(a)); secondly, *alkali soils* which are characterised by the dominance of sodium on the exchange complex, some of which have been called *solonetzes* (USDA *Natrargids*) (Fig. 26.5(d)). The former may not be appreciably alkaline nor the

latter saline but where both conditions are present, the soils are called *saline–alkali*.

All water contains soluble salts and in humid climates these are usually limited in amount and are readily lost in leaching. However, in dry climates they tend to remain in the soil profile and the drier the climate, the stronger is this tendency. Wherever this is accompanied by a supply of continually evaporating water (i.e. from a shallow water table or periodic inundation), soluble salts will tend progressively to accumulate and the soil becomes salinised. Sodium salts are now usually dominant, mainly chloride and sulphate, and furthermore, there is a tendency for salts deposited in this way to be sorted in order of solubility with the alkaline earth carbonates deposited first, then their sulphates and alkali salts last. Therefore, they tend to be found in this sequence upwards in soils and towards the centre of dischargeless basins.

Alkali soils have a pH over about 8.5 and a relatively high percentage (over about 15) of sodium ions on the exchange complex. They originate from the leaching of soils under conditions where sodium becomes more concentrated than calcium. Where organic matter is present, sodium carbonate (Na_2CO_3) and bicarbonate ($NaHCO_3$) are formed and these alkalise the soil solution, causing it to dissolve part of the organic matter and disperse other organic particles and clays. In wet periods, these sodic clays are massive but when dry, their contraction leads to the formation of a prismatic structure and the precipitation of sodium carbonate and organic matter at the surface. If leaching occurs, the profile becomes differentiated, first forming a somewhat less alkaline E horizon over a strongly alkaline columnar B horizon. This is followed (after continued leaching) by the exhaustion of the sodium and its replacement by hydrogen and aluminium ions, acidifying the surface and forming a *solodi* soil. This is often accompanied by hydromorphic conditions and so leads to some reduction and mobilisation of iron compounds.

(c) Intrazonal soils on particular rocks and topography

The other main causes for the formation of intrazonal soils are the occurrence of certain parent materials or slopes which dominate profile formation. An important example is the *rendzina soil* (USDA *Rendoll*) (Fig. 26.5(e) and Plate V(b)) formed on calcareous parent materials, especially where they contain few impurities (i.e. are *calci-*

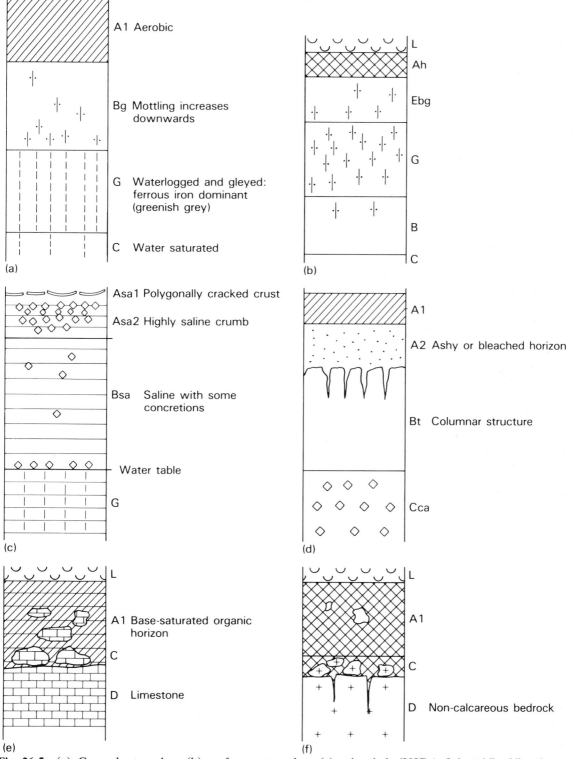

Fig. 26.5 (a) Groundwater gley; (b) surface water gley; (c) solonchak (USDA Salorthid); (d) solonetz (USDA Natrargid); (e) rendzina (USDA Rendoll) and (f) ranker (USDA Inceptisol) soils. Key in Fig. 24.4.

morphic). Rendzina soil is characterised by a high proportion of free carbonates and hence of base saturation in the A horizon. With a relatively high level of biotic activity, a stable mull forms which directly overlies the parent material often without a B horizon. Another intrazonal soil type is the *ranker* (Fig. 26.5(f)). This is relatively shallow soil with an A horizon which has developed directly on non-calcareous rock and it often occurs on slopes whose steepness is such that there is continual loss of material by lateral creep.

26.4 Retrospect

Soil profiles are a palimpsest from which their causative environmental processes can often be interpreted. These processes on normal sites broadly reflect the earth's climatic zones expressed through their effects on vegetal cover and the action of soil water. Thus, a world soil map resembles a climatic one and since these effects derive largely from differences in temperature and rainfall regime, relatively consistent soil variations can be seen along a temperature transect from Arctic to equatorial zones and along rainfall transects across both temperate and tropical zones. At a local scale, site characteristics are influenced by the topographic situation, parent materials, the legacy of past conditions and human activities which either prohibit soil formation, as on bare rock uplands, or dominate it as in hydromorphic lowlands. In humid climates, basin soils tend to be peaty and gleyed but in dry climates the high evaporation causes them to be salinised, with often a repetitive toposequence or *catena* from higher to lower ground (Plate V(c)). Particular rock types also give important variations in the zonal pattern, especially when they cause overlying soils to be shallow due to topographic reasons.

Key topics

1. The relevance of the zonal concept to pedology.
2. The significance of cation exchange in soil processes.
3. The influence of climate on soil-forming processes.
4. The modification of climatic influences by terrain characteristics.

Further reading

Brady, N. C. (1990) *The Nature and Properties of Soils* (9th edn). Macmillan; London and New York.
Bridges, E. M. (1979) *World Soils*. Cambridge University Press; Cambridge.
Fenwick, I. M. and Knapp, B. J. (1982) *Soils: Process and Response*. Duckworth; London.
Wild, A. (ed.) (1988) *Russell's Soil Conditions and Plant Growth*. Longman; London and New York.
Strahler, A. N. (1969) *Physical Geography*. John Wiley and Sons; New York and London.

Chapter 27 Soils and plant growth

This chapter considers the relation between the soil and the growth of higher plants, with particular reference to its role in supplying mineral nutrients. Soil–plant interactions are complex and may be studied in a number of different ways. Firstly, it is possible to group the nutrient elements together and consider their movements in a series of stages, namely through the soil solution which represents the immediate environment of the root surfaces known as the *rhizosphere*, within the plants themselves and in the form of organic residues which are broken down and returned to the soil. Secondly, the process may be viewed in terms of the energy flows from their solar source through the soil–vegetation system. Thirdly, it is possible to consider the activities of each of the main chemical elements separately. Since this latter approach has the advantage of being an established method which is relevant to practical applications, it is adopted here and the factors controlling the growth of higher plants are set in the context of the limitations imposed by acidity, salinity and alkalinity. This is followed by a consideration of the role and movements of the main nutrient elements and of their cycling through certain soil–vegetation systems.

27.1 Soil factors controlling the growth of higher plants

Soil–vegetation relationships are cyclical, since soils provide mechanical support and supply plants with water, nutrient elements and simple compounds during the growing period. Most of the required nutrients are derived from the underlying rocks but an appreciable quantity (especially of nitrogen and sulphur) can be added to the soil in rainwater. Nutrients taken up are returned in litter fall and through the death of organisms. The cycle is completed when they are transformed into compounds and elements which can again be absorbed by plants. The soil adds further nutrients in weathering. In humid climates, some are lost by leaching although some of the leached materials are brought back to upper levels by the soil biota. The cycle can be broadly represented as in Fig. 27.1.

Eighteen elements appear to be essential for plant growth. Carbon and oxygen are obtained from atmospheric carbon dioxide by photosynthesis (section 29.1). Oxygen is also acquired from soil air and hydrogen is available in soil water. These three elements generally make up between 94 and 99.5% of plant tissues whereas the remainder are the *mineral nutrients* which although less abundant, are equally essential to life. They can be subdivided into *macronutrients* and *micronutrients*. The macronutrients are nitrogen, phosphorus, potassium, calcium, magnesium, sulphur and silicon

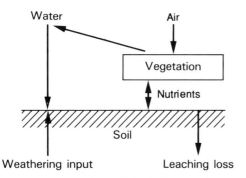

Fig. 27.1 Simple model of plant nutrient cycling.

and some of their usual ranges of abundance in mineral soils are given in Table 27.1 The micronutrients are iron, manganese, boron, copper, zinc, molybdenum, chlorine and cobalt. Organic matter is sometimes regarded as a complex nutrient and it is important as a source of nitrogen, phosphorus and sulphur. It also influences soil structure which facilitates root access to other nutrients, notably exchangeable calcium, potassium and magnesium ions.

Plant growth is limited by the element which is least abundant in relation to requirements and crops sometimes give low yields because of the lack of a single element when all others are present. Furthermore, nutrient elements should be present in the correct proportions so that an excess of one does not cause deficiencies of others. Too much calcium may, for instance, interfere with plant uptake of phosphorus and boron. In general, plants take only the nutrients they need but where these are abundant in the soil, some can accumulate unnecessary concentrations, such as soluble salts in some desert plants.

Table 27.1 Total amounts of organic matter and nutrients in temperate region mineral surface soils. (After Brady 1974)

Constituents	Ranges that may ordinarily be expected (%)	Representative analyses	
		Humid region soil (%)	Arid region soil (%)
Organic matter	0.40–10.00	4.00	3.25
Nitrogen	0.02–0.50	0.15	0.12
Phosphorus	0.01–0.20	0.04	0.07
Potassium	0.17–3.30	1.70	2.00
Calcium	0.07–3.60	0.40	1.00
Magnesium	0.12–1.50	0.30	0.60
Sulphur	0.01–0.20	0.04	0.08

Plants obtain their nutrients through the *osmosis* (i.e. the flow of a solvent through semipermeable membranes) of soil water into roots (section 4.3). The soil solution reaches the root either through capillary movements of the soil water in the profile or from the elongation of their roots. The former depend on the *tension gradient* which results from the effects of a difference in moisture content between neighbouring parts of the soil and from three different forces, namely gravity, capillary attraction and the osmotic suction of plant roots. The rates of movement depend on the *capillary conductivity* or hydraulic conductivity (K) of the soil (section 6.1) which can be defined as the readiness with which water flows through it in response to a given suction or gradient. Water will move fastest through a saturated soil, for example, where all the pores are filled with water. When the soil dries and air enters, the rate decreases because the first pores to empty are the widest so, as drying proceeds, the flow becomes increasingly restricted to the fine pores. Since the rate of water flow through capillaries is proportional to the fourth power of their diameter, the reduction of flow can be abrupt. Although the two terms capillary conductivity and hydraulic conductivity are interchangeable, some authors reserve capillary conductivity for unsaturated and use hydraulic conductivity for saturated soils and this practice is followed here. In the field, K is measured by means of infiltration tests in unsaturated soils and by measuring the rate of rise of water tables in auger borings of known dimensions in saturated soils (van Beers 1958).

The rate of flow in a saturated subsoil (between two points at different levels) can be illustrated by the example in Fig. 27.2 and expressed in the following equation:

$$Q = Ka\frac{hL}{L} \qquad [27.1]$$

where Q is water flow across a (m^3s^{-1}), K is hydraulic conductivity ($m\ s^{-1}$), a is cross-sectional area perpendicular to flow direction, hL is head loss (m) and L is distance (m). Assuming that flow is continuous and the relative levels of A and B remain unchanged, water will flow from A to B. If K is 0.01 m s^{-1}, hL is 1 m, L is 10 m and a is 20 m^2, the quantity of flow (Q) is determined as follows:

$$Q = 0.01 \times 20 \times \frac{1}{10} = 0.02 \ m^3 \ s^{-1}$$

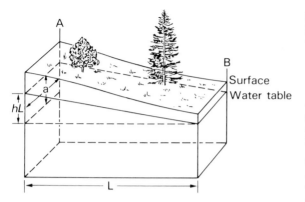

Fig. 27.2 Flow of water in a saturated soil.

K values tend to be a close reflection of soil texture and generally vary between extremes of less than 1 cm day^{-1} in impermeable clays to more than 2 m day^{-1} in sands. Where drought limits the growth of plants, roots can often extend rapidly enough to outpace the recession of the soil water, as long as the plant remains above the *permanent wilting point* (i.e. the percentage moisture content of the soil at which the irreversible wilting of plants occurs; section 4.3). When a part of the soil profile suffers permanent or temporary waterlogging, oxygen will no longer be able to reach the plant roots. This will decrease or arrest both their growth and their ability to absorb water and nutrients from the soil solution. Only plants with some protective or adaptive mechanism in their root surface, allowing them to extract water and nutrients from the soil, can live. Paddy rice, for instance, has internal air spaces which allow oxygen to diffuse down through the root into the immediately surrounding soil and thus permit it to take up ions aerobically.

(a) Acidity and base status

Acidity is an important control on the soil–plant system but has widely varying effects on different plants because of its influence on the concentrations and availability of the various ions. Acid soils usually result from the leaching of calcium and other cations from the exchange complex (section 26.1) and their replacement by hydrogen. They are characterised by a relatively high concentration of aluminium and manganese ions and a low concentration of calcium ions and are defined as being of low *base status*. In general, soils are most productive when at least 80% of the exchange complex is occupied by calcium ions.

Calcium is readily lost from the soils of humid regions and, for example, in the UK, a normal cereal crop will take up approximately 7–10 kg ha^{-1} yr^{-1} of calcium compared with 59 kg with some root crops and 72 kg with legumes (Table 27.2). Leaching will normally remove between 300 and 400 kg ha^{-1} yr^{-1} but this can rise to as much as 1000 kg ha^{-1} yr^{-1} on bare soil after the application of heavy dressings of lime. This deprives plants of the required calcium and hinders the uptake of other nutrients and also results in a decline in the activity of micro-organisms, particularly bacteria and earthworms, which are important in the maintenance of soil structure and in recycling nutrients. Low pH in agricultural soils is sometimes corrected by the addition of lime although

Table 27.2 Amounts of various elements taken up by common agricultural crops of England. (After Russell 1973)

Crop	Dry matter (t ha^{-1})	kg ha^{-1}								
		N	P	K	Na	Ca	Mg	S	Si	Cl
Cereals, etc.										
Wheat	4.69	56	10.4	26.8	2.3	7.4	4.9	8.7	50.7	2.8
Oats	4.47	59	9.5	42.8	4.5	9.3	6.0	9.0	44.7	7.4
Meadow hay	3.16	55	6.1	47.5	7.6	25.7	9.8	6.4	29.8	16.4
Legumes										
Beans	3.89	120	14.2	62.4	1.9	23.5	6.5	10.4	3.8	6.0
Red clover hay	4.21	110	12.2	79.5	4.3	72.1	19.1	10.5	3.7	11.0
Roots										
Potatoes	3.76	52	10.5	71.1	3.1	2.7	4.3	3.0	1.3	4.9
Turnips	5.23	123	16.3	138.5	20.4	59.2	6.4	23.4	4.0	24.8
Mangolds	8.49	167	25.8	280.0	98.7	34.3	28.7	15.7	9.5	93.1

if this addition exceeds requirements, it causes an undesirable rise in pH. This, as shown on Fig. 27.3 reduces the availability of some nutrients and can also cause them to become deficient by encouraging their rapid exhaustion through improved plant growth.

(b) Salinity and alkalinity

Saline soils do not normally have a pH over about 8.5 but they inhibit plant growth and at high concentrations, they can even prevent it. When a water solution containing a relatively large amount of dissolved salts is brought into contact with a plant cell, it causes shrinkage and collapse of the protoplasmic lining. At high concentrations, this kills the plant due to the reversal of osmotic pressure causing water to pass outwards from the root cell to the more concentrated soil solution. Although plants vary widely in their tolerance to salinity, it is possible practically to grade the risk to the majority of those of economic importance in terms of the total salt concentration in the soil solution. This is normally measured in terms of its *electrical conductivity*, expressed in milliSiemens (mS) cm^{-1} at 25 °C. Some sensitive plants will be affected at values of over 4, most will be inhibited at levels of over 8, whereas only highly salt-tolerant plants and halophytes will survive values of over 15.

Alkaline or *sodic* soils are dominated by active sodium. This harms plants directly through the effects of carbonate, bicarbonate and sodium ions on their metabolism and nutrition and indirectly through the check on water movement imposed by the reduced soil permeability, which results from the deflocculation of clays. The growth of most plants is seriously restricted where sodium composes over about 15% of the exchangeable cations or where the pH is over 9. Saline-alkali soils, with both high salinity and pH, combine their harmful effects.

27.2 The main mineral nutrients

(a) Nitrogen

Nitrogen is essential in the enzyme systems of plants which enlarge cell sizes and therefore increases leaf areas, making plants more succulent and darker green in colour. Excess nitrogen may, however, cause plants (and especially cereals) to develop leaf at the expense of stalk and grain

and thus to lose firmness and become more vulnerable to disease. Table 27.1 indicates the amount of nitrogen present in mineral surface soils. Soil nitrogen, apart from that applied in fertilisers, is mostly derived from organic wastes and from atmospheric fixation by soil micro-organisms. Some nitrogen is also added in rainwater which in humid temperate climates may amount to 4–10 kg ha^{-1} yr^{-1}, all of which is in the ammoniacal or nitrate form usable by plants.

Most soil nitrogen is associated with the soil humus but, since only 2–3% of this is mineralised annually under normal conditions, the yield is relatively slow. Up to 8% exists as ammonium ions on the exchange complex and 1–2% occurs as soluble ammonium and nitrate compounds, which are the forms most accessible to plant roots. Most of the uptake is of nitrate because it is normally present in larger quantities than ammonium, due to the rapid oxidation of the latter in most soils. Nitrogen compounds are highly soluble and as a result, they tend to be relatively easily lost in leaching. Some loss also results from *denitrification*, which represents the reduction of nitrogen compounds to gases (especially under anaerobic conditions) which are lost by vaporisation.

(b) Phosphorus

Phosphorus is a constituent of the cell nuclei of plants and is essential for their division. Its presence is especially critical at early stages of growth and cannot be made up later. It thus tends to be taken up from the soil in larger quantities early in the growing season. Phosphate-deficient plants tend to be stunted, to lack new shoots, to have a dull grey-green colour or to show red pigmentation at growth points. Excess phosphate, on the other hand, may speed maturation and thus depress the final yield of crops.

The supply of phosphorus to plant roots is restricted less by the overall quantities shown on Table 27.1, than by the fact that most is usually 'fixed' and unavailable because of chemical combination and absorption on to soil surfaces. The degree to which this occurs depends to a large extent on its ionic form which is in turn determined by the pH of the surrounding solution.

In general, the dominant form where the pH is below 6.5 is $H_2PO_4^-$, but above this level HPO_4^{2-} dominates, the former being somewhat more readily assimilated by plants than the latter. Under very acid conditions, however, iron oxides and soluble aluminium and aluminium ions (and

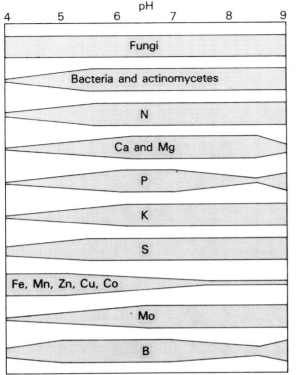

Fig. 27.3 Diagram showing the relationships in mineral soils between pH on the one hand and the activity of micro-organisms and the availability of plant nutrients on the other. The width of the bands and the degree of shading indicate the zones of greatest microbial activity and most ready availability of nutrients. (After Brady 1974)

at high pH values, calcium ions) will reduce the availability of the phosphorus by combining with it to form complex insoluble compounds. Its maximum availability thus practically lies within the range shown on Fig. 27.3. Although the proportion of total phosphorus in organic combination varies widely between different soils, the element is most continuously available where organic matter is abundant. In this situation, organic phosphorus provides a continuing source through mineralisation, is less easily fixed than inorganic phosphorus and generates acids which dissolve some fixed phosphorus.

(c) Potassium

Potassium differs from nitrogen and phosphorus in that it is concerned in plant metabolism rather than the cell make-up. Plants continue to absorb it throughout their 'lives' as the K^+ ion but as they ripen, they return some to the soil. Furthermore, potassium is subject to 'luxury' consumption and so more than necessary is taken up if it is available. While about half the phosphorus and almost all the nitrogen in the topsoil is in organic combination, only a small fraction of the potassium is available in this form.

Table 27.1 shows that it is relatively abundant in soils, especially in arid areas. It is held in three main forms. Firstly, between 90 and 98% of the potassium is retained as a fixed constituent of rocks (such as feldspars and micas) and is only made available to plants after weathering releases it from the rock. Secondly, between 1 and 10% occurs in the lattice structures of clays (such as montmorillonite and illite) and is only released at a slow rate. Finally, some 1 or 2% is held as exchangeable cations or in soil solution and is the only potassium that is readily available to plants, although it is highly vulnerable to leaching.

(d) Sulphur

Sulphur is an essential constituent of many proteins and of some vitamins and plant oils, although it is required in relatively small quantities. The main features of its cycle are given in Fig. 27.4. It is taken up as the sulphate ion, which is derived largely from the oxidation of organic matter. Furthermore, some is added to the soil in precipitation and some absorbed by plants directly from the atmosphere, notably in areas near to sources of the combustion of fossil fuels, especially coal and oil. The oxidation of sulphur leads to the formation of sulphuric acid and this may be serious on land where drainage follows a period of waterlogging, during which substantial quantities of sulphides have formed.

(e) Magnesium and silicon

Magnesium is needed by all plants since it is a constituent of chlorophyll and contributes to the transportation of phosphorus in plant tissues. It derives originally from minerals such as mica, hornblende, dolomite and serpentine, which also contain calcium, and its behaviour in soils is closely similar with deficiencies only occurring under acid conditions. Silicon is taken up by plants in substantial amounts and is especially important in the stems of cereals and in the leaves of tropical trees (Rodin and Bazilevich, 1967). It is taken up

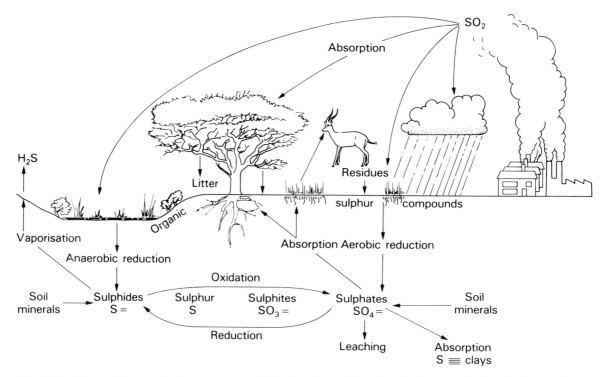

Fig. 27.4 The sulphur cycle, showing some of the transformations in soil, plants and animals. Most sulphur is usually in organic combinations (After Brady 1974). This diagram can be compared with those for the oxygen, carbon and nitrogen cycles (Figs 30.2–30.4).

by plants as silicic acid and although it can be highly variable in quantity from as much as 40% in many soils to less than 5% in those in tropical rainforests, it is generally abundant and has not been recorded as limiting to plant growth.

(f) Micronutrients

All micronutrients are required in very small amounts and are harmful when the available forms are present in high concentrations. For example, molybdenum may be beneficial if added to agricultural land at rates as small as 35–70 g ha^{-1}, while applications of over 1.5–2 kg ha^{-1} may be toxic to most plants. Several micronutrients appear to be necessary to the enzyme systems of plants concerned with oxidation–reduction and in the formation of proteins. Iron, copper and manganese are needed in the synthesis of chloroyphyll, boron is used in fruit and seed formation and zinc in seed and grain maturation.

The micronutrient content of soils depends on their parent materials and both deficiencies and excesses can usually be traced to this source. All micronutrients, with the exception of molybdenum, become more available with increasing acidity. Weathering and soil formation release oxides and sometimes sulphides of such metals as iron, manganese and zinc and their cations enter the exchange complex in the same way as the more common ones. Borate and molybdate anions may undergo reactions similar to those of the phosphates. Chlorine is the most soluble of the group and is easily lost in leaching, but is supplemented by annual additions in atmospheric precipitation.

27.3 The cycling of nutrients

It is possible to obtain some idea of the annual cycling of plant nutrients through soil and vegetation by measuring them in the incremental growth of plants and in the litter fall. This section considers some examples which show first the uptake of nutrients by field crops and secondly, their distribution between soil and vegetation in various

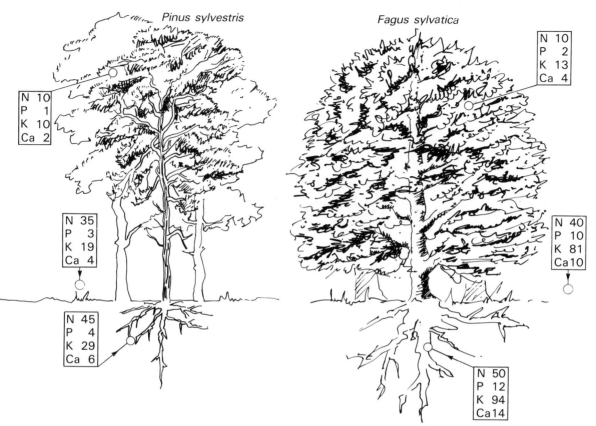

Fig. 27.5 Annual uptake (values below the soil line), retention (values in the crowns) and return (values above the soil line) of macronutrient elements (in kg ha⁻¹) by Scots pine and European beech. (After Duvigneaud and Denaeyer-De Smet 1970)

woodland species. Reference is also made to the proportions of the nutrient uptake which are retained as wood increment and returned as litter and to the zonal distribution of nutrient balance.

Table 27.2 shows the nutrient uptake of some common agricultural crops in England and it is apparent that they all take relatively large amounts of nitrogen and potassium. Cereals have the lowest uptake of mineral nutrients in relation to the amount of dry matter produced, except for the high values for silicon. Root crops have a relatively large demand (especially for potassium) and beans are to some extent intermediate. Sodium is not regarded as a necessary plant nutrient, but can to some extent substitute for potassium, and this view is supported by the relatively high values for sodium in turnips and mangolds.

Figure 27.5 shows the annual uptake of some macronutrients and compares their distribution between annual wood increment and litter fall in

deciduous and coniferous woodlands. The totals are substantial (especially for nitrogen and potassium) but, at all stages, the deciduous trees have higher values than conifers and also differ from them in the greater importance of both potassium and calcium relative to nitrogen. When the content of nutrients in the soil is also considered, it is possible to assess the importance of the fraction taken up by plants. Figure 27.6 shows their amounts in the vegetation, litter and humus and soil in four deciduous-forest ecosystems.

The vegetation contains a substantial proportion of the total nutrients having, in general, several times as much as the litter and humus. There is considerable variation between the tree species and, for example, the beech gives the richest and the birch the poorest litter and humus, although the latter reflects the high level of lime and the low level of nitrogen in its underlying soil. The chestnut forest is considerably lower in nitrogen, phos-

247

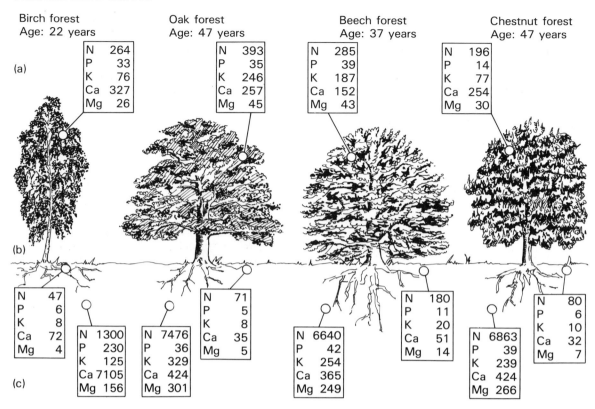

Birch forest
Age: 22 years

Oak forest
Age: 47 years

Beech forest
Age: 37 years

Chestnut forest
Age: 47 years

(a)

Birch	
N	264
P	33
K	76
Ca	327
Mg	26

Oak	
N	393
P	35
K	246
Ca	257
Mg	45

Beech	
N	285
P	39
K	187
Ca	152
Mg	43

Chestnut	
N	196
P	14
K	77
Ca	254
Mg	30

(b)

N	47
P	6
K	8
Ca	72
Mg	4

N	71
P	5
K	8
Ca	35
Mg	5

N	180
P	11
K	20
Ca	51
Mg	14

N	80
P	6
K	10
Ca	32
Mg	7

N	1300
P	230
K	125
Ca	7105
Mg	156

N	7476
P	36
K	329
Ca	424
Mg	301

N	6640
P	42
K	254
Ca	365
Mg	249

N	6863
P	39
K	239
Ca	424
Mg	266

(c)

Fig. 27.6 Distribution of micronutrients in deciduous forest ecosystems in the UK (kg ha^{-1}). The boxes represent, from top to bottom, total nutrients in: (a) biotic components of the ecosystem, including leaves, branches, fruits, boles and ground flora, both living and dead; (b) litter and humus; (c) soil to a depth of 50 cm. (After Duvigneaud and Denaeyer-De Smet 1970)

phorus and potassium than the others, but this difference is not derived from the soil or reflected in the litter and the humus. The amounts and proportions of nutrients cycled thus depend both on the nature of the soils and the vegetation.

At a wider scale, soil–vegetation systems are related to climatic zones. For example, Table 27.3 presents an estimate of the amounts of mineral nutrients taken up annually by different kinds of plant cover and shows the relatively large uptake in the tropical rainforest and steppe zones. The relative importance of different nutrients also varies between zones. Nitrogen, potassium and calcium are important everywhere and, as has been shown, they are dominant in temperate field crops and forests. However, silicon dominates the nutrient contribution in steppes, savannas and tropical rainforests (Rodin and Bazilevich 1967).

Table 27.3 Approximate annual mineral nutrient uptake by vegetation in different ecological zones in kg ha^{-1} yr^{-1}, arranged approximately according to latitude (including N, P, K, Ca, Mg, Na, S, Cl, Si, Al, Mn, Fe). (After Rodin and Bazilevich 1967)

Ecological zone	Uptake	Return in litter fall
Forest and swamp communities in taiga zone	60–300	20–150
Deciduous and broad-leaved forests	300–500	200–400
Steppe land	250–1800	350–700
Deciduous subtropics	1000	800
Tropical rainforest	2000	Over 1500

27.4 Retrospect

Soils are the part of the ecosystem which supplies plants with nutrients and receives organic residues. The processes are cyclical and involve a complex of biotic activities and both organic and inorganic reactions, controlled ultimately by the input of solar energy. Their physical expression can be seen in the movements and combinations of a relatively small number of nutrient elements. These are derived from rocks and the atmosphere and progress through the cycle is via a number of states, namely, the exchange complex, the soil solution, the tissues of living plants and organic residues. Both local ecosystems and broad ecological zones can be approximately characterised by the amounts and proportions in which they cycle these nutrients.

Key topics

1. Conditions in the rhizosphere which control plant growth.
2. Nutrient availability and transfer from soil to plant.
3. The factors limiting nutrient availability and uptake.
4. The variation of nutrient uptake under different climatic regimes.

Further reading

Black, C. A. (1968) *Soil–Plant Relationships*. John Wiley and Sons; New York and London.

Eyre, S. R. (1966) *Vegetation and Soils: a World Picture*. Edward Arnold; London.

Martin, J. K. (1983) Biology of the rhizosphere, CSIRO (Australia), *Soils: an Australian Viewpoint*. CSIRO/Academic Press; Melbourne and London.

Sanchez, P. A. (1976) *Properties and Management of Soils in the Tropics*. John Wiley and Sons; New York and London.

Trudgill, S. T. (1979) *Soil and Vegetation Systems*. Clarendon Press; Oxford.

Part V
Vegetation systems
Chapters 28–32

The following five chapters introduce the basic concepts of biogeography and ecology and emphasise their significance for assessing those earth surface processes which concern vegetation systems. The ecosystem concept is reviewed in the context of the systems approach in physical geography and its relevance to resource management. Of ecosystem characteristics, energy flow and biogeochemical cycling are discussed as the most important ways whereby processes operate in the biosphere. The reciprocal relationship between vegetation systems and the environmental systems discussed in previous chapters are emphasised, as well as the significance of anthropogenic activities in ecosystem function. Finally, vegetation succession is considered as an important process in landscape development and the controversial climax concept is re-examined.

The Californian Redwoods, USA.

Chapter 28 Biogeography, ecology and ecosystems

The purpose of this chapter is primarily to show how biogeography, ecology and ecosystems relate to systems theory in physical geography, as detailed in Chapter 1. Its second function is to provide some basic insight, in a conceptual framework, into the components and processes which are characteristic of ecosystems, themes which will be discussed in greater detail in the following chapters. Definitions of some basic terms in current usage in biogeography and ecology are also given and a flexible approach to ecosystem classification is adopted in order to provide as wide a view as possible of biogeographical and ecosystem theory in relation to geography.

28.1 Biogeography and ecology: development and scope

As the word suggests, *biogeography* is concerned with both biology and geography and represents the study of plant and animal distributions and their environmental relationships. It is biological because it is concerned with plants and animals which occupy the *biosphere* (Fig. 28.1(a)) and geographical because it is concerned with the distribution of these organisms and the environmental factors controlling these distributions. However, given the location of the biosphere at the interface between the atmosphere, hydrosphere and lithosphere, it should not be overlooked that there are also numerous overlaps with other disciplines such as climatology, geology and hydrology. Classical biogeography developed chiefly from the biological sciences, via the general field of natural history, utilising the foundations of biological taxonomy and nomenclature established by Linnaeus (1707–78) and the wealth of vegetation and environmental data collected by late eighteenth- and early nineteenth-century explorers such as Humboldt (1769–1859). Such observations on the variations of plant and animal life led Darwin (1809–82) to formulate his theory of evolution and the origin of species. From then on, biogeography became an intrinsic part of the newly emerging biological sciences within which two distinct, but related, branches of study developed.

One such approach is taxonomic, which is characteristic of plant (*phytogeography*) and animal (*zoogeography*) biogeography and is concerned with explaining the origin, evolution and dispersal of organisms on the basis of their distributional patterns. This type of approach, though modified, has been adopted by Dansereau (1957) and, more recently, by Simmons (1979) with emphasis on the role of ancient and modern societies in affecting plant distributions. Points of focus have been domestication, the inception and development of agriculture, the effects of industrialisation and biotic resource use. The second

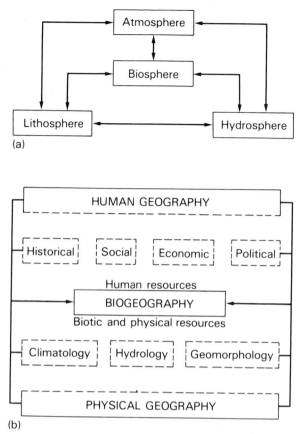

Fig. 28.1 (a) The position of the biosphere in relation to the atmosphere, lithosphere and hydrosphere; and (b) the relationship of biogeography to aspects of human geography and physical geography.

approach, which developed in tandem with phytogeography and zoogeography, led to the establishment of the subject known today as *ecology*. The origin of the word ecology is the Greek *oikos*, meaning 'household', 'home' or 'place to live', in association with the root *logy*, which means the 'science of' or the 'study of'. The term ecology, as in present-day usage, is derived from *okologie* or *oecology*, a word coined by the German zoologist Ernst Haeckel in 1869. He defined the subject as 'the entire science of the relations of the organisms to the surrounding exterior world, to which relations we can count in the broader sense all the conditions of existence. These are partly of organic, partly of inorganic nature' (quoted in Friederichs 1958).

Early development of the subject was slow, due largely to much interest at that time being given

to the (then) contentious proposals of Darwin and Wallace on evolution, but it was a major turning-point in scientific thought involving a more holistic and process-orientated approach to the natural world. By the 1900s, ecology had become well established but with some debate as to its precise meaning, since many workers (such as Elton 1927) equated it with scientific natural history. However, others continued the tradition of Friederichs (1958) who defined ecology as 'the science of the living beings as members of the whole of nature', with emphasis on the 'living beings' rather than on their reciprocal relationships with their environment. Modern ecologists, however, have tended to adopt ecology as 'the science of organisms in relation to their total environment, and the interrelationships of organisms interspecifically and between themselves' (Fraser Darling 1963).

As a result, there is now a division of ecology into *synecology*, which is concerned with the study of plant communities in relation to their habitats, and *autecology* which is the study of the ecological relations of individual species. The former approach is the one most widely adopted by modern geographers studying the composition, structure, function and development of plant communities. Through lack of a better alternative, such workers are called biogeographers and the field of study within the geography curriculum is known as biogeography. This should not be confused with the above description of classical biogeography, the scope of which is more clearly defined and from which geographers have incorporated many principles, in combination with those of the ecological approach, into their own discipline.

Figure 28.1(b) illustrates the position of biogeography within the greater field of geography as it stands today. Biogeography's concern with the habitat and environment provides a unifying link at the interface between the atmospheric and earth process or geomorphic systems. At the same time, human communities have long been regarded as major factors in ecosystem changes, both historically and presently. As such, they warrant the attention of biogeography which therefore provides a unifying link between human and physical geography. However, it is regrettable that biogeography did not become a major focus of attention within geography until the 1960s and 1970s, with the demise of regional geography and its instatement beside other systematic approaches, such as climatology and geomorphology (see section 32.1). In addition, little attention has been paid to animal communities due, in part, to the emergence of the classical biogeography and ecology from botany and the difficulties of studying often highly mobile and elusive animals.

Having documented a brief history of biogeography and ecology, what of their scope? Both are concerned with living organisms, their interrelationships and dependence upon, or control of, other environmental characteristics (such as the hydrological, atmospheric, soil and sediment transfer systems) as well as anthropogenic activity. This latter factor must, of necessity, be considered as an ever-increasing input into ecosystem functioning. On the one hand, human communities are entirely dependent upon ecosystems for their fuel and food energy requirements while, on the other hand, they are ever developing both new requirements and ways of manipulating ecosystem resources. Indisputably, therefore, vegetation and its associated fauna represent the surface expression of the interactions of the components of the above-mentioned systems, e.g. micro- and macroclimate, water balance, geology, soil type, geomorphological processes and social requirements.

As a consequence, the scope of biogeography and ecology is almost boundless. It is concerned with autecology, synecology, energy flows and movement of materials (the biogeochemical cycles) in the terrestrial environments, which are as diverse as the tundra and tropical rainforests, and in the aquatic environments, which are as diverse as the Arctic Ocean and tropical lakes. In addition, it is also concerned with the demands and effects of human communities on these ecosystems, an aspect of increasing importance for geography and the natural sciences concerned with resource management.

28.2 Ecosystems and their relationship with general systems

The word ecosystem was first used by A. G. Tansley (1935) as follows:

'The more fundamental conception is ... the whole system ... including not only the organism-complex, but also the whole complex of physical factors forming what we call the environment. ... We cannot separate them [the organisms] from their special environment with which they form one physical system. ... It is the system so formed which provides the basic units of nature

on the face of the earth. ... These ecosystems as we may call them, are of the most various kinds and sizes.'

Thus, the term ecosystem is self-defining: *eco* indicates environmental and *system* indicates a group of coordinated parts. This brings us back conveniently to the description of a system given in Chapter 1 and defined as '... a set of elements together with relations between the elements and among their states' by Hall and Fagen (1956). Ecosystems, therefore, represent one type of general system or group of general systems wherein living organisms play a fundamental role and represent one of the many variables which may be characteristic of any system. The relevance of the ecosystem concept to geography cannot be overstated since it provides not only a means of describing, analysing and understanding a wide variety of problems in physical geography, but also provides a theoretical framework and approach which can be validly used in human geography. Stoddart (1965) pointed out the following chief qualities, which make the ecosystem concept a useful one in geographical investigation:

(a) The concept is monistic, fusing environment, people, plant and animal life into one framework within which the interrelationships between the components can be analysed.

(b) Ecosystems have an organisation and are structured, which is conducive to study and rationalisation.

(c) All ecosystems function, involving the components and the processes in operation between them. This makes quantification possible for at least part of an ecosystem, in the case of large ecosystems, or possibly the entire ecosystem in the case of small examples.

(d) Ecosystems are a type of general system as evidenced by the definitions given above.

(e) Ecosystems may be defined at any scale.

28.3 The components and processes within ecosystems

It has been stated in Chapter 1 that all systems have three major components, viz. elements, states and relations between elements or states. The elements are objects (e.g. sediments) and each element has properties or states such as number, size, colour, etc. This is also true of ecosystems, elements of which comprise, for example, plant and animal species which have a variety of states (e.g. the population levels of individual species). In addition, all systems have inputs and outputs and, in ecosystems (Fig. 28.2), both inputs and outputs closely relate to the flow of energy in the form of organic matter. The chief input is energy from sunlight, which is utilised by green plants in the process of photosynthesis to manufacture food, and output may be in the form of harvest either by animals or people. In general terms, ecosystems have both *biotic* and *abiotic* elements with the former being defined as plants, animals and (in most contemporary ecosystems) human beings, whereas the latter are the non-living elements of the ecosystem, such as sediments, soils, nutrients and water. These biotic and abiotic attributes represent the means through which the ecosystem functions and also represent a storage element.

The driving force of the ecosystem is solar energy, which is responsible for the circulation of all materials in the ecosystem and the operation of processes (section 2.2). The input of solar energy is used in the processes of photosynthesis by green plants, which thereby make energy available in a usable form to the plants themselves (the primary producers) and to the consumers. Intricately associated with this process of energy transference (to be considered in detail in Ch. 29) is the circulation of nutrients, gases and water from soils and the atmosphere, via the producers to the consumers and eventually to human beings or out of the ecosystem entirely. This again serves to illustrate that processes within the ecosystem are inextricably linked with other environmental systems including nutrient transfer. This latter process is termed biogeochemical cycling and will be dealt with in Chapters 30 and 31.

28.4 Ecosystem stability or ecosystem homeostasis

One aspect of systems which has not so far been discussed, is the question of balance or equilibrium, since all systems (including ecosystems) develop towards achieving a balance between the inputs, outputs, elements and processes which are in operation. Subsequently, the ecosystem can operate at its optimum without major changes in its characteristics and thereby maintain its 'personality'. To achieve this, ecosystems have self-regulatory controls whereby they respond to changes or perturbations often through the populations that make up the ecosystem.

(a) Feedback in ecosystems

Feedback is characteristic of all systems and ecosystems and is a property whereby output influ-

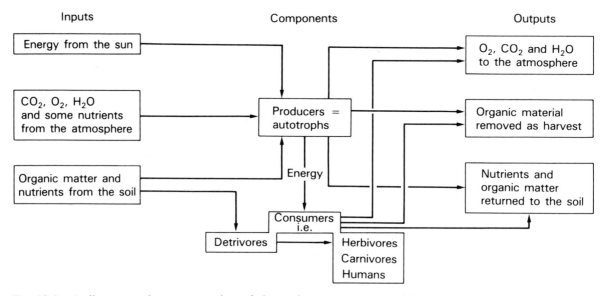

Fig. 28.2 A diagrammatic representation of the major components and interrelationships in a terrestrial ecosystem.

ences input. There are two types of feedback operative in ecosystems. Firstly, negative feedback has a stabilising ecosystem-perpetuating influence in so far as it prevents major changes in any one of the ecosystem components. That is to say, the effects of the component act to inhibit the very factors which were responsible for that component originally flourishing and therefore has a dampening effect. For example, an animal population within an ecosystem may increase, resulting in the depletion of the food supply. Animals will consequently die through lack of food and the population will be reduced, ultimately to restore equilibrium between predator and prey and restore the original ecosystem characteristics. This type of mechanism, which often amounts to population control, may occur at both producer and consumer levels or be restricted to consumer levels only in the ecosystem.

Secondly, there may be positive feedback which is damaging to ecosystem equilibrium because, when it occurs, one component (or set of components) becomes dominant, usually at the expense of other components. This implies that the ecosystem changes from its original characteristics to a different ecosystem, with a different array of characteristics. For example, fire is an important means whereby the characteristics of heathland communities are maintained in most temperate parts of the world. If this is eliminated, the ecosystem will gradually be invaded by non-heathland species and may develop into scrub and, eventually, woodland.

(b) Thresholds in ecosystems

A threshold can be delimited as the condition which marks the transition from one state to another and, in ecosystems, the ideal state is one of dynamic equilibrium. If, however, impetus for change from outside the ecosystem is so great that control within the ecosystem cannot accommodate it, a state of imbalance will occur followed by the establishment of a new ecosystem. The threshold, therefore, is the 'point of no return' when internal ecosystem mechanisms operate, but fail to cope with the change stimulated or initiated by external factors. Such changes may be gradual, as in the succession of seral stages of vegetation development (to be discussed in Ch. 32) or rapid, as in large-scale clearance of areas of tropical rainforest which, in recent years, has resulted in nutrient depletion, soil erosion and general degradation of the ecosystem. Thresholds in ecosystems are not always easy to delimit (even on a short time-scale covering recent geological time) when the only evidence is provided from micro- and macroscopic fossil-bearing deposits and limited historical records, which do not directly provide details of the vegetation communities. Present-day and very recent thresholds, however, are well documented due to large-scale changes in vegetation in many areas of the world, such as Malaysia and the Amazon Basin.

(c) Lagged response in ecosystems

Not all responses to change are immediate and, indeed, in something as complex as an ecosystem, the response is likely to be slow or delayed, due largely to the immense amount of internal control within the ecosystem. Thus, changes in inputs of energy and nutrients often take a considerable time to pass through the ecosystem; this is lag time. A lagged response is usually a much more common reaction in ecosystems than is rapid change and makes the detection of thresholds, as explained above, difficult. Furthermore, it is also clear from the fossil record of the Pleistocene period that some animals, especially the Coleoptera (beetles), tend to react much more rapidly to environmental change than do plants.

The lagged response characteristic of ecosystems, and the difference of response between plants and Coleoptera, is particularly well exemplified by studies of the British Late Devensian period. Fossil Coleopteran evidence (Coope 1977) shows that after the retreat of the last major ice advance in the UK (about 12 000 years ago), average summer temperatures were sufficient to support a deciduous forest environment, including such species as oak (*Quercus*), elm (*Ulmus*), alder (*Alnus*), and lime (*Tilia*). On the other hand, however, palaeobotanical evidence (Pennington 1977) indicates the presence of a park tundra environment with open habitat species, such as grasses (Gramineae) and sedges (Cyperaceae) with inspreading pioneer species of birch (both *Betula nana* – a dwarf birch, and *Betula pubescens* – a tree birch). One explanation of the evidence, therefore, involves a lagged response of the vegetation during this period to a rapid climatic amelioration, in comparison with the more rapid response of the Coleoptera. The lagged response of vegetation to changing inputs into ecosystems is also illustrated by the heathland example quoted in section 28.4(a). When fire is eliminated, the ecosystem changes in character but the response

257

is not immediate and it can take 20–50 years before woodland is established.

28.5 The classification of ecosystems

Classification as an entity in itself is tedious yet it is necessary in order to break down reality into meaningful, quantifiable and understandable units, and this applies equally to ecosystems as to systems in general. The classification of ecosystems has been undertaken in a variety of ways and some of these are documented below, in order to acquaint the reader with the variety available and to emphasise the flexibility of ecosystem classification schemes.

Ecosystems can be classified according to habitat and a simple classification is twofold, namely terrestrial ecosystems and aquatic ecosystems. However, this general classification may be further broken down in a variety of ways and examples of this are given in Tables 28.1 and 28.2 and Fig. 28.3. Broadly, aquatic ecosystems may be subdivided further, chiefly on the basis of habitat (i.e. freshwater or marine ecosystems) and ecosystems which are marginal between terrestrial and aquatic ecosystems (i.e. estuaries and sea shores). The terrestrial biomes (Fig. 28.3), however, have been classified in another way using the life zone system, devised by Holdridge (1947), which is based on two major assumptions. Firstly, that mature stable plant formations represent physiognomically (physiognomy meaning the form and structure) discrete vegetation types which are recognisable on a world-wide basis. Secondly, that geographical boundaries of vegetation correspond closely with the boundaries between climatic zones, which may be determined chiefly by the interaction of temperature and rainfall, associated with the prevailing circulation regimes (section 11.5 and Table 11.1).

Tables 28.1 and 28.2 and Fig. 28.3 represent the majority of ways in which ecosystems have been classified in most biological, environmental and geographical textbooks in the last two decades. However, a further type of classification of ecosystems (Odum 1975) is of particular interest to geographers because it details ecosystems as entities in the context of energy flow, which is the most fundamental attribute of all ecosystems (Table 28.3). This type of classification may be considered to be, on the one hand, more useful than those described above, which are based on

Table 28.1 A classification of aquatic ecosystems and ecotones*

A. Freshwater ecosystems
Lakes and ponds
Marshes and swamps
Springs
Streams and rivers

B. Marine ecosystems
Open sea

C. Aquatic ecotones or ecosystems which have inputs from both terrestrial and aquatic ecosystems
Estuaries
The sea shore
Tidal salt-marshes
Coral reefs
Deltas
Mangrove swamps

* An ecotone may be defined as a junction zone between neighbouring ecosystems.

Table 28.2 A classification of terrestrial ecosystems

Tundra	
Forests	Northern coniferous forest Moist temperate (mesothermal) coniferous forest Temperate deciduous forest Broad-leaved evergreen subtropical forest Tropical rainforest
Temperate grassland	
Savanna	
Chaparral	
Desert	

either habitat or climatic control, because it utilises a truly common denominator (i.e. energy flow) to classify ecosystems. On the other hand, something is lost in this classification, namely climatic inputs into ecosystems which inevitably have a major role in determining ecosystem or land-use characteristics. In many ways, however, this latter approach provides a means whereby ecosystems may be considered as inclusive of anthropogenic influence or as control systems. In view of geographical philosophy, this is a much more possibilistic than deterministic approach and has considerable application in the real world, given that more ecosystems will become control systems in the future.

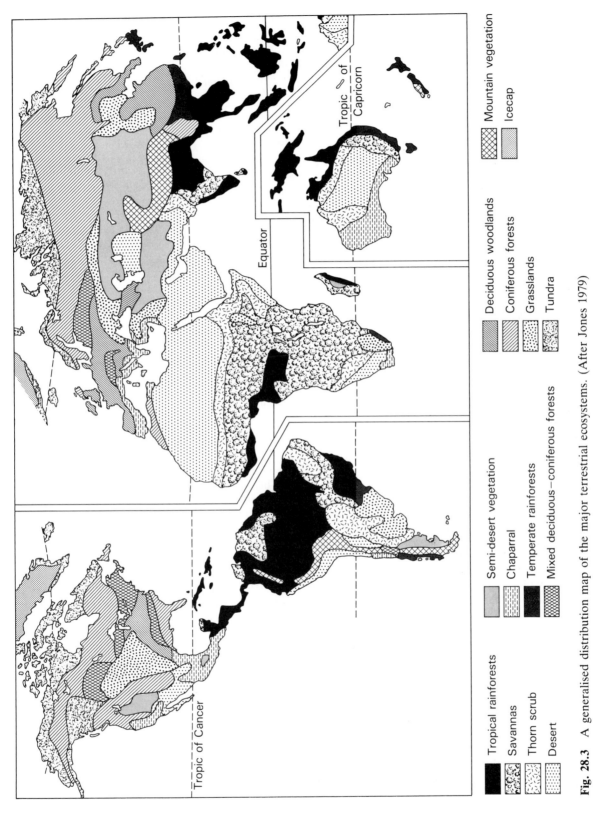

Fig. 28.3 A generalised distribution map of the major terrestrial ecosystems. (After Jones 1979)

Tropic of Capricorn

Equator

Tropic of Cancer

Mountain vegetation

Icecap

Deciduous woodlands

Coniferous forests

Grasslands

Tundra

Semi-desert vegetation

Chaparral

Temperate rainforests

Mixed deciduous–coniferous forests

Tropical rainforests

Savannas

Thorn scrub

Desert

Table 28.3 The classification of ecosystems based on energy characteristics. (Based on Odum 1975; Simmons 1979)

	Annual energy flow (kJ m⁻²)	
	Range	Estimated average
1. Unsubsidised natural solar-powered ecosystems, e.g. open oceans, upland forests. Anthropogenic factor: hunter-gatherer, shifting cultivation	4 130–41 800	8 360
2. Naturally subsidised solar-powered ecosystems e.g. tidal estuary, coral reef, some rainforests. In these ecosystems natural processes augment solar energy input. Tides, waves, for example, cause an import of organic matter and/or recycling of nutrients, while energy from the sun is used in the production of organic matter. Anthropogenic factor: fishing, hunter-gatherer	41 800–167 200	83 600
3. Human-subsidised solar-powered ecosystems, e.g. agriculture, aquaculture, silviculture. These are food- and fibre-producing ecosystems which are subsidised by fuel or energy provided by human communities, e.g. mechanised farming, use of pesticides, fertilisers, etc.	41 800–167 200	83 600
4. Fuel-powered urban–industrial systems, e.g. cities, suburbs, industrial estates. These are the wealth-generating systems (as well as pollution-generating) in which the sun has been replaced as the chief energy source by fossil fuel. These ecosystems (including socio-economic systems) are totally dependent upon types 1, 2 and 3 for life support, including fuel and food	418 000–12 540 000	8 360 000

28.6 Ecosystems in time

Time is an important factor in ecosystem development since inputs, components, processes and outputs are dynamic and change over time. The magnitude of such changes, and the operation of feedbacks and controls, have determined present-day ecosystem characteristics. *Palaeoecology* (which is the study of the relationships of past floras and faunas with their environment) affords the opportunity to trace ecosystem development over time, by examining the contained fossils of various types of deposits. For example, much research since the 1920s has been directed at reconstructing the Pleistocene interglacial and postglacial development of ecosystems in North America and Northwest Europe. Studies have used sub-fossil Coleoptera and Mollusca, the remains of higher plants (e.g. fruits, nuts, seeds, pollen), algae (e.g. diatoms) and the remains of lower plants (e.g. spores), which are often preserved in lake and peat deposits and buried soils.

This work has shown that, in addition to climatic change, one of the major causes of ecosystem change has been the impact of human communities. Such work is of considerable importance on two counts. Firstly, it provides insight into the development of ecosystems during interglacial periods, when the activities of human communities had little or no marked effect, and is therefore a record of truly natural ecosystem development. Secondly, studies of the postglacial period (the last 12 000 years) have shown that, over the last 5000 years, there has been a considerable change from one type of ecosystem to another, from the relatively simple ecosystems of the open or cascading type to the controlled type.

In other words, human populations have ceased to be a merely integral or symbiotic part of ecosystems and have become dominants in determining ecosystem development. It is also clear that human populations have attained this pre-eminence in parallel with increasing technological ability, i.e. from stone-using to metal-using to plastic-using to nuclear-power-using communities. This trend is likely to continue in the future with further exploitation of ecosystems for food production and their conversion to fully controlled land-use systems. The consequences have been, and will continue to be, seen in areas other than ecology or biogeography, such as social, economic and legal considerations. These, alongside ecology, have already played a major role in the establishment of, for example, national Parks in both the developed and developing parts of the world and there is an increasing role for environmentalists in politics. Furthermore, society has also begun to realise that extractive industries (such as coal and metal mining, sand and gravel extraction) as well as domestic waste disposal have led to the creation of new ecosystems in the form of spoil tips, etc. which can be put to use, in terms of food production or amenity provision, if properly treated (Ch. 31).

28.7 Retrospect

It remains only to be emphasised that biogeographers, whether trained in biology, chemistry, physics or geography, have a major role to play in ensuring that ecosystems and land-use are at their optimum. To this end they must consider the input of palaeoecological and historical data which provide an insight into long-term ecosystem development, the mechanisms of ecosystem function and the needs of human communities, in terms of food, mineral and recreational requirements. This will provide a broad and flexible framework which can be used as a predictive model for present and future conservation and management. 'Conservation is often presented as if it represented a cling-ing to the past: not so – what we are engaged in preserving is opportunities for the future' (Evans 1976). The resources of the future will be determined by present ecosystem management which can be effective only if ecosystem attributes and their interrelationships are understood.

Key topics

1. The relationship between ecosystems and other earth surface systems.
2. The interplay between components and processes in ecosystems.
3. The regulation of ecosystems via positive and negative feedback and changes effected by crossing thresholds.
4. Changes in ecosystems over time.

Further reading

Chorley, R. J. and Kennedy, B. A. (1971) *Physical Geography – a Systems Approach.* Prentice-Hall; London and New Jersey.

Huggett, R. (1980) *Systems Analysis in Geography.* Clarendon Press; Oxford.

Shugart, H. H. and O'Neill, R. V. (eds) (1979) *Systems Ecology.* Dowden, Hutchinson and Ross Inc.; Stroudsburg, Pennsylvania.

Stoddart, D. R. (1965) Geography and the ecological approach: the ecosystem as a geographic principle and model, *Geography*, **50**, 242–51.

Taylor, J. A. (ed.) (1984) *Themes in Biogeography.* Croom Helm; London and Sydney.

Tivy, J. and O'Hare, G. (1981) *Human Impact in the Ecosystem.* Oliver and Boyd; Edinburgh.

Whittaker, R. H. (1975) *Communities and Ecosystems.* Collier-Macmillan; London and New York.

Chapter 29   Energy flow in ecosystems

As stated in the previous chapter, energy flow in ecosystems is of vital importance and the energy input from the sun is, ultimately, the most important input into all systems. Solar radiation reaching the earth's surface provides two forms of energy, viz. *heat energy* and *photochemical energy*. The former is responsible for driving the hydrological cycle (section 2.1) and atmospheric systems, while the latter is used by plants in the process of photosynthesis. The importance of photosynthesis cannot be overstated since it is the source of almost all energy in both ecosystems and all living systems, and is essential to life. Thus, it sustains ecosystems themselves and their characteristics and is vital to the maintenance of human communities and their activities.

29.1 The process of photosynthesis

The process of photosynthesis consists of a complex series of biochemical stages and is chiefly carried out by chlorophyll-containing terrestrial and aquatic plants. The process can be summarised in a simplified form by the equation:

$$6CO_2 + 6H_2O \xrightarrow[\text{chlorophyll}]{\text{energy}} C_6H_{12}O_6 + 6O_2 \quad [29.1]$$

and consists essentially of two parts. Initially, light energy is absorbed by chlorophyll and splits a molecule of water, releasing oxygen. Then, the energy is used in several stages to synthesise carbohydrate from carbon dioxide. In addition, the rate of photosynthesis in an ecosystem is limited by the internal character of the ecosystem itself as well as external environmental conditions. These constraints may include varying light intensity, temperature, water availability, soil nutrients, leaf area, physiognomy of the community and the presence of layering, as well as the photosynthetic productive capacity of plants. This again serves to illustrate the interaction between ecosystems and other environmental systems.

However, once carbohydrate has been produced, it can be put to a variety of uses. For example, it may be converted into starch for storage or combined with sugar to produce cellulose which is the chief structural material of plants. It can also combine with elements essential to plant growth and metabolism (such as nitrogen, sulphur and phosphorus) to produce amino acids and proteins. In addition, some of the sugar which is produced is used as an energy source for growth and metabolic processes by the plants themselves. The energy is liberated by the process of respiration, summarised by the following equation:

$$C_6H_{12}O_6 + 6O_2 \rightarrow 6CO_2 + 6H_2O + energy \quad [29.2]$$

In this process, chemical energy is converted to heat which is a much more dispersed low-grade energy than chemical energy and consequently, it is lost from the ecosystem. This means that the amount of energy available to consumers (animals) in the ecosystem is less than the total amount of energy 'fixed' by the producers (the plants). To take it a step further, animals which feed on other animals have less energy again available to them for the same reason, i.e. energy is utilised and dissipated as heat energy by each successive consumer. This topic of energy transfer will be considered in later sections of this chapter but this generalised statement emphasises the one-way flow of energy.

29.2 Energy and its relationships with the laws of thermodynamics

Two kinds of energy exist. Potential energy, which is energy at rest, is capable of and available for work, while kinetic energy is a result of motion and results in work (section 4.2). Work that occurs in this way can bring about energy storage (as potential energy) or can arrange matter without storing energy. The one-way flow of energy, described in section 29.1, is related to this use and storage of energy and is the result of the operation of the laws of thermodynamics. The first law states that energy is neither created nor destroyed but may be transformed from one type (e.g. light) into another (e.g. the potential energy of food such as sugar or starch). This occurs in the process of photosynthesis where the molecules of the products (carbohydrates) store more energy than the reactants (water and carbon dioxide) with the extra energy being derived from sunlight. This is an *endothermic* reaction, when energy from outside is put into a system to raise it to a higher energy state. However, in the process of respiration, much of the potential energy stored in the food (i.e. in carbohydrate or sugar) is degraded into a form (heat energy) incapable of further work, which randomly disperses the molecules involved. This is known as *entropy* and relates to the second law of thermodynamics which states that, when energy is transferred or transformed, at least part of the energy takes a form that cannot be passed on any further. This is exactly what happens in an ecosystem in the process of energy transference from producer to consumer, when much of the energy is degraded as heat and the remainder is stored in living organic material.

In conclusion to this section, ecosystems main-

tain themselves in their highly organised and low-entropy state because they can transform energy from high-order to low-order states. As a corollary, if the quantity or quality of energy flow through an ecosystem is reduced, that ecosystem begins to cross thresholds (section 28.4) involving an inefficient use of energy. This results in greater entropy, until it reaches a new equilibrium and a new ecosystem therefore emerges as a response to failure of negative feedback controls.

29.3 Gross and net primary productivity

Energy enters the ecosystem via *autotrophs* which are the chlorophyll-bearing organisms (*autotroph* means self-nourishing whereas *heterotroph* means dependent on autotrophs for nourishment). The energy, which is accumulated by green plants via the process of photosynthesis, is called *primary production* and represents the first energy storage in an ecosystem. The total amount of organic matter synthesised per unit time is *gross primary productivity* (GPP). However, as explained in section 29.1, in order to carry out photosynthesis and all their other metabolic processes, the plants must use up some of the energy they have trapped by releasing it in the exothermic and entropic process of respiration. The energy which is left after respiration, and stored as organic matter, is the *net primary productivity* (NPP) which is essentially plant growth. Productivity can be expressed quantitatively in two ways, either as energy per unit area per unit time (e.g. kJ m^{-2} yr^{-1}) or as the weight of dry organic matter per unit area per unit time (e.g. g m^{-2} yr^{-1}).

Closely related to NPP is *biomass* which is the NPP that has accumulated with time. On a seasonal basis, some of the NPP (e.g. plant litter) will decompose and enter the detrital food chain, as described in the next section. However, the remainder is stored in the living components of the ecosystem and constitutes the *standing crop biomass*, usually expressed as dry weight of organic matter per unit area (i.e. g m^{-2}). The essential difference between productivity and biomass is that the former represents the rate at which organic matter is produced by photosynthesis, while the latter represents a measure of the storage of organic matter within an ecosystem. This storage is contained in the above-ground and below-ground plant parts (e.g. leaves, branches, roots, seeds, etc.) and its distribution varies according to the species present in the ecosystem. In a forest environment, for example, most of the biomass is stored in the above-ground plant parts, especially the woody tissues of the trunk and branches.

This means that there is a low root to shoot ratio (R/S) and ecosystems where this occurs can usually assimilate more light and thus have a higher productivity than ecosystems with a high R/S. In these (e.g. tundra ecosystems), most of the biomass is stored in the root systems, ensuring access to necessary water and nutrients, which is essential under such harsh environmental conditions. In addition, the distribution of biomass between above- and below-ground plant parts may change as a plant community develops on a new and previously unvegetated substrate, such as open water or bare rock. Taking an example in a temperate environment in the early stages of colonisation, the R/S will be high, gradually changing with each successive stage until finally

temperate forest with a low R/S develops. This process of succession will be considered in Chapter 32.

In section 28.5 it was stated that, on a global basis, the distribution of the major vegetation communities (biomes), illustrated in Fig. 28.4, corresponds closely with that of the earth's atmospheric circulation zones (section 11.5). The nature of these biomes is determined by the biomass characteristics which, in turn, are a function of productivity. Some of the variables, which determine the rate of photosynthesis and hence productivity, are given in section 29.1. Of the climatic variables mentioned, temperature and precipitation most strongly influence productivity which, as a result, shows wide variations over the earth's surface. Examples are given in Figs 29.1 and 29.2 and Table 29.1 and, in general, productivity is greatest where mean annual rainfall and temperatures are highest. This occurs in the tropical rainforests, which are among the most productive

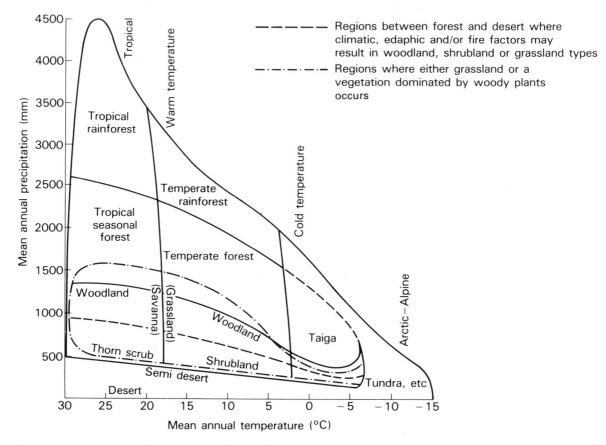

Fig. 29.1 The distribution of world biomes in relation to mean annual temperature and mean annual precipitation. (Adapted from Whittaker 1975)

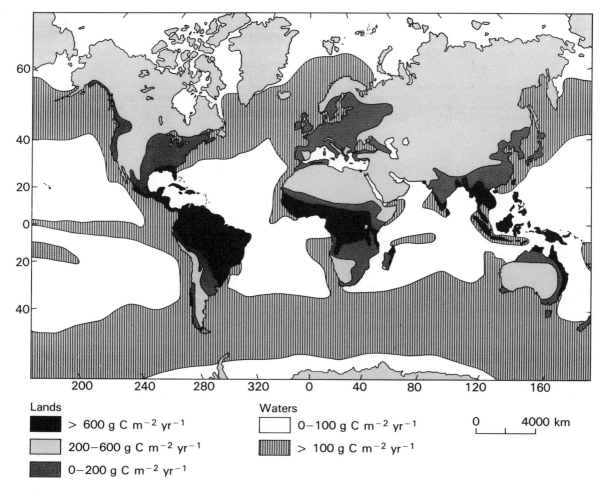

Lands

⬛ > 600 g C m⁻² yr⁻¹

▨ 200–600 g C m⁻² yr⁻¹

▨ 0–200 g C m⁻² yr⁻¹

Waters

☐ 0–100 g C m⁻² yr⁻¹

▥ > 100 g C m⁻² yr⁻¹

0   4000 km

Fig. 29.2 The distribution of world primary productivity. (Adapted from Leith and Whittaker 1975)

ecosystems in the world, with productivity in the range of 1000–3500 g m⁻² yr⁻¹.

The least productive terrestrial ecosystems are the cold (tundra) and hot deserts where either temperature or precipitation are the major limiting factors to productivity, which ranges between 100 and 250 g m⁻² yr⁻¹. These are figures for some of the unsubsidised natural solar-powered ecosystems described in Odum's ecosystem classification (section 28.5). Where there is an addition of energy to an ecosystem, high productivity may result. This subsidy may be due to natural processes, as in the case of estuaries, or to inputs of fuel or energy provided by human communities (as in agriculture) which represent the naturally-subsidised and human-subsidised solar-powered ecosystems of Odum's classification. In the case of estuaries, the subsidy is in the form of tide and wave energy

which causes import of organic matter and nutrient recycling. In the case of agriculture, fuel energy is added to the ecosystem in the form of mechanisation and the application of fertilisers and pesticides.

29.4 Secondary productivity

The amount of secondary productivity which can occur in an ecosystem is entirely dependent on NPP since this is the amount of energy available to the heterotrophs (consumers). However, some of the NPP may be exported from the ecosystem by agencies such as wind, water or humans. The amount exported will vary considerably between ecosystems; for example, almost all of the net

265

Table 29.1 Net primary production and plant biomass for the earth. (After Whittaker 1975)

Ecosystem type	Area (10⁶ km²)	Net primary productivity per unit area (g m⁻² yr⁻¹)		World net primary production (10⁹ t yr⁻¹)	Biomass per unit area (kg m⁻²)		World biomass (10⁹ t)
		Normal range	Mean		Normal range	Mean	
Tropical rainforest	17.0	1000–3500	2200	37.4	6–80	45	765
Tropical seasonal forest	7.5	100–2500	1600	12.0	6–60	35	260
Temperate evergreen forest	5.0	600–2500	1300	6.5	6–200	35	175
Temperate deciduous forest	7.0	600–2500	1200	8.4	6–60	30	210
Boreal forest	12.0	400–2000	800	9.6	6–40	20	240
Woodland and shrubland	8.5	250–1200	700	6.0	2–20	6	50
Savanna	15.0	200–2000	900	13.5	0.2–15	4	60
Temperate grassland	9.0	200–1500	600	5.4	0.2–5	1.6	14
Tundra and Alpine	8.0	10–400	140	1.1	0.1–3	0.6	5
Desert and semi-desert scrub	18.0	10–250	90	1.6	0.1–4	0.7	13
Extreme desert, rock, sand and ice	24.0	0–10	3	0.07	0–0.2	0.02	0.5
Cultivated land	14.0	100–3500	650	9.1	0.4–12	1	14
Swamp and marsh	2.0	800–3500	2000	4.0	3–50	15	30
Lake and stream	2.0	100–1500	250	0.5	0–0.1	0.02	0.05
Total continental	149		773	115		12.3	1837
Open ocean	332.0	2–400	125	41.5	0–0.005	0.003	1.0
Upwelling zones	0.4	400–1000	500	0.2	0.005–0.1	0.02	0.008
Continental shelf	26.6	200–600	360	9.6	0.001–0.04	0.01	0.27
Algal beds and reefs	0.6	500–4000	2500	1.6	0.04–4	2	1.2
Estuaries	1.4	200–3500	1500	2.1	0.01–6	1	1.4
Total marine	361		152	55.0		0.01	3.9
Full total	510		333	170		3.6	1841

productivity of human-subsidised (agricultural) ecosystems will be exported to fuel-powered urban-industrial systems and, in some salt-marsh ecosystems, about 45% of the NPP may be exported to adjoining estuaries (Teal 1962). Furthermore, some of the NPP is unavailable to consumers for a variety of reasons. For instance, not all plants are accessible since living plant material is unavailable to decomposer organisms, and consumers often have preferences. In addition, a proportion of the food ingested by consumers may not be digestible and will pass through the animal's body. For example, Drozdz (1967) has shown that bank voles individually consume between 54.4 and 62.8 kJ day⁻¹ of which 43.4–53.4 kJ are assimilated. Also, for yellow-necked fieldmice, the intake is between 50.7 and 70.9 kJ day^{-1} with 39.4–64.7 kJ assimilated. The remaining food is egested unused to be either stored, exported, decomposed by micro-organisms (the detritus food chain, which is examined in the following section) or consumed by other heterotrophs.

Both producers and consumers must respire in order to maintain their structure and function. Consequently, part of the energy extracted from the producers is used in respiration and lost as heat energy, although assimilated energy not respired is available for production of new tissue, growth and new individuals. This is secondary productivity and unlike primary productivity, there is no analogous gross and net productivity at secondary level. However, the actual assimilation of energy is closest to *gross secondary productivity* while that remaining after respiratory losses are taken into account is closest to *net secondary productivity*. This is usually greatest when animal populations have high birth and growth rates. The animals which feed on plants are the *herbivores* or *primary consumers* which may, in turn, be fed upon by *carnivores* or *secondary consumers*. The process of energy transfer between these is similar to that described above and the net result is that, at each transfer, there is less energy available for the next consumer level. The following equation summarises the energy budget of a consumer population:

$$C = A + Fu \qquad [29.3]$$

where C is the energy ingested, A is the energy assimilated and Fu is the energy lost through faeces, urine and gas. The term A can be further refined as:

$$A = P + R \qquad [29.4]$$

where P is the secondary production and R is energy lost through respiration. Thus:

$$C = P + R + Fu \qquad [29.5]$$

or secondary production:

$$P = C - Fu - R \qquad [29.6]$$

Both primary and secondary production are limited by internal and external variables in the ecosystem. Furthermore, NPP itself (in terms of the quantity, quality and accessibility of energy) represents a major constraint on secondary productivity, as do the degree to, and efficiency with, which both primary and secondary consumers utilise available productivity.

29.5 The movement of energy through ecosystems: food chains and food webs

Green plants, and the energy they provide as NPP, are essential to sustaining life on earth since few other organisms have the ability to trap energy. When energy is transferred from one organism to another (i.e. from plant to herbivore to carnivore, in stages of eating and being eaten), the transference is known as a *food chain*. An example is given in Fig. 29.3, where plants are fed upon by ptarmigan which in turn are fed upon by Arctic fox. However, the situation is not usually as simple as this in most ecosystems since, to return to Fig. 29.3, the plants are eaten by a variety of other organisms (such as Collembola, Diptera, mites and Hymenoptera), and a variety of birds, which in turn are consumed by a variety of predators. As a result, the linear food chains become interlinked to form a *food web* or an energy exchange network. The example given in Fig. 29.3 is a relatively simple one from the Arctic tundra ecosystem of Bear Island, Spitsbergen. In other ecosystems (such as temperate and tropical forests, where temperature is not such a limiting factor in ecosystem development and function as it is in the tundra), food webs are much more complex. In general, the more complex the food webs, the more stable the ecosystem with more negative feedback controls. Furthermore, since energy is used at every stage or level in the food chain or web, the number of exchanges is usually limited to four or five.

The components of the food web fall into four main categories. The primary consumers are the herbivores which have the ability to convert energy, stored in plant tissue, to energy stored in animal tissue. They are essential in the energy exchange network, because of their physiognomic and physiological adaptations to consuming a diet which is rich in cellulose, and without them other consumers could not exist. Examples of herbivores include ruminants such as cattle, sheep and deer and the lagomorphs, which include rabbits and hares. The secondary consumers are the carnivores which utilise the herbivores as their energy source and provide a source of energy for second-level carnivores. Omnivores represent the third category but they may be either primary or secondary consumers, since they eat both plants and animals (for example, the badger is an omnivore feeding on both plants and insects). The final

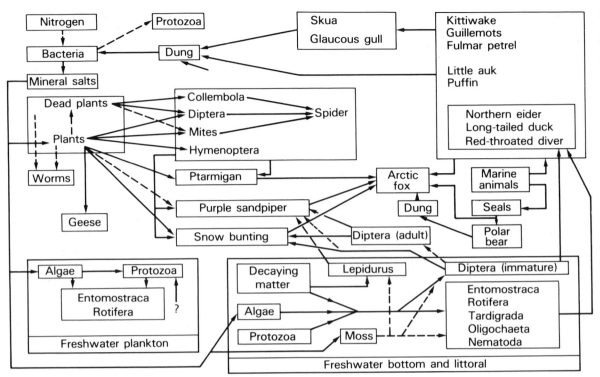

Fig. 29.3 A diagrammatic representation of a food web for Bear Island, Spitsbergen. (Adapted from Summerhayes and Elton 1923)

category of consumers in the food web are *the decomposers* who also play a vital role in nutrient cycling, by releasing nutrients as they break down plant and animal tissue in order to obtain their energy. Two groups of decomposer organisms occur. One group consists of macro-organisms or *detrivores* (e.g. earthworms, slugs and molluscs) which are reducer-decomposers that ingest organic matter and break it down into smaller pieces to condition the material for subsequent breakdown by the second group, the micro-organisms. These are bacteria and fungi which bring about the final breakdown of detritus and transform organic matter into nutrients available for plants.

29.6 Energy transfer networks: the grazing and detrital pathways

The NPP of an ecosystem supports all consumer organisms by providing their energy needs. Consumer organisms may be, for convenience, divided into two groups, viz. the *macroconsumers* (the herbivores, carnivores and omnivores described in

the previous section) and the *decomposers* (the detrivores and micro-organisms). In the former group, the herbivores and omnivores feed directly on living plant material and are therefore *biotrophic* providing an energy source for carnivores, while the decomposers obtain their energy from dead plant (and animal) tissues and are *saprotrophic*. This gives rise to the two major routes along which energy passes, i.e. the *grazing and detrital pathways*, both of which occur in all ecosystems and between which energy is often exchanged. The importance of one pathway relative to the other will vary between ecosystems, in parallel with the amount of energy which passes along each of the routes.

(a) The grazing pathway

This appears to be the most obvious route for energy flow in most ecosystems because it is the most easily observed and quantifiable route. However, more often than not, it actually plays a subordinate role to the detrital pathway, especially in terrestrial ecosystems. Referring back to Fig. 29.3, the grazing pathway of the food web is represented by the sequence plants → ptarmigan → Arctic fox, although a more familiar example of the grazing pathway is that which occurs in the pastoral agricultural system (i.e. grass → sheep → humans). However, in most terrestrial ecosystems, only a relatively small proportion of energy transfer occurs via the grazing food chain. For example, Table 29.2 illustrates the relatively minor role of herbivores and the grazing food chain in a grassland ecosystem, both in terms of ingestion of primary productivity and secondary productivity. The productivity passing through the grazing food chain is only 1.6%, compared with 98.4% passing through the detrital food chain.

Similarly, Andrews *et al.* (1974) found that cattle grazing on the prairie accounted for only 15% of the above-ground NPP or 3% of total NPP, despite the fact that these animals may consume between 30 and 50% of the above-ground NPP. However, in this situation, research on ungrazed, lightly grazed and heavily grazed plots illustrated that grazing pressure (i.e. herbivore population numbers) had a direct effect on the ecosystem. Where grazing pressure was greatest, more of the net production was confined to the below-ground root systems of the plants; where it was light, the allocation to shoots was higher while in ungrazed plots, only a small proportion went into shoot pro-

duction. This illustrates that biotrophically-based energy transfer has a direct effect on the autotrophs themselves and the distribution of biomass. In the case quoted, light grazing actually encouraged the above-ground distribution of NPP and such findings relate to the statement made in section 28.7, in the context of resource management.

However, returning to this example, Dean *et al.* (1975) showed that in both lightly and heavily grazed plots, between 40 and 50% of energy consumed by the cattle was returned to the ecosystem, largely as faeces. Furthermore, this was utilised in the detrital pathway of energy transfer, thereby illustrating the interrelationships between the two pathways. The examples given above illustrate that the grazing pathway of energy transfer (albeit a very important food source for human communities) plays a minor role in energy flow in most terrestrial ecosystems, when compared with the detrital pathway. In many deep-water aquatic ecosystems, however, the reverse is true with most energy passing along the grazing pathway. These kinds of ecosystems generally have a low biomass (Table 29.1) and a rapid turnover of organisms and harvest. In Long Island Sound, USA, for example, Riley (1956) has shown that between 50 and 75% of the primary producers (i.e. the phytoplankton or unicellular organisms, mostly algae,

Table 29.2 Calculated ingestion, production, respiration and egestion by heterotrophs ($kJ\ m^{-2}\ yr^{-1}$) per 100 $kJ\ m^{-2}$ net annual primary production in a grassland ecosystem. (After Heal and Maclean 1975)

	Ingestion	*Production*	*Respiration*	*Egestion*
Herbivore system				
Herbivores:				
Vertebrate (H_v)	25.000	0.250	12.250	12.500
Invertebrate (H_i)	4.000	0.640	0.960	2.400
Carnivores:				
Vertebrate (C_v)	0.160	0.003	0.123	0.031
Invertebrate (C_i)	0.170	0.040	0.095	0.034
Saprotroph (decomposer) system				
Saprotrophs:				
Invertebrate (S_i)	15.153	1.212	1.818	12.122
Microbial (S_m)	136.377	54.551	81.826	–
Microbivores:				
Invertebrate (M_i)	10.910	1.309	1.964	7.637
Carnivores:				
Vertebrate	0.041	0.001	0.032	0.008
Invertebrate	0.648	0.155	0.363	0.130
Total				
% passing through:	192	58	99	35
Herbivore system	15.2	1.6	13.5	42.9
Saprotroph system	84.8	98.4	86.5	57.1

which can photosynthesise) are grazed by zoo-plankton, which are free-swimming, grazing heterotrophs (e.g. crustaceans). As a consequence, more energy is transferred along the grazing pathway than along the detrital pathway.

(b) The detrital pathway

The grassland examples quoted in the previous section illustrate that only a relatively small propor-tion of energy flow occurs via the grazing pathway. This is true of almost all terrestrial and *littoral* (i.e. the ecotone between terrestrial and marine ecosystems) ecosystems, where the detrital pathway predominates because only a small proportion of the available NPP is directly grazed by herbivores, who themselves return energy in the form of faeces to the detrital pathway. This is a much less obvious and a much more complicated passage of energy, compared with the grazing path-

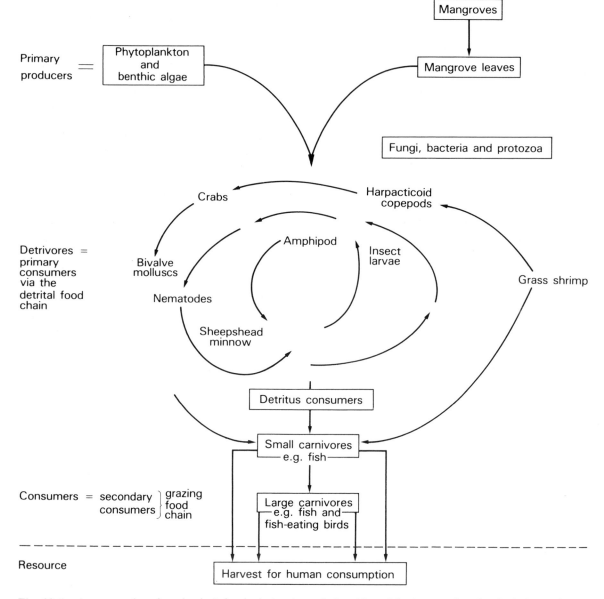

Fig. 29.4 An example of a detrital food chain, its relationship with the grazing food chain and as a component of an economic resource.

way, and is also more difficult to quantify. Figure 29.4 shows a detrital energy flow network for a mangrove ecosystem in Florida, USA (Odum and Heald 1972) and in this littoral ecosystem, only a very small proportion of the NPP (as partly represented by mangrove leaves) is used directly by herbivores.

In this example of a naturally subsidised solar-powered ecosystem (Section 28.5 and Table 28.3) the majority of leaves, etc. are transported by tides and currents from the Florida coast south to Central and South America. Micro-organisms, such as bacteria and fungi, begin to decompose the organic material to which algae add enrichment. It is then eaten and re-eaten by a group of detrivores (Section 29.5) during which process the micro-organisms are removed and the substrate egested to be recolonised by a further set of micro-organisms. This means that the original mangrove leaves provide energy to the decomposer micro-organisms which then provide the energy source for the detrivores. These in turn enter the grazing pathway by becoming the major source of food for fish, which are harvested by humans. As a consequence, the mangroves are an important, although indirect, economic resource and represent a feature which again can only serve to emphasise the relevance of the ecosystem concept to resource management (Section 28.7).

29.7 Trophic levels, ecological pyramids and models of energy flow

So far, the major biological components of ecosystems have been referred to as producers and consumers, reflecting both functional and feeding interrelationships which are also known as *trophic relationships*. Each step in energy transfer is a trophic level (section 25.3) with producers at the first trophic level, herbivores at the second trophic level and first-order carnivores at the third and so on, with some animals (e.g. omnivores) capable of participating in several trophic levels. Trophic levels should include organisms participating in the detrital pathway as well as those of the grazing pathway. For example, decomposers which feed on dead plant material may be described as functional herbivores while those feeding on dead animal material may be considered as functional carnivores.

As a result of this trophic layering in ecosystems and because there is less energy available at each level (section 29.5), there is a steep stepwise decrease in productivity up the sequence, giving rise to the *pyramid of productivity*. Two further pyramids occur as a consequence. The *pyramid of numbers* (Fig. 29.5) reflects the decrease in numbers of individual organisms involved at each successive level. This pyramid, however, may become reversed if many small organisms feed on one large organism at a lower trophic level. For example, many thousands of insects may feed on one tree or many thousands of bacteria may decompose one leaf or one animal carcass. As a result of energy transfer, the biomass for each trophic level also decreases up the sequence giving rise to the *pyramid of biomass* (Fig. 29.5). This is usually measured as weight of living material (e.g. dry g m^{-2}) and represents the standing crop or the total amount of fixed energy extant at any one time. Occasionally, biomass pyramids can also be reversed as in some aquatic ecosystems, where primary production by phytoplankton is low and these are heavily grazed by zooplankton. As a result, the biomass represented by the base of the

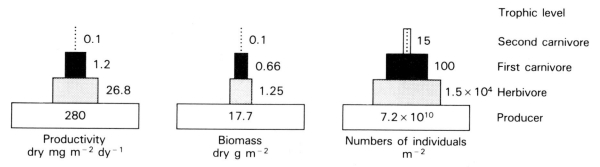

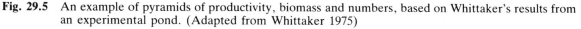

Fig. 29.5 An example of pyramids of productivity, biomass and numbers, based on Whittaker's results from an experimental pond. (Adapted from Whittaker 1975)

pyramid is smaller than the biomass of the consumers in the ecosystem.

The transfer of energy is dependent on the *efficiency of transfer* from one trophic level to another. The most important aspect of energy transfer efficiency is *consumption efficiency*, generalised values for which are given in the energy transfer example illustrated in Fig. 29.6. This is a model of both the grazing and detrital flow pathways and illustrates, in a theoretical form, the interrelationships between the two in a terrestrial ecosystem. The consumption efficiency (*CE*) can be expressed by the formula:

$$CE = (C_n/P_{n-1}) \qquad [29.7]$$

where C_n is the consumption by trophic level n and P_{n-1} is the production of the previous trophic level $(n-1)$.

Heal and Maclean (1975) have based their numerical values (which can only be considered as generalisations) on energy flow in a temperate grassland ecosystem. Here, the grazing pathway is dominated by herbivorous insects, which consume less than 5% of the NPP and vertebrate herbivores, which consume about 20%. The detrital pathway is dominated by saprophytic invertebrates consuming 10% of the litter fall, while the remaining 90% is utilised by micro-organisms. Clearly, the figures would vary between one ecosystem and another but, in general, the model is very useful and illustrates some other important ecosystem characteristics. For example, it shows the direct and inverse relationship between primary producers and primary herbivore consumers (section 29.6a), while the decomposer subsystem has no direct effect on the rate at which energy enters the system. In addition, the model shows that energy passing through the grazing pathway can take two routes. It can be respired and dissipated as heat or is passed on to the detrital pathway, although energy entering the detrital pathway can be lost only through respiration. This emphasises the interrelationships between the two energy transfer pathways and, as Heal and Maclean (1975) state, it shows that the detrital pathway is energy-conserving through recycling. Because of this, and because in most ecosystems

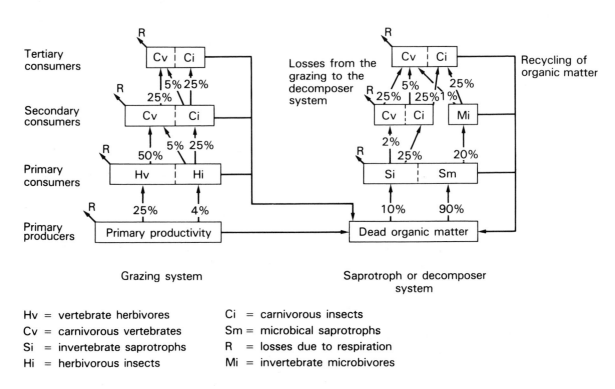

Hv = vertebrate herbivores   Ci = carnivorous insects
Cv = carnivorous vertebrates   Sm = microbical saprotrophs
Si = invertebrate saprotrophs   R = losses due to respiration
Hi = herbivorous insects   Mi = invertebrate microbivores

Fig. 29.6 A generalised trophic structure for terrestrial ecosystems and consumption efficiencies (see equation [29.7] in text) between trophic levels in a temperate grassland ecosystem. (Adapted from Heal and Maclean 1975)

the majority of the NPP is absorbed via the detrital route, it is a vital regulatory mechanism and consequently a determinant of ecosystem characteristics.

Figure 29.6 essentially represents a model of energy flow but it should be emphasised that this concept is not new in ecological thinking. A model of energy flow was first developed by Lindeman (1942) utilising some of the principles already discussed in section 29.2, especially the relevance of the laws of thermodynamics and trophic structure. In its simplest form, the biomass of any one trophic level is designated by $\wedge$ along with a subscript (e.g. $\wedge_1$, $\wedge_2$, $\wedge_3$) to denote trophic level. In addition, it is possible to describe the amount of energy which is transferred from one level to another, designated by λ. Thus, the amount of energy which is transferred from $\wedge_n$ to $\wedge_{n+1}$ is shown by λ_{n+1}. Similarly, energy which is dissipated as heat in the process of respiration is designated as R_1, R_2, etc. The loss of heat (R) plus the transferring energy λ_{n+1} between $\wedge_n$ to $\wedge_{n+1}$, is symbolised as $\wedge_{n1}$, representing energy which, at any one trophic level, does not go into biomass production. Consequently, it is possible to write a generalised formula for energy flow from one trophic level to another, as follows:

$$\frac{\triangle \wedge_n}{\triangle t} = \lambda_n - \lambda_{n^1} \qquad [29.8]$$

This equation gives the rate at which energy is assimilated at a given trophic level minus the rate at which energy is lost from it.

29.8 Retrospect

In conclusion to this section, it is apparent that the analyses of community structure (expressed as trophic levels, the pyramids of productivity/biomass/numbers and the Lindeman energy flow concept) all illustrate the relevance of ecosystem theory to the general systems approach described in Chapters 1 and 28. All of these attributes can be recognised in any ecosystem, allowing quantification and providing a model framework in which to relate components and to compare within and between ecosystems. It also provides a mechanism for prediction and especially, in the case of both ecosystems and agricultural systems, a mechanism for determining effective management policies. As stated in section 28.5 (Table 28.3), the fuel-powered urban-industrial ecosystems are totally dependent on the unsubsidised, naturally and human-subsidised ecosystems for their food energy. Furthermore, '. . . an acre of a city requires not only many acres of agro-ecosystems to feed it, but even more acres of general life support, natural or semi-natural environment to take care of the carbon dioxide and other large routine wastes and to supply it with huge volumes of water and other materials' (Odum 1975). Thus, such ecosystems place a great deal of stress on the solar-powered ecosystems and in these days of high population levels (with attendant poverty and food scarcity in many developing countries), the emphasis must be on efficient resource management. In this chapter, the relevance of the principles of energy transfer to resource management has been emphasised. A thorough understanding of energy flow is also a prerequisite for efficient conservation of natural and semi-natural communities, and the resources and opportunities they provide for the future. Taking a wider view, however, other inputs into ecosystems must also be considered in these contexts. These include soil degradation and erosion, land and air pollution, water quality and ecosystem use for non-productive purposes (such as recreation), and some of these factors are dealt with elsewhere in this book.

Key topics

1. The significance of energy transfer processes in ecosystems.
2. Productivity as a major ecosystem characteristic.
3. The various pathways of energy transfer through trophic levels.
4. The quantification of ecosystem inputs and outputs and their relevance to management/conservation.

Further reading

Anderson, J. M. (1981) *Ecology for Environmental Sciences*. Edward Arnold; London.

Collier, B. D., Cox, G. W., Johnson, A. W. and Miller, C. P. (1973) *Dynamic Ecology*. Prentice-Hall; London and New Jersey.

Odum, E. P. (1975) *Ecology*. Holt, Rinehart and Winston; London and New York.

Pimm, S. L. (1982) *Food Webs*. Chapman and Hall; London.

Smith, R. L. (1980) *Ecology and Field Biology*. Harper and Row; London and New York.

Swift, M. J., Heal, O. W. and Anderson, J. M. (1979) *Decomposition in Terrestrial Ecosystems*. Blackwell; Oxford.

Whittaker, R. H. (1975) *Communities and Ecosystems*. Macmillan; London and New York.

Chapter 30 The flow of materials within ecosystems: biogeochemical cycles 1

All ecosystems, including the fuel-powered ecosystems described in section 28.5, function and are maintained by energy flow (Ch. 29). Equally important, however, and closely related to energy flow, is the *circulation of materials* within ecosystems. While the earth has a continuous source of energy from the sun, its source of materials (apart from minor inputs from the solar system in the form of meteorites, etc.) is constant and finite from the earth itself. Since the earth originated, these materials (such as carbon, nitrogen, oxygen, phosphorus and sulphur) have been cycled and recycled via the climatological, hydrological, soil and sediment transfer systems described elsewhere in this book and, since the evolution of life, transfer has also been effected via ecosystems. The processes of material or nutrient cycling are called *biogeochemical cycles*, a term which illustrates the fact that interchanges of the elements and compounds in materials occur between both the biotic or organic components (e.g. plants and animals) and abiotic or inorganic components (e.g. bedrock, soils and atmosphere) of ecosystems.

The relationship between the flow of energy and the biogeochemical cycles is complex since each influences the other and both determine ecosystem characteristics, such as the abundance of organisms, the plant and animal species present and productivity (sections 29.3 and 29.4). In particular, specific elements and compounds are essential for energy fixation and its subsequent flow through an ecosystem (e.g. carbon dioxide and water). On the other hand, the transport of elements and compounds (e.g. nitrates and water) from the soil to plants cannot be brought about without the expenditure of energy on the part of plants. This two-way relationship is an example of feedback within ecosystems (section 28.4) and, if the situation is one of positive feedback, the more nutrients that are available the greater the energy flow, which in turn results in increased mineral cycling. In this case, during the process of vegetation succession (Ch. 32) the ecosystem will change from one type, having a set of specific characteristics and functions, to another type of ecosystem, with a different set of characteristics and functions. In section 28.4, the example of heathland maintenance by fire was given. If fire is eliminated, resulting in a different distribution of materials from the biotic (the plants) to the abiotic (soil) components of the ecosystem, a different ecosystem will emerge. Conversely, when negative feedback exists in an ecosystem, available nutrients may be limited by the nature of such factors as bedrock, overlying soils and atmospheric input. This limits the abundance of organisms, their capacity to fix energy and sustain higher trophic levels and consequently their ability to circulate nutrients. The negative feedback mechanism maintains the character of the ecosystem while the positive feedback mechanism stimulates change.

Furthermore, there has been a recent increase in studies of biogeochemical cycles providing a wealth of information on material fluxes at levels of investigation ranging from the global cycling of nutrients to the role of individual species. This has come about due to the development of radioactive tracers, the advent of sophisticated numerical and computer techniques for modelling and, in particular, because an understanding of such processes provides the key to many contemporary agricultural and pollution problems. It is impossible, in this text, to provide a comprehensive review of the available data but this and the following chapter summarise recent progress and applications.

30.1 Model of nutrient flow

Figure 30.1 represents a general model of *nutrient flow* and illustrates the relationship between the flow of materials and components of the ecosystem responsible for energy flow. The autotrophic components depend not only on the supply of solar energy but also on the availability of nutrients which, in turn, is related to inputs into the detritus and the rate at which nutrients are released by decomposition processes. In addition, the storage component of the ecosystem is important since this represents the reserve of organic material and nutrients present in the soil, litter and standing crop. Usually, the larger the storage component the more stable the ecosystem is, counteracting positive feedback. The field characteristics of any ecosystem are a reflection of the distribution and movement of organic matter and nutrients between these various components.

Examples of energy flow in different ecosystems are given in section 29.6. In general, the flow of energy through aquatic ecosystems, where the grazing pathway is predominant, is very different from that in terrestrial environments where the detrital pathway dominates. Similarly, nutrient utilisation is different since the flow of energy is intricately linked with the flow of materials. For example, in aquatic ecosystems there is a variety of autotroph populations usually with low biomass and with different responses to environmental conditions, such as temperature. Given optimal environmental conditions, individual populations respond rapidly with high reproductive rates and associated increased utilisation of nutrients. On an annual basis, different populations will respond as conditions change, resulting overall in a continuous supply of energy and nutrient uptake. In a terrestrial ecosystem, such as a forest, the situation is different. Here, the autotroph population consists of large individuals which, in comparison to the algae of the aquatic ecosystem, reproduce slowly. Furthermore, they contain a large store of energy and nutrients in the biomass, enabling them to survive unfavourable conditions within certain limits.

Both types of ecosystems, however, contain reserves of organic matter which are very important for nutrient conservation and hence ecosystem maintenance. In the aquatic ecosystem, this reserve occurs in particulate and dissolved forms while in the terrestrial ecosystem, it occurs as soil organic matter and in the standing crop of the autotrophs themselves. This is a very important attribute of all ecosystems since the organic component has a vital role in recycling nutrients and prevents rapid large-scale losses out of the system. This is because large amounts of nutrients form part of all types of organic matter and, consequently, they are not readily available unless released by leaching or decomposer organisms. This

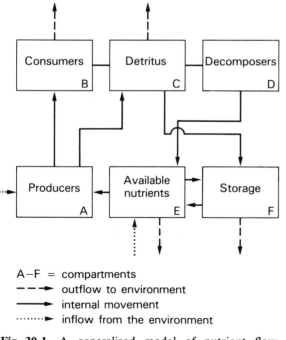

A–F = compartments
– – → outflow to environment
——→ internal movement
········► inflow from the environment

Fig. 30.1 A generalised model of nutrient flow within an ecosystem. (Adapted from Webster *et al* 1975)

again illustrates the close interrelationships which exist between the components of any ecosystem. If the standing crop is limited by any factor, for example water supply, then nutrient recycling is also limited.

Furthermore, most ecosystems have time-dependent nutrient-recycling mechanisms, as exemplified in the forest ecosystem where leaves, constituting the short-term cycle, are recycled much more rapidly than wood which represents the long-term cycle. Nutrient cycling is also influenced by the species present in an ecosystem; for example, deciduous species are more active in mineral cycling than conifers (especially pines) under similar environmental conditions. Duvigneaud and Denaeyer – De Smet (1970) have shown that in deciduous forests, the annual return of K, Ca, N and P to the litter is respectively 10, 81, 40 and 10 kg ha^{-1} while in coniferous forest, the values are 4, 19, 35 and 3 kg ha^{-1}. This topic will be discussed further in section 31.1.

30.2 General characteristics of biogeochemical cycles

Biogeochemical cycles have two major attributes, viz. *pools* and *flux rates*. A biogeochemical pool is a quantity of any chemical substance in either a biotic or abiotic component of an ecosystem. The atmosphere, for example, contains a large pool of carbon as the gas carbon dioxide while the ocean waters contain carbon as dissolved bicarbonate. Furthermore, calcareous rocks contain carbon as particulate carbonate and living organisms contain it in the form of organic matter. Exchange between these pools is brought about by processes such as erosion and photosynthesis and the rate at which this transfer is effected is the flux rate, which is defined as the quantity of material passing from one pool to another (per unit time and per unit area or volume of the system). *Turnover rates* and *turnover times* are also terms used to describe the relative importance of a flux process in relation to the pools involved. The former is the flux rate into or out of a pool, divided by the quantity of nutrient in the pool, and allows the importance of the flux process to be determined in relation to pool size. The quantity of nutrient in the pool, divided by the flux rate, is the turnover time and provides an indication of the time necessary for movement of a quantity of nutrient equal to that in the pool.

Thus, the transfer of nutrients from one pool to another is cyclical and, as such, is quite distinct from energy flow which is a one-way process whereby energy enters an ecosystem as solar energy, is converted to chemical or food energy and ultimately leaves as heat energy unavailable for use by living organisms. In contrast, the same atoms of an element may be constantly fluxed from one pool to another using kinetic energy, as in the case of an inorganic flux process such as erosion, or food energy if the flux is brought about by living organisms. A further characteristic of biogeochemical cycles is also important, especially if one is to consider the effect that modern technology and industrialisation must have had on the flow of materials on a global scale. Many workers consider that the major biogeochemical cycles, prior to the introduction of an industrial society, must have been in a dynamic but balanced or steady-state condition. This is in keeping with natural processes and the systems approach in general whereby inputs balance outputs and, as a result, the general characteristics of the ecosystem are maintained (see climax vegetation, section 32.3).

30.3 The classification of nutrients and biogeochemical cycles

Table 30.1 lists the elements necessary for life and classifies them into three groups, according to their importance as structural parts of organic matter. Other workers, however, identify only *macronutrients* and *micronutrients* (e.g. Odum 1975). The former, which includes the major constituent group of Anderson (1981), are those elements and compounds that are required in large quantities by living organisms and are major constituents of protoplasm. The micronutrients include elements and compounds essential to the functioning of living systems, but which are needed in only small quantities. In addition, there are two further groups of substances which circulate between the biotic and abiotic parts of ecosystems. The first group are elements and their compounds which have no known biological function, whereas the second group are man-made compounds, such as pesticides and radionuclides (for discussion see section 31.4).

As far as the biogeochemical cycles of these nutrients are concerned, they have been classified in two ways. Most workers classify them according

Table 30.1 Elements necessary for life, and some of their functions. (After Anderson 1981)

Category	Element	Symbol	Some known functions
Major constituents (20–60 atoms %)	Hydrogen Carbon Oxygen	H C O	Universally required for organic compounds of cell
	Nitrogen	N	Essential constituents of proteins and amino acids
	Sodium	Na	Important counter-ion involved in nerve action potentials
	Magnesium Phosphorus	Mg P	Cofactor of many enzymes, e.g. chlorophyll Universally involved in energy transfer reactions and in nucleic acids
Macronutrients (0.02–2 atoms %)	Sulphur	S	Found in proteins and other important substances
	Chlorine	Cl	One of major anions
	Potassium	K	Important counter-ion involved in nerve conduction, muscle contraction, etc.
	Calcium	Ca	Cofactor in enzymes, important constituent of membranes and regulator of membrane activity
	Boron	B	Important in plants, probably as cofactor of enzymes
	Silicon	Si	Found abundantly in many lower forms such as diatoms
	Vanadium	V	Found in respiratory pigments of some lower animals
	Manganese	Mn	Cofactor of many enzymes
Micronutrients (trace elements) (< 0.001 atoms %)	Iron	Fe	Cofactor of many oxidative enzymes, e.g. haemoglobin
	Cobalt	Co	Constituent of vitamin B_{12}, required for N fixation
	Copper	Cu	Cofactor of many oxidative enzymes
	Zinc	Zn	Cofactor of many enzymes, e.g. insulin
	Molybdenum	Mo	Cofactor of a few enzymes, particularly nitrogenase (N fixation)
	Iodine	I	Constituent of thyroid hormones which regulate growth and development in vertebrates

to whether they are *gaseous*, associated chiefly with the atmosphere and ocean, or *sedimentary*, where their main pools are associated with the soil and rocks of the lithosphere. On a more functional basis, Collier *et al.* (1973) have described a spectrum between *perfect* and *imperfect biogeochemical cycles*, relating to the degree of regulation or the efficiency of negative feedback controls. At one end of the spectrum, the perfect biogeochemical cycles have a large abiotic pool of the substance and many negative feedback controls. The carbon (C), nitrogen (N) and oxygen (O) cycles are examples where the large abiotic pool is present in the atmosphere and metabolic activity is required to flux the substance from the abiotic

to the biotic pool. In all three cases, the flux is often mediated via organisms and the negative feedback control is manifested in changing population numbers. Imperfect biogeochemical cycles include many which are sedimentary in nature and frequently lack the direct control of populations of organisms acting in a negative feedback manner (e.g. the phosphorus cycle, section 30.5).

30.4 Gaseous biogeochemical cycles

The cycles of carbon, oxygen, nitrogen and hydrogen are fundamental to all life on earth. The first three are detailed below, but the hydrogen cycle is excluded since it is intricately associated with the hydrological cycle discussed in section 2.1.

(a) The oxygen cycle

Oxygen is present almost everywhere on earth. It is chemically very reactive, being able to combine with a wide range of other elements e.g. with hydrogen to form water (H_2O); with carbon to form carbon monoxide (CO) and carbon dioxide (CO_2); with sulphur to form sulphur dioxide (SO_2); with nitrogen to form nitrites and nitrates (NO_2^-, NO_3^-) and with phosphorus to form phosphates (PO_4^{3-}). It is also an important constituent of organic substances such as sugars, starches and cellulose and is present in almost all rocks and minerals. Because of the high degree of chemical reactivity of

oxygen and its presence in a wide range of biotic and abiotic ecosystem components, its biogeochemical cycling is extremely complex. It is intricately associated with hydrogen in the hydrological cycle (section 2.1) and with carbon and the carbon cycle (see below) and is also important in weathering processes (section 16.2) and sedimentary cycles. Some of these relationships are given in Fig. 30.2.

The two main pools of free oxygen (i.e. oxygen in its elemental state) are in the atmosphere, which contains about 0.8×10^{15} t (Johnson 1970), with the percentage composition averaging 20.946% for dry pure air, and in the oceans which contain 0.2×10^{15} t. These pools have been produced and are being renewed by two important processes. Firstly, photodissociation of water vapour in the upper atmosphere which results in oxygen release as water molecules are split by

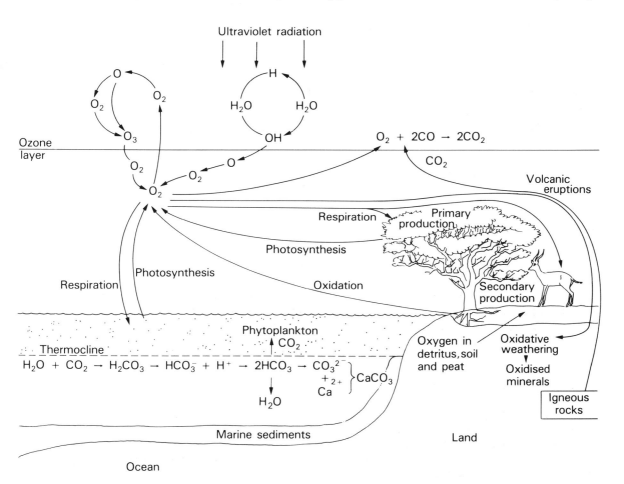

Fig. 30.2 A simplified diagram of the oxygen cycle. Note the interplay between the oxygen cycle and the carbon cycle. (Adapted from Cloud and Gibor 1970)

light, although Johnson (1970) has suggested that only some 10^{15} t of molecular oxygen have been produced in this way during the history of the earth. Secondly, the process of photosynthesis (section 29.1), whereby energy is fixed and oxygen released, is considered to be the major contributor of molecular oxygen as life evolved over geological time. More oxygen was produced by past ecosystems than was used in respiration, the decay of organic matter or in the oxidation of rocks. Lack of oxidation and the decay of some types of ancient ecosystems are attested by the presence of coal, oil, natural gas, etc., which represent reduced carbon withdrawn from the biogeochemical cycles.

While it is generally difficult to assess global biogeochemical cycles, it is thought that the oxygen cycle is a perfect cycle in a state of dynamic equilibrium, despite anthropogenic activity involving fossil-fuel burning. It has been estimated, for example, that such burning releases 2000 times more carbon per year than is stored, resulting in rapid removal of oxygen from the atmosphere. However, Machta and Hughes (1970) have indicated that for the period 1910–70, this has had no discernible effect on oxygen concentrations. Negative feedback mechanisms in this cycle may be the reason why it has maintained a steady-state condition. Redfield (1958) has suggested that sulphate-reducing bacteria, which occur in anaerobic environments, may operate in such a way.

These chemosynthetic bacteria have the ability to use the sulphate ion as an oxygen source as follows:

$$\underset{\substack{\text{sulphate} \\ \text{ion}}}{SO_4^{2-}} + \underset{\text{carbon}}{2C} \rightarrow \underset{\substack{\text{carbon} \\ \text{dioxide}}}{2CO_2} + \underset{\substack{\text{sulphide} \\ \text{ion}}}{S^{2-}} \quad [30.1]$$

Free oxygen is not released in this process but organic matter is decomposed and nutrients are released, such as nitrates, phosphates and carbon dioxide. The latter may be subsequently fluxed to a situation where it may be utilised in photosynthesis and thus result in oxygen release. Since these bacteria survive only in anaerobic habitats, any change in atmospheric oxygen content would reflect the abundance of such habitats, and consequently bacteria populations, which in turn would regulate oxygen supply. Similar regulatory controls may be brought about by increased carbon dioxide concentrations in the atmosphere which may promote increased photosynthesis and, again, an increased production of molecular oxygen. However, anthropogenic activity in the form of large-scale deforestation and pollution may reduce photosynthesis and lead to an imbalance in both the carbon and oxygen cycles.

(b) The carbon cycle

As stated earlier in this chapter, the flow of energy and the cycling of materials in ecosystems are very closely linked and this interrelationship is particularly well illustrated by the carbon cycle. Nearly 50% of organic matter (by dry weight) consists of carbon which, in reduced chemical form, is the major means of energy transfer through living systems (Ch. 29), predominantly through the processes of photosynthesis, respiration and decomposition. On a longer time-scale, geological factors are also important in the carbon cycle. For example, the storage of large quantities of carbon (i.e. coal, oil and gas) in sedimentary rocks or as calcium and magnesium carbonates, represents carbon fixed by past ecosystems and subsequently withdrawn from the actively circulating part of the biosphere. The main features of the carbon cycle are given in Fig. 30.3 which shows that the main readily available pools of carbon are in the oceans (i.e. the largest pool, where carbon dioxide is in solution or occurs as carbonate and bicarbonate ions), in the atmosphere (where it occurs as carbon dioxide) and in living and dead organic matter. It is considered that the atmospheric and oceanic pools are in equilibrium with each other, with an annual exchange of about 100×10^9 t of carbon (Bolin 1970) and with a 3-year mean residence time for a gaseous carbon atom in either pool (Etherington 1982), compared with a 15–20-year residence in living terrestrial vegetation (Anderson 1981). Carbon cycling in the oceans is almost a closed system with exchange occurring at the water surface. It is determined by the reaction between the atmospheric carbon dioxide with the carbonate ions present in sea water to form bicarbonates, as follows:

$$\underset{\substack{\text{carbon} \\ \text{dioxide}}}{CO_2} + \underset{\substack{\text{carbonate} \\ \text{ion}}}{CO_3^-} + \underset{\text{water}}{H_2O} \rightleftharpoons \underset{\substack{\text{bicarbonate} \\ \text{ion}}}{2HCO_3^-} \quad [30.2]$$

Once in sea water, the circulation of carbon occurs physically (via currents) and biologically via assimilation by phytoplankton and subsequent passage along food webs. Figure 30.3 shows a variation of carbon concentrations between the ocean surface and the ocean deeps, separated by the thermocline (Ch. 34) which is a layer between the upper warmer and lower colder zones. Some mix

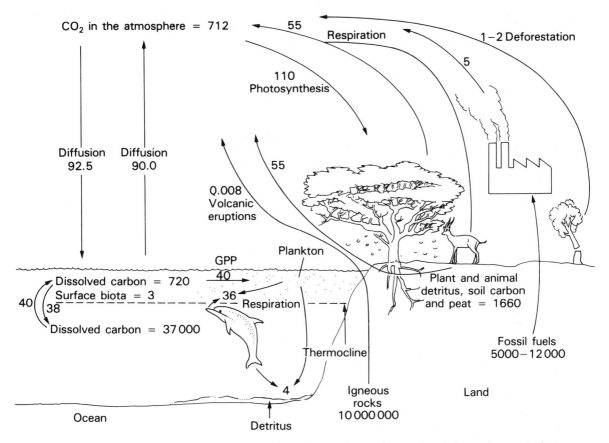

Fig. 30.3 The global carbon cycle. The major fluxes (arrows) are 10^9 t yr^{-1} and the major pools (compartments) are 10^9 t. (Data based on Baes *et al.* 1977; Bowen 1979). Note: many values are estimates

by eddying occurs between the two zones and between 10 and 20% particulate matter sinks to the ocean bottom, where about 15% of the calcium carbonate is incorporated into deep sediments. Carbon incorporated into sediments has a residence time of about 10^8 years while that in the water is about 10^5 years. Thus in 100 000 years, there is complete replacement (Baes *et al.* 1977).

In terrestrial environments, there are similar slow and fast fluxes of carbon occurring. Some 1456×10^9 t in humus and recent peat, and 600×10^9 t in large stems/roots, form the slowly fluxing carbon pools, while 160×10^9 t are rapidly fluxing. It is generally considered that fluxes between terrestrial carbon pools and those in the atmosphere are in equilibrium with the amount of carbon (approximately 113×10^9 t) removed by photosynthesis, equalling the amount replaced by plant decomposition and respiration. The main users of atmospheric carbon dioxide are the most productive biomes (i.e. the forests), which are also

the chief terrestrial pool of organic carbon. The other main terrestrial pool of carbon is detrital carbon contained on, and in, the soil. Schlesinger (1977) has estimated that this latter value is 1456×10^9 t compared with 826×10^9 t for total world terrestrial biomass. These figures illustrate the importance of the detrital pathway of energy exchange detailed in section 29.6.

On a global scale, carbon exchange (especially that reflected in productivity and respiration) will vary seasonally and this is illustrated by seasonal changes in the carbon dioxide content of the atmosphere in the northern hemisphere. The mean concentration of atmospheric carbon dioxide is approximately 320 ppm, but fixation of carbon dioxide during the summer exceeds its return to the atmosphere by respiration and decomposition, while the reverse is true in winter. This seasonal variation may be as much as 20 ppm (Bolin 1970) and similar variations may occur on a diurnal basis. For example, at daylight, photosynthesis begins removing carbon dioxide from the air so that the concentration falls sharply. However, in the afternoon, the process begins to be reversed as respiration increases and at sunset, photosynthesis decreases and carbon dioxide concentrations sharply increase.

Despite these variations on a diurnal and seasonal time-scale, the overall production to respiration ratio is approximately 1, reflecting the equilibrium state mentioned earlier in this section, and is partly maintained by the slow flux of carbon which occurs between the hydrosphere (especially the oceans) and the atmosphere. The equation quoted above is a reversible one and can be expanded, as follows:

$$CO_2 + H_2O \rightleftharpoons H_2CO_3 \rightleftharpoons H^+ +$$

carbon water carbonic hydrogen
dioxide acid ions

$$HCO_3^- \rightleftharpoons 2H^+ + CO_3^{2-} \qquad [30.3]$$

bicarbonate hydrogen carbonate
ions

The direction of the reaction depends essentially on the relative concentration of the components. For example, if atmospheric carbon dioxide is depleted by terrestrial autotrophs (or if bicarbonate levels in the ocean are reduced by phytoplankton), a series of regulatory or homeostatic reactions will occur over a time-span of several years, which restores equilibrium.

There is now, however, evidence to show that anthropogenic activities, especially fossil-fuel burning and large-scale deforestation, may have resulted in some instability within this self-regulating system. Raiswell et al. (1980) have shown that, between 1958 and 1976, the mean annual concentration of carbon dioxide in the atmosphere has risen from 316 to 332 ppm. Of the 5×10^9 t of carbon injected into the atmosphere by fossil-fuel burning, approximately 50% is retained in the atmosphere (Stuiver 1978), almost half dissolves in the oceans and a small percentage is added to land biomass (Bacastow and Keeling 1973). As far as the oceanic 'sink' is concerned, 34% of excess carbon dioxide is stored in surface waters and 13% is carried to the deeps. Computer models, however, suggest that in a century from now, between 46 and 80% of the excess carbon dioxide will remain in the atmosphere due to saturation of ocean waters (Siegenthaler and Oeschger 1978) and their consequent impairment as a negative feedback control in the carbon biogeochemical cycle. Further complications arise if the terrestrial 'sink' is considered. Autotrophs do remove carbon

dioxide from the atmosphere but in doing so, they respire and thus inject carbon dioxide into the atmosphere. In addition, the problem is compounded by the rapid rate of clearance of the world's forests, especially the tropical rainforests which contain 72% of the world's organic matter in the standing crop and soil organic matter. These forests are being cleared at the rate of 11 million ha yr^{-1} and it has been estimated that the consequent burning of timber and soil aeration has resulted in a release of carbon which is double that from fossil fuels (Woodwell et al. 1978).

These trends, especially that of increased atmospheric content of carbon dioxide, may have considerable climatic implications. Carbon dioxide allows incoming short-wave solar radiation to penetrate through the atmosphere but absorbs outgoing long-wave radiation, which is returned to earth in the so-called 'greenhouse effect' (section 2.2). Thus, increased carbon dioxide levels would be expected to result in an increase in global temperature and a rise in sea-level, as polar ice caps melt (section 36.1).

(c) The nitrogen cycle

Nitrogen is an important constituent of organic matter, especially proteins which are present in all living organisms. Furthermore, the nitrogen cycle is one of the more complex nutrient cycles because so many of the fluxes between pools are brought about by specific organisms in complicated chemical reactions. The various components and estimates of pool and flux size are given in Fig. 30.4. The atmosphere consists of approximately 79% nitrogen and this constitutes the main pool of the nutrient although, in its gaseous elemental form, it is unavailable to the majority of organisms. The processes of converting this gaseous nitrogen to an available form for assimilation in the biosphere constitutes a major part of the nitrogen cycle. The first stage in nitrogen flux from the atmospheric pool is via a fixation process and, apart from a small amount (about 10%) fixed by lightning, cosmic radiation or meteorite trails, most fixation of nitrogen occurs biologically. In this latter process, nitrogen (N_2) is split into two atoms thus: $N_2 \rightarrow 2N$ and for every 28 g of nitrogen, this step requires 668.8 kJ of energy. The free nitrogen atoms produced are then combined with hydrogen to form ammonia, with a release of about 54.3 kJ of energy, as follows:

$$2N + 3H_2 \rightarrow 2NH_3 \qquad [30.4]$$

(nitrogen) (hydrogen) (ammonia)

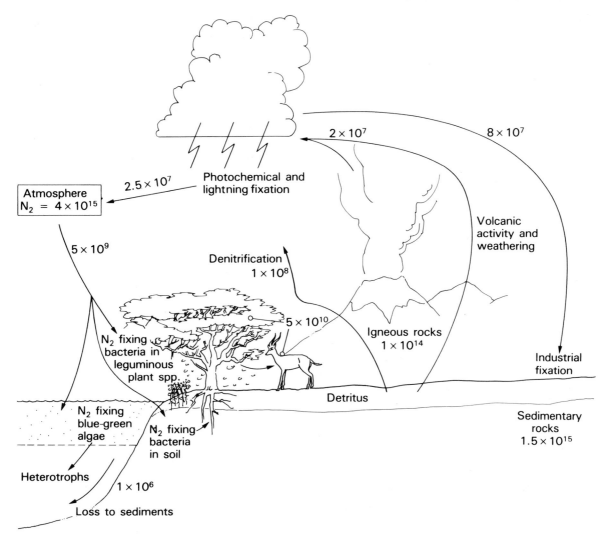

Fig. 30.4 A simplified diagram of the global nitrogen cycle. The major fluxes (arrows) are 10^9 t yr^{-1} and the major pools (compartments) are 10^9 t (Based on data in Holland 1978; Bowen 1979). Note: many values are estimates

There are three groups of species responsible for this fixation, viz. bacteria living in a symbiotic relationship with leguminous plants and in the root nodules of non-leguminous plants, free-living aerobic bacteria and blue-green algae. According to Smith (1980), there are some 200 species of nodulated legumes which, via their symbiotic bacteria, are the chief nitrogen fixers in agricultural systems. Furthermore, there are 12 000 species of nodule-living and free-living bacteria and blue-green algae which are responsible for nitrogen fixation in non-agricultural ecosystems. In symbiotic systems (e.g. those which occur in leguminous plants, such as clover), the bacterial genus re-

sponsible for nitrogen fixation is *Rhizobium* which inhabits the rhizosphere, the immediate surroundings of the plant roots. It eventually enters the plant roots in response to secretions, forming nodules and undertaking nitrogen fixation, which it makes available to the host as ammonia. A similar process occurs with non-leguminous nodule-bearing plants such as alder (*Alnus*). As far as free-living soil bacteria are concerned, there are 15 genera known to fix nitrogen of which the most common are Azotobacteriaceae (e.g. the aerobic *Azotobacter*), *Clostridium* and *Desulphovibrio* spp. plus some *Bacillus* spp. and *Enterobacter* spp. which fix nitrogen anaerobically. Of the blue-green algae which fix nitrogen, the genera *Nostoc* and *Calothrix* are the most common, occurring in soil and aquatic habitats often as pioneer species. Once nitrogen has been fixed to produce ammonia, it can be utilised by the plants to form proteins which can then be passed along food webs to herbivores and carnivores, or to the decomposers.

This latter group of organisms can break down organic matter by ammonification, nitrification and denitrification, all of which also make nitrogen compounds available to autotrophs. Excretory products and dead organic matter are broken down to amino acids by heterotrophic bacteria and fungi and in ammonification, these amino acids are oxidised by decomposer organisms to produce ammonia and energy, as follows:

$$CH_2NH_2COOH + 1\tfrac{1}{2}O_2 \rightarrow 2CO_2 +$$
amino acid $\quad\quad$ oxygen $\quad\quad$ carbon
$\quad\quad\quad\quad\quad\quad\quad\quad\quad\quad\quad\quad\quad$ dioxide

$$H_2O + NH_3 + 735 \text{ kJ} \quad\quad\quad [30.5]$$
water $\quad$ ammonia $\quad$ energy

The ammonia produced may then be utilised by plants to synthesise proteins or alternatively, it may undergo nitrification which is a process whereby ammonia is oxidised to nitrite and nitrate by micro-organisms, and energy is released. Two groups of bacteria are responsible for these processes. Firstly, *Nitrosomonas*, which oxidise ammonia to nitrate in order to produce energy, thus:

$$NH_3 + 1\tfrac{1}{2}O_2 \rightarrow HNO_2$$
ammonia $\quad$ oxygen $\quad$ nitrous
$\quad\quad\quad\quad\quad\quad\quad\quad\quad$ acid

$$+ H_2O + 689.7 \text{ kJ}$$
$\quad$ water $\quad\quad$ energy $\quad\quad\quad\quad [30.6]$
and $HNO_2 \rightarrow H^+ + NO_2^-$

Secondly, the *Nitrobacter* group of bacteria can then oxidise the nitrate ions in another energy-releasing reaction to form nitrate, as follows:

$$NO_2^- + 1\tfrac{1}{2}O_2 \rightarrow NO_3^- + 75.24 \text{ kJ}$$
nitrite $\quad$ oxygen $\quad$ nitrate $\quad$ energy
ion $\quad\quad\quad\quad\quad\quad$ ion $\quad\quad\quad\quad [30.7]$

In general, these processes are beneficial to ecosystems by making nitrogen compounds available to plants and subsequently to animals, but they can also lead to losses of nitrogen since nitrates are more soluble and thus more easily leached out of the system. In an extreme situation, where large amounts of excess nitrates are produced, they may be leached from the soil to concentrate in lakes and rivers causing pollution and this situation will be discussed below. Nitrates, however, provide the substrate for the process of denitrification whereby nitrates are reduced to gaseous nitrogen by fungi and the bacteria *Pseudomonas*. In an oxygen-limited environment, they can break down nitrate in the following reaction:

$$C_6H_{12}O_6 + 4NO_3^- \rightarrow 6CO_2 +$$
carbohydrate $\quad$ nitrate $\quad\quad$ carbon
$\quad\quad\quad\quad\quad\quad$ ion $\quad\quad\quad$ dioxide

$$6H_2O + 2N_2$$
water $\quad$ nitrogen $\quad\quad\quad\quad\quad [30.8]$

As a consequence, nitrogen may be lost from the ecosystem as it passes back into the atmosphere.

On a global basis, the quantities of nitrogen in the various components of the biosphere can only be estimated (Fig. 30.4). Postgate and Hill (1979), for example, estimate that 126 million t yr^{-1} are biologically fixed with a further 26 million t yr^{-1} fixed by atmospheric processes. Ehrlich *et al.* (1977), however, have estimated that the annual uptake of nitrogen is 35 times higher than fixation rates. This illustrates a high degree of nitrogen conservation, regulation of its cycle and recycling which occurs in natural systems. However, cultural influence is also extant within the global nitrogen cycle. For example, large amounts of nitrogen gases are injected into the atmosphere by burning fossil fuel in various industrial processes and motor-fuel consumption. This has led to increased concentrations of nitric acid in the atmosphere, which eventually reaches the earth's surface in precipitation. It causes the phenomenon known as acid rain which is also associated with sulphur dioxide emissions from fossil-fuel burning (section 5.1).

Probably by far the most important cultural input for the nitrogen cycle is the application of nitrogenous fertilisers (produced by industrial fixation of atmospheric nitrogen) in agricultural systems. Only about one-third of the fertiliser applied to soils is actually assimilated by the crop, the remainder is leached from the system into lakes and rivers causing pollution. This is associated with high phosphate levels (section 30.5) which also contribute to the process of *cultural eutrophication*. Excessive nitrate levels in freshwater bodies have deleterious effects on the ecosystem, and nitrate levels in domestic water supply, especially in agricultural areas, are often in excess of World Health Organisation maximum limits and may cause gastric cancer and methaemoglobinaemia (i.e. a serious blood disorder) in babies. Essentially, the homeostatic balance between nitrogen fixation and denitrification has been disrupted and the problem is accentuated by the animal and human wastes containing large amounts of nitrates, which may also contaminate water bodies.

30.5 Sedimentary biogeochemical cycles

Living organisms require a variety of nutrients of an inorganic nature (Table 30.1) which are derived from abiotic sources, usually originating from rocks. Mineral salts are derived from rocks by weathering (section 16.2) and those which are soluble enter the hydrological cycle. Autotrophs obtain their minerals in solution from salt, fresh or soil water while heterotrophs obtain their minerals either from solution, autotrophs or other heterotrophs. These minerals are returned to the soil and water by decomposer organisms. Every element has its own biogeochemical cycle, some wholly sedimentary, such as the phosphorus cycle, while others involve a gaseous phase.

The phosphorus cycle

Phosphorus is an essential nutrient for all living organisms and, in many ecosystems, it is often a limiting factor to growth chiefly because it lacks an atmospheric pool and its biogeochemical cycle is linked closely with sediment systems (Ch. 16) and long-term geological cycles. Figure 30.5 illustrates the global phosphorus cycle. It shows that the main pools of phosphorus occur in rocks from which it is released by weathering, leaching, erosion and mining. A small pool also occurs in dead organic matter from which it is released by decomposer organisms. Phosphorus, in terrestrial ecosystems, is made available to higher plants as inorganic phosphate, is incorporated into plant tissues as organic phosphates and is subsequently passed on to herbivores and carnivores. On passing to the decomposer system, organic phosphates are reduced to inorganic phosphates which again become available for autotroph assimilation. In the marine environment, phosphorus compounds are contained in both biotic and abiotic components. These include the phytoplankton, inorganic phosphates, organic phosphorus compounds and a soluble macromolecular colloidal phosphorus. However, the most important flux of phosphorus

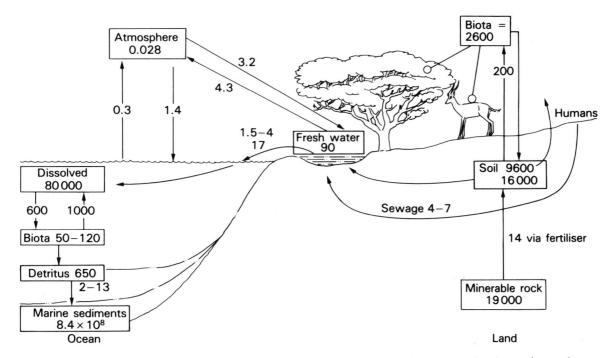

Fig. 30.5 A diagrammatic representation of the global phosphorus cycle. The major fluxes (arrows) are 10^6 t yr^{-1} and the major pools (compartments) are 10^6 t. (Adapted from Richey 1983)

occurs between the inorganic phosphate and the particulate material. The organic fraction readily converts to the colloidal type, both releasing phosphorus to the soluble inorganic component and a rapid recycling occurs through the plankton.

Thus, in both terrestrial and aquatic ecosystems, phosphorus is efficiently conserved and recycled but overall there is a gradual loss of phosphorus from the system, as particulate matter sediments from the terrestrial environment to the ocean floor. On a human time-scale, it is effectively lost from the system because diagenetic processes (incorporating it into sedimentary rock) may take many thousands or millions of years and an even greater time may elapse before uplift occurs. As a consequence, phosphorus becomes progressively scarcer in the biosphere. In common with many other biogeochemical cycles, human activities have caused disequilibrium to occur in the phosphorus cycle. In particular, large quantities of phosphate fertilisers are applied to agricultural systems (in common with nitrate fertilisers, described in section 30.4(c)) to increase productivity. Much of this is immobilised as insoluble salts, as the phosphate combines with calcium, ammonium and iron in the soil, and thus becomes unavailable for biological assimilation. In addition, organic phosphates (derived from fertilisers) are removed from the system by harvesting and are eventually released as human or animal sewage or as waste from food processing plants. Primary sewage treatment removes only 10% of this phosphorus and secondary treatment removes only a further 30%. Most of the remaining 60% enters the rivers and lakes, causing cultural eutrophication (section 30.4) which leads to an imbalance in these ecosystems and excessive plant growth, especially unicellular algae. These can multiply rapidly, forming blooms over water surfaces which deplete both the oxygen and nitrogen supplies within the water, causing death to many aquatic plants and animals. In addition, once the nitrogen supply has been depleted, nitrogen-fixing blue-green algae proliferate and produce metabolites that are toxic to a variety of animals, such as cattle, fish and water-birds.

Thus, the disruption of the phosphorus biogeochemical cycle by human activity has two major effects. Firstly, cultural eutrophication, which is occurring on a large scale in densely populated parts of the world, causes major changes in freshwater aquatic ecosystems. Secondly, at the same time, mining of phosphate rocks for fertilisers is leading to an increased loss of phosphorus from the biosphere to non-recoverable concentrations in marine sediments. Essentially, the flux of phosphorus from the biosphere to the ocean has been greatly accelerated and, furthermore, the flux in the reverse direction (which takes the form of guano and harvested fish) in no way counteracts it.

Key topics

1. The relationship between energy flows and biogeochemical cycles.
2. Transfer processes and storage in gaseous and sedimentary biogeochemical cycles.
3. The significance of pools, fluxes and controls in the major biogeochemical cycles.
4. The effect of anthropogenic intrusions on biogeochemical cycles.

Further reading

Biosphere (1970) Scientific American Books, W. H Freeman & Co.; San Francisco.

Bolin, B. and Cook, R. B. (eds) (1983) *The Major Biogeochemical Cycles and their Interactions.* John Wiley and Sons; New York and London.

Furley, P. A. and Newey, W. W. (1982) *The Geography of the Biosphere.* Cambridge University Press; Cambridge.

Hallberg, R. (ed.) (1983) Environmental biogeochemistry, *Ecol. Bull.* (Swedish Res. Council), **35**.

Post, W. M., Emmanuel, W. R., Sinka, P. J. and Stangenberger, A. G. (1982) Soil carbon pools and world life zones, *Nature*, **298**, 156–9.

Revelle, R. (1982) Carbon dioxide and world climate, *Scient. Amer.*, **247**, 33–41.

Spiedel, D. H. and Agnew, A. F. (1982) *The Natural Geochemistry of Our Environment.* W. Dawson and Sons Ltd, Folkestone; Westview Press, Boulder, Colorado.

Chapter 31 The flow of materials within ecosystems: biogeochemical cycles 2

In the previous chapter, the biogeochemical cycling of some of the most important elements was described at the global level. This chapter is concerned with biogeochemical cycling at the ecosystem, biome, plant community and species levels. In addition, the biogeochemical cycling of some pollutants and radionuclides is included, to illustrate some of the effects of industrial and high-technology societies on the environment.

31.1 Biogeochemical cycling at the ecosystem level

An alternative approach to examining biogeochemical cycling at the level of the biosphere, in terms of individual cycles, is to examine the nutrient budgets of individual ecosystems. Such a study requires an examination of the processes affecting the quantity of nutrients in circulation or in available pools. The *nutrient capital* of an ecosystem is the total quantity of nutrients in both biotic and available abiotic pools. In the case of a mature ecosystem, the nutrient capital will be relatively stable and in a state of dynamic equilibrium. However, in a developing ecosystem undergoing succession (Ch. 32) towards maturity, the nutrient capital will be constantly changing. The same is true of the *nutrient budget* of an ecosystem, which is a measure of the input and output of materials via the various ecosystem components. The major inputs and outputs of nutrients in any ecosystem are summarised in Table 31.1. While the inputs and outputs due to direct human intervention have been retained as a separate category, it should be borne in mind that both direct and indirect human activity may also affect inputs and outputs ascribed to natural routes. In addition, an understanding of biogeochemical cycling must involve an analysis of the processes of transfer between components within the ecosystem, as well as the fluxes across the boundaries.

Three major techniques have been used to quantify flux rates of nutrients in current studies. These include direct measurement of biotic and abiotic processes and characteristics, for example precipitation (Ch. 5), inflow and outflow of water (Ch. 7) and productivity and biomass (Ch. 29), combined with measurement of nutrient content. Secondly, measurement by difference may also be undertaken to estimate flux rates. For example, if inputs and outputs across the boundaries of a terrestrial ecosystem are known, together with the rate of change in total quantity of the nutrient in active pools, then it is possible to estimate the rate of addition of the nutrient by weathering processes. Finally, flux rates may be determined by using *radioactive tracers* but only if radioisotopes of the nutrient are available or when, as in the case of ^{137}Cs, it behaves in an almost identical way

Table 31.1 The major routes of nutrient inputs and outputs in ecosystems. (Based partly on Collier *et al.* 1973)

	Inputs	*Outputs*
Natural routes	Fixation from the atmosphere	Release to the atmosphere
	Weathering of substrate	Loss by leaching or erosion
	Precipitation	Runoff
	Particulate fall-out from the atmosphere	Wind erosion
	Immigration of animals	Emigration of animals
Routes caused by human activity	Fertilisation	Harvest
	Energy due to mechanisation	Harvest
	Pesticides	Harvest

to a particular nutrient (in this case potassium). In order to use this technique, a quantity of the radioisotope is introduced into a particular nutrient pool. Subsequent measurements of its rate of disappearance from this pool, and its appearance in others, facilitate the calculation of both turnover rates and turnover times (section 30.2).

(a) The Hubbard Brook ecosystem

In the last three decades, a great deal of research has been undertaken in a variety of ecosystems using the approaches mentioned above. However, probably the most well-known work is that of Bormann, Likens and co-workers, who have examined a series of small forested ecosystems in the Hubbard Brook Experimental Forest in central New Hampshire, USA. The work is summarised in Likens *et al.* (1977) and Bormann and Likens (1979) and involved an analysis of biogeochemical cycling under natural conditions, as well as that which followed experimental deforestation as the ecosystem recovered. As an example from this work, Table 31.2 gives average annual input

Table 31.2 Average annual input and output (losses) of nutrients for a 55-year-old aggrading forested ecosystem at Hubbard Brook with values in kg ha^{-1} yr^{-1} during the period 1963–74. (After Bormann and Likens 1979)

	Meteorologic input		*Geologic output-streamflow*			
	Bulk precipitation	*Net gas and aerosol*	*Dissolved substances**	*Organic particulate matter**	*Inorganic particulate matter**	*Net loss*
Aluminium	†	†	2.0	†	1.4	3.4
Silicon	†	†	17.6	0.1	6.1	23.8
Calcium	2.2	‡	13.7	0.06	0.17	11.7
Magnesium	0.6	‡	3.1	0.05	0.14	2.7
Potassium	0.9	‡	1.9	0.01	0.51	1.5
Sodium	1.6	‡	7.2	<0.01	0.25	5.9
Iron	†	†	†	0.01	0.63	0.64
Hydrogen ion	0.96	†	0.10	–	–	– 0.86
Phosphorus	0.04	‡	0.01	<0.01	0.01	– 0.02
Nitrogen	6.5	14.2₁§	3.9	0.11	†	−16.7
Sulphur	12.7	6.1§	17.6	0.02	0.01	– 1.2
Chloride	6.2	‡	4.6	†	†	– 1.6

* Dissolved substances and organic particulate matter constitute net output of elements that occur or have occurred in ionic form within the ecosystem: inorganic particulate matter is largely composed of unweathered primary or secondary minerals.

† Nearly zero.

‡ Relatively small.

§ Partly due to biological activity within the ecosystem, i.e. nitrogen fixation, gaseous absorption, and impaction on plant surface.

and output data for nutrients in a 55-year-old aggrading forest ecosystem. These data, for an 11-year period, show that overall nutrient losses were small and that for some elements (such as magnesium and potassium), the geological outputs are considerably offset by meteorological inputs. In addition, other elements (e.g. hydrogen ions, phosphorus and nitrogen) are increasing within the system. For example, there is an annual input of nitrogen of $20 \, kg \, ha^{-1}$ and an output of $4.0 \, kg \, ha^{-1} \, yr^{-1}$, leaving $16.0 \, kg \, ha^{-1} \, yr^{-1}$ to accumulate within the system. Since the weathering bedrock contains only traces of nitrogen, and input from precipitation is $6.5 \, kg \, ha^{-1} \, yr^{-1}$, much of the remainder must be supplied from the net gain of nitrogen fixed biologically over that released

in denitrification. All the elements showing net losses (e.g. aluminium, silicon, calcium, magnesium, potassium and sodium) are probably balanced by inputs from weathering bedrock.

The *calcium cycle* (Fig. 31.1) of the same ecosystem illustrates a sedimentary biogeochemical cycle where $2.2 \, kg \, ha^{-1} \, yr^{-1}$ are contributed in bulk precipitation (Table 31.2), $9.5 \, kg \, ha^{-1} \, yr^{-1}$ are stored in the biomass (Table 31.3) and a further $21.1 \, kg \, ha^{-1} \, yr^{-1}$ are released into the ecosystem by weathering. The remainder ($13.9 \, kg \, ha^{-1} \, yr^{-1}$) is lost from the system in drainage water, with 98.6% of this loss occurring in solution. The annual net loss from the ecosystem of $11.7 \, kg \, ha^{-1} \, yr^{-1}$ is only 2% of the total available calcium within the ecosystem and about

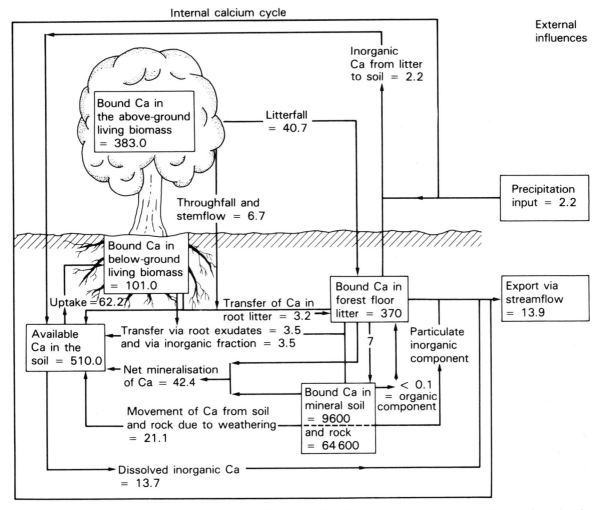

Fig. 31.1 The annual calcium budget for the Hubbard Brook ecosystem, an example of a northern hardwood forest in the USA. The major fluxes are kg ha yr^{-1} and the major pools (compartments) are kg ha^{-1}. (Adapted from Likens *et al.* 1977)

Table 31.3 Average annual net accretion of nutrients in biomass of the aggrading forest ecosystem 55 years after clear-cutting.* (After Bormann and Likens 1979)

| | Nutrients in biomass (kg ha^{-1} yr^{-1}) | | | |
Nutrient	Living biomass†	Dead wood	Dead biomass in forest floor	Total biomass accretion
Calcium	8.1	0	1.4	9.5
Magnesium	0.7	0	0.2	0.9
Sodium	0.15	0	0.02	0.2
Potassium	5.8	0	0.3	6.1
Phosphorus	2.3	0	0.5	2.8
Sulphur	1.2	0	0.8	2.0
Nitrogen	9.0	0	7.7	16.7
Iron	1.5?	0	1.2	2.7?

* Total biomass accretion is the sum of net nutrient accumulation in three biomass subcompartments: living biomass, dead wood and forest floor.
† Living biomass is taken as an average of two pentads (1955–60 and 1960–65).

19% of the amount which is circulated annually by the vegetation. Of its distribution within the ecosystem, more than 99% is present in the soil complex and only 0.5% is present in the vegetation. Therefore, the cycling of the calcium is predominantly within the forest–soil system, indicating a tight intra-system cycle for retaining calcium within the plant–soil components.

Part of the Hubbard Brook experiment also involved an analysis of biogeochemical cycling before and after deforestation and subsequent ecosystem recovery. Data for this are given in Fig. 31.2 and, with Table 31.4, illustrate that during the devegetated period (when transpiration was almost nil) runoff greatly increased, as did the concentrations of dissolved substances. This was due to accelerated decomposition of organic matter on the forest floor and the much reduced uptake of nutrients by vegetation. From 1970, the forest showed a rapid recovery with consequent reductions in erosion and transport of particulate matter but, even after seven years of recovery,

Table 31.4 Annual weighed concentrations of nitrate, calcium, and potassium ions in stream water from reference watershed 6 (W6) and the experimentally devegetated watershed 2 (W2), in mg l^{-1}. (After Bormann and Likens 1979)

| | Nitrate | | Calcium | | Potassium | |
Year	W6	W2	W6	W2	W6	W2
Pre-cutting period						
1963–64	– *	–*	1.5	2.1	0.3	0.3
1964–65	1.0	1.3	1.0	1.4	0.2	0.2
1965–66	0.9	0.9	1.4	1.8	0.2	0.2
Devegetated period						
1966–67	0.7	38.4	1.3	6.5	0.2	1.9
1967–68	1.3	52.9	1.3	7.6	0.3	3.0
1968–69	1.3	40.4	1.3	6.0	0.3	2.9
Regrowth period						
1969–70	3.2	37.6	1.6	6.2	0.3	2.8
1970–71	2.4	18.8	1.5	4.0	0.2	1.6
1971–72	2.3	2.7	1.4	2.2	0.2	0.7
1972–73	1.8	0.3	1.4	1.8	0.2	0.5
1973–74	2.3	0.1	1.4	1.7	0.2	0.5
1974–75	2.4	0.1	1.4	1.7	0.2	0.4
1975–76	2.2	0.1	1.3	1.6	0.2	0.4

* No data.

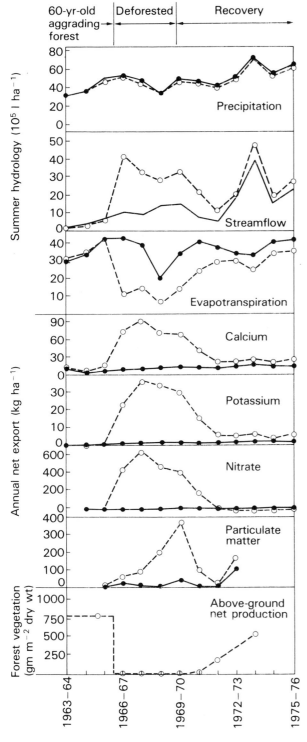

streamflow and the net export of calcium and potassium were still greater than pre-cutting levels. On the basis of these results, Likens *et al.* (1978) have estimated that the time necessary to compensate for losses during the deforested period (and for biomass to develop to pre-cutting levels) may be 60–80 years.

(b) Similar studies elsewhere

A variety of studies similar to that of the Hubbard Brook experiment have been undertaken in diverse parts of the world as part of the International Biological Programme (IBP), established in 1964 and terminated in 1974, to promote a better understanding of environment and the management of natural resources. The structure and function of forest ecosystems constituted a major part of this programme and the results are synthesised in Reichle (1981). Some of these data, summarised by Cole and Rapp (Reichle 1981) are given in Table 31.5, from which certain trends emerge. There is a greater above-ground accumulation of biomass and nutrients in temperate coniferous and deciduous forests than in those of the boreal zone, with greater accumulation in temperate coniferous than in temperate deciduous forests. As far as nitrogen is concerned, only 7% on average is found in the above-ground parts with the lowest amounts occurring in the boreal forests. In terms of the forest floor organic matter, the greatest accumulation occurs in the boreal coniferous forests, followed by temperate coniferous, boreal deciduous, temperate deciduous and Mediterranean forests and a similar trend occurs for the nutrient elements listed in Table 31.5.

The largest return of organic matter and nutrients in litter occurs in the temperate deciduous forest. For example, 60 kg ha^{-1} of nitrogen are returned annually in a temperate deciduous forest compared with only 36 kg ha^{-1} yr^{-1} in a temperate coniferous forest, reflecting longer foliage retention in the latter. However, not all of the nutrients are returned to the forest floor via litter fall since Cole and Rapp's data show that 85% of the

Fig. 31.2 Effects of deforestation on various parameters of the Hubbard Brook forest. The open circles represent the experimentally deforested watershed, the black circles a control area. Hydrological effects were minor during the winter period because of continuous snow cover. (After Likens *et al.* 1978)

Table 31.5 Data for IBP study sites: average nutrient accumulation in above-ground organic matter, forest-floor organic matter and annual production and nutrient cycling. (Adapted from Cole and Rapp; in Reichle 1981)

Forest region			Boreal coniferous	Boreal* deciduous	Temperate coniferous	Temperate deciduous	Mediterranean*	Overall averages
Above-ground organic matter and nutrient accumulation (kg ha⁻¹)	Biomass		51 268	97 343	307 341	151 900	269 000	207 526
	Nitrogen		116	221	479	442	745	429
	Potassium		44	104	340	224	626	263
	Calcium		258	164	480	557	3 853	589
	Magnesium		26	38	65	57	151	60
	Phosphorus		16	20	68	35	224	52
Forest-floor organic matter and nutrient accumulation (kg ha⁻¹)	Biomass		113 720	68 772	74 881	21 625	11 400	52 343
	Nitrogen		617	548	681	377	125	515
	Potassium		109	99	70	53	10	65
	Calcium		360	489	206	205	361	236
	Magnesium		140	139	53	28	20	55
	Phosphorus		115	79	60	25	4	50
Annual production and nutrient cycling (kg ha⁻¹ yr⁻¹)	Total above-ground production biomass		1 206	5 164	8 354	10 050	7 100	8 287
	Nitrogen	Uptake	5.1	25.0	47.4	75.4	47.7	55.0
		Requirement	4.7	55.9	47.5	97.9	77.2	66.3
	Potassium	Uptake	2.1	12.8	32.6	50.7	52.9	38.4
		Requirement	2.4	25.7	27.9	47.8	36.9	35.4
	Calcium	Uptake	6.1	38.9	44.6	85.0	120.7	63.3
		Requirement	3.1	18.3	19.7	55.6	63.7	37.4
	Phosphorus	Uptake	1.1	5.7	5.6	5.6	7.3	5.2
		Requirement	0.6	5.7	5.5	7.2	8.6	5.9
	Magnesium	Uptake	0.6	10.6	7.1	13.2	11.2	9.4
		Requirement	0.6	9.0	4.6	10.4	8.2	7.0

* Data based on one sample only.

nitrogen, 85% of phosphorus, 71% of calcium, 60% of magnesium and only 41% of potassium are returned in this way. The remainder is returned by throughflow and stemflow (section 7.1). Turnover times for organic matter and nutrients have also been calculated for the various forest types, by comparing the rate of nutrient return with total accumulation within the forest floor. The more northerly boreal forests show a much slower turnover time for both organic matter and nutrients. For example, boreal coniferous forests retain nitrogen 13 times longer than the temperate coniferous forest and 42 times longer than the temperate deciduous forest.

The data for annual elemental uptake and requirement (Table 31.5) also show some general trends. For example, the average annual uptake of nitrogen for all the sites was only 55 kg ha^{-1}, with a maximum uptake of 75 kg ha^{-1} yr^{-1} occurring in the temperate deciduous zone. In addition,

the amount of nitrogen necessary for a growth increment is generally low and again, the requirement is highest for the temperate deciduous forest. Uptake and requirement of nitrogen for growth is greatest in the temperate deciduous forest followed by the temperate coniferous and then boreal coniferous forests. A similar trend is apparent for potassium and calcium.

The Cole and Rapp data (Reichle 1981) do not include information on tropical forests which have also been the subject of several intensive studies on biogeochemical cycling. Some of the results are given in a UNESCO/UNEP/FAO report (1978) and, in addition, Herrera et al. (1978) have summarised the nutrient cycling characteristics of a range of Amazon ecosystems. In this region, there are a number of ecosystems of which the lowland rainforest is the most extensive. The climate is more or less uniform throughout the region and the various ecosystems are considered to be the

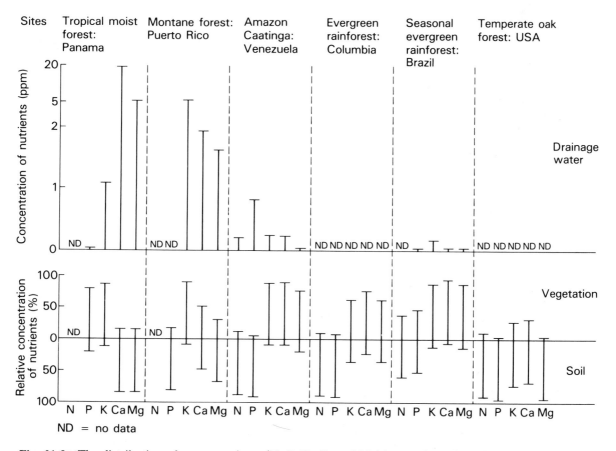

Fig. 31.3 The distribution of macronutrients (N, P, K, Ca and Mg) in a variety of tropical ecosystems. Note the comparison between the distribution of nutrients in these ecosystems and that of the temperate oak forest. (Adapted from Herrera et al. 1978)

result of edaphic conditions. Despite variations, however, all soils are acidic (being heavily leached and low in nutrients) and, in terms of chemical elements, Amazonian waters are among the poorest on earth. Summarising the research of a variety of workers, Herrera *et al* have shown that nutrient concentrations in Amazonian vegetation, soils and waters are lower than in many other forest ecosystems (Fig. 31.3).

These results are in accord with the early hypothesis of Walter (1936) that Amazon rainforests are not as reliant on soil nutrients as are most other types of vegetation. Alternatively, they depend on nutrients contained in their own biomass which become available as litter accumulates and subsequently decays and it appears that this tight intra-system control is effected through mechanisms for nutrient conservation. These include: the establishment of a dense root mat on the soil with a high nutrient retention capacity; direct cycling of nutrients from litter to roots via mycorrhiza, a root–fungi symbiosis; nutrient conservation by plant components such as nutrient recovery before leaf fall and the multi-layered physiognomy of the forest, which ensures efficient extraction of nutrients from throughfall water. The major implication of these results is in the context of conservation and survival of these ecosystems. Their maintenance is dependent on their tight intra-system nutrient budgets, which appear to be relatively unaffected by native slash and burn techniques of clearance for cultivation. However, large-scale destruction will result in a rapid loss of nutrients out of the ecosystem and a consequent rapid decline in productivity. Such is the problem associated with current large-scale deforestation in the Amazon Basin today.

A recent ongoing study of nutrient cycles has also been undertaken in a lower montane rainforest in New Guinea by Edwards and Grubb (Edwards 1982) and some of their results are given in Fig. 31.4. They have distinguished between the short-term cycle, which involves nutrient cycling in leaf litter and throughfall, and the long-term cycle which they define as the availability of nutrients for wood growth (section 30.1) and use the former as an indication of the latter. The quantities returned annually in litter fall and throughfall, as a percentage of the quantities in the aboveground standing crop, were nitrogen 18%, phosphorus 21%, potassium 15%, calcium 9% and magnesium 16%. This suggests that the short-term cycling of minerals is relatively rapid, in comparison with the store of nutrients in the biomass,

and that only a small proportion of the annual uptake of nutrients is incorporated into new wood. Comparing these data with similar data from other tropical forest regions, Edwards states that the short-term cycling of nitrogen and phosphorus are high relative to the amounts in the vegetation, although the short-term cycles of potassium and magnesium are similar to those of lowland rainforests. However, the total amount of nitrogen cycling annually is low, compared with lowland forests, and may indicate that long-term cycles of nitrogen, and possibly phosphorus, are slower in the New Guinea montane forest than in most lowland forests. Values for nitrogen and phosphorus in the soil (Fig. 31.4) are much higher than those in the vegetation, since the soil pools may be largely unavailable for plant growth.

In general, the data indicate that nitrogen is a limiting factor in the ecosystem since its concentration in the biomass is considerably lower than for lowland rainforests, and more nitrogen is withdrawn from the leaves as they mature before they fall. In addition, values for calcium and magnesium show that these elements are abundantly available. However, the potassium supply in the soil is only a little more than four times that in the annual litter and throughfall and its cycling is a tight intra-system budget. On much more local scales, nutrient budget studies have been undertaken on a wide variety of ecosystems. For example, Gray (1983) and Gray and Schlesinger (1983) have examined nutrient budgets in evergreen and deciduous shrubs in southern California, and Ernst (1983) details the element nutrition of dune plants in coastal sand dunes of the Netherlands.

31.2 Biogeochemical cycling at the biome level

As a corollary to the Hubbard Brook study, Gersmehl (1976) has proposed an open *three-compartment model* for nutrient cycling, which can also be used to characterise the nutrient circulation in *zonal ecosystems* or *biomes*. The three linked compartments of ecosystems are the soil, the biota and the litter. If the system was closed, the simplified movements of nutrients would consist of uptake from the soil by plants and, via plants to animals, the return of nutrients from the biota into the litter, which decays re-

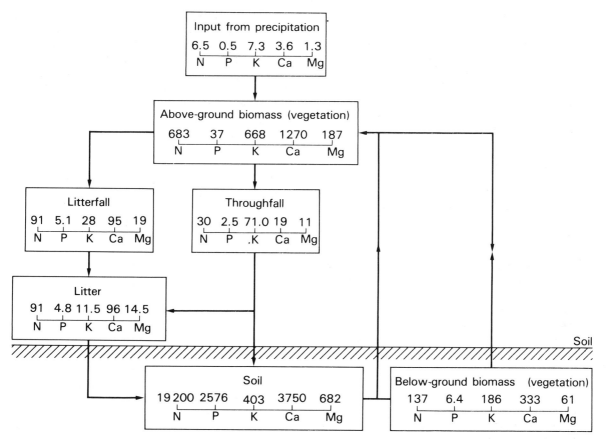

Fig. 31.4 A diagrammatic representation of mineral cycling in a montane rainforest in New Guinea (kg ha^{-1}). Arrows represent the major transfer pathways. (Adapted from Edwards 1982)

leasing nutrients back to the soil. However, ecosystems of a closed type rarely, if ever, occur and the significance of inputs and outputs of nutrients (as well as internal cycling) is well attested by the Hubbard Brook study (section 31.1). A combination of inputs and outputs with internal ecosystem compartments gives rise to a model of nutrient flow, as shown in Fig. 31.5. The importance of each flux route will vary according to the ecosystem's characteristics and, for example, the Hubbard Brook study (Fig. 31.5) illustrates the changes in nutrient flow patterns which occur as a result of deforestation. Similar representations can be made for other types of ecosystems, such as agricultural systems and fire-dependent ecosystems, where inputs and outputs are made more complex by anthropogenic factors.

Furthermore, Gersmehl emphasises the role of this simplistic model as a useful synthesising tool and a means of expressing the characteristics of the large-scale life zones or biomes of the world

(see Fig. 28.4). Since energy and moisture balances (manifested at least in part in biotic characteristics) are the primary regulators of nutrient fluxes, varying climatic regimes give rise to distinctive patterns of nutrient circulation (Table 31.6 and Fig. 31.5). For example, a typical steppe is frozen for half the year and does not receive high amounts of precipitation, so nutrients circulate slowly and there is little surplus water to cause high losses by runoff and leaching. In equilibrium, such an ecosystem stores few nutrients in the biomass compared to the litter and, especially, the soil. This is due to the predominance of mineral flux routes by litter fall, aided by the death of annual plants and regular firing of dried grass, and its incorporation into the soil. The implications for agriculture in such regions should be obvious. Repetitive harvesting of annual crops may, without careful management, cause a reduction in litter fall, depleting the mineral reservoir in the soil, which results in further losses by

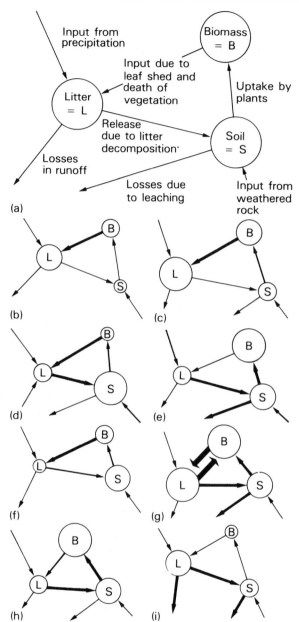

(a)

(b)

(c)

(d)

(e)

(f)

(g)

(h)

(i)

Fig. 31.5 A general model of mineral cycling, its application in a variety of ecosystem types and in the Hubbard Brook ecosystem. Arrows indicating nutrient transfers are drawn proportionately. (Adapted from Gersmehl 1976) (a) general circulation of nutrients within any terrestrial ecosystem; (b) tundra; (c) boreal forest; (d) steppe; (e) temperate deciduous forest; (f) desert; (g) tropical rainforest; (h) Hubbard Brook ecosystem before deforestation; (i) Hubbard Brook ecosystem during deforestation.

soil erosion (as in the Dust Bowl problems of the 1930s, in the USA).

Applying the model in a similar way to other zonal ecosystems, the varying significance of the litter, biomass and soil compartments (and the degree of exchange between them) can be usefully illustrated. In particular, the model illustrates interrelationships in the natural world (sections 28.2 and 28.3) and also helps to explain regional differences in agricultural practice. However, it is limited as a research tool on a more local scale since it represents a generalisation of reality, as does any model. Nutrient fluxes may deviate from this theoretical pattern in four ways: firstly, relationships are non-linear and doubling the nutrient input does not always give rise to a proportionate response; secondly, similar changes may cause different responses at different times; thirdly, nutrient mobility varies with chemical status, such as soil acidity or alkalinity, as well as with the climatic environment; finally, biotic effects will also vary because different plants create different conditions, which affect nutrient storage and movement. Nevertheless, the model provides a useful framework for an analysis of large-scale ecosystems. It provides insight into the way anthropogenic activity can disrupt the natural equilibria of ecosystems (resulting in degradation) and it emphasises the constraints that nutrient cycles place on exploitation of ecosystem resources.

31.3 Biogeochemical cycling at the plant community and species levels

Sections 31.1 and 31.2 detail large-scale and theoretical approaches to biogeochemical cycling, the applications of which are obvious for conservation practices and resource management. In parallel to these studies, aspects of biogeochemical cycling have been pursued at the more regional and local scales of the plant community and individual species. Applications of the results of these researches are numerous but the most prominent are in the fields of *biogeochemical prospecting* and the *rehabilitation of derelict land* (especially mine waste), some examples of which are given below.

(a) Biogeochemical prospecting

The Russian geologist Karpinsky (1841) was one of the first workers to realise that different plant

Table 31.6 Measured mineral characteristics of twelve zonal ecosystems, in kg ha^{-1}. (After Gersmehl 1976)

	Minerals stored in:		Annual mineral movement		Nitrogen as % of total mineral flow
	Biomass	Litter	From soil to biomass	From biomass to litter	
Arctic tundra	159	280	38	37	53
Spruce forest	3 350	2 100	178	145	35
Sphagnum bog	609	–	100	73	37
Birch forest	2 100	1 600	385	290	35
Oak forest	5 800	800	340	255	23
Wet grassland	1 183	800	683	682	25
Dry grassland	345	70	162	161	27
Shrub desert	185	–	60	59	27
Salt desert	143	–	85	84	13
Subtropical forest	5 283	600	992	795	28
Savanna grassland	978	–	319	312	26
Equatorial forest	11 081	178	2 028	1 540	22

communities occur on different bedrock types and may often characterise the geology of specific areas. Subsequently, biogeochemical prospecting techniques have been developed, using mineral-specific plant communities and species, often in conjunction with soil chemical analysis, to locate certain types of rocks and mineral deposits. This is made possible because the composition of plant communities is determined in large part by soil pH, the presence and availability of mineral nutrients and their biogeochemical cycling. Table 31.7 provides information on some of the most well-documented communities.

Calciphilous floras have a world-wide distribution and are so called because they thrive on calcareous soils and benefit either by the presence of abundant calcium ions or from lime-rich soils, which are usually well aerated and good conductors of heat and water. Such a flora usually contains *calcicole species* (which require above-average amounts of calcium for plant growth) and excludes *calcifuge species*, which require only a

Table 31.7 Examples of plants which indicate the presence of mineral deposits. (Adapted from Brooks 1983)

Mineral	Species	Family	Locality
Cobalt	*Crassula alba* (L)	Crassulaceae	Zaire
	Silene cobalticola (U)	Caryophyllaceae	Zaire
Copper	*Bulbostylis barbata* (U)	Cyperaceae	Australia
	Cyanotis cupricola (U)	Commelinaceae	Zaire
	Elsholtzia haichowensis (L)	Labiatae	China
	Impatiens balsamina (L)	Balsaminaceae	India
	Minuartia verna (L)	Caryophyllaceae	UK
	Tephrosia s. nov. (L)	Gramineae	Australia
Iron	*Acacia patens* (L)	Leguminoseae	Australia
	Eriachne dominii (L)	Gramineae	Australia
Lead	*Alyssum wulfenianum* (U)	Cruciferae	Austria/Italy
Nickel	*Alyssum pintodasilvae* (U)	Cruciferae	Portugal
	Lychnis aplina var. *serpentinicola*	Caryophyllaceae	Fennoscandinavia
Selenium	*Astragalus albulus* (L)	Leguminosae	USA
Uranium	*Astragalus argillosus* (L)	Leguminosae	USA
Zinc	*Hutchinsia alpina* (L)	Cruciferae	Pyrenees
	Thlaspi calaminare (U)	Cruciferae	W. Europe

low pH and low concentrations of calcium ions. As a result of such conditions, many species which ordinarily prosper in warm dry climates tend to concentrate increasingly on calcareous soils at higher altitudes and latitudes. The beech (*Fagus sylvatica*), for example, is successful on almost any soil type in southern Europe but in England it is, in its natural state, mainly confined to chalk soils.

Saline soils may also give rise to specific plant communities, consisting of *halophyte floras*, and are usually found near the sea or in areas of arid and semi-arid climates. Such communities develop and survive because they are adapted physiologically to tolerate and accumulate large quantities of salts. One important application is their use in *hydrogeochemical studies* in the USSR, as an indication of water quality (Chikishev 1965). Similarly, in the USA, Australia and Canada, certain plant communities have been used to delimit selenium-rich soils because the plant species have the ability to substitute selenium for sulphur in their metabolism. Cannon (1960) has successfully used similar communities (especially those dominated by *Astragalus* spp.) to locate uranium deposits on the Colorado Plateau, USA, where the abundance of carnotite results in greater availability of selenium to plants.

As well as variations in the species composition of indicator communities, the presence of certain minerals in the substrate can cause distinct morphological changes. Areas of serpentine rocks, for example, tend to carry a sparse vegetation as well as a limited range of species. This is because soils in these areas are rich in chromium, cobalt, iron, magnesium and nickel but are lacking in calcium, molybdenum, nitrogen, phosphorus and potassium and only a limited range of plants can tolerate such conditions. Similar sparse communities may occur over soils with high concentrations of copper, lead or zinc and the metal content of certain species has been successfully used to determine the nature of the underlying bedrock. Warren (1972), for example, found that the ash of blueberry (*Vaccinium* spp.) leaves at a Canadian site contained 2000 ppm of copper as compared with only 570 ppm in ash of leaves from a neighbouring uncontaminated area, while values for green alder (*Alnus sinuata*) were 25 650 ppm and 310 ppm respectively. Similarly, Tiagi and Aery (1982) have shown that *Impatiens balsamina* has exceedingly high concentrations of zinc (up to 12 141 ppm) and grows exclusively on metal-rich ground, with highest densities on mineral dumps in the Zawar mines area of Udaipur, India.

(b) The rehabilitation of derelict land

The examples which follow illustrate how plant species, and a knowledge of their role in biogeochemical cycling, can be used for effective *land restoration*. Heavy metals (such as copper, lead and zinc) play a vital role in modern society but their extraction results in the production of extensive deposits of waste materials, the rehabilitation of which is a major problem. The metals remaining in the wastes are often toxic to plants, animals and humans and they may be dispersed in drainage water to contaminate river systems far beyond the area of mineral extraction. Furthermore, finely ground wastes may be subject to wind erosion, resulting in contamination of wide areas. Such materials are also deficient in elements (especially nitrogen and phosphorus) which are essential for plant growth, and where there is more than 1000 ppm of any metal present, it is usually toxic to plants so that many areas of soil remain bare and liable to erosion for many years.

However, on some older mine wastes, some plants can be found including species such as violet (*Viola calaminaria*) and alpine pennycress (*Thlaspi alpestre*) which are rarely found on normal soils and which have been used as indicators of the presence of metals. In many cases, it appears that the species found on mine wastes have evolved metal-tolerant populations and when species growing on uncontaminated areas are planted on mine wastes, they do not survive. Furthermore, the tolerance is specific for individual metals and Smith and Bradshaw (1979), for example, have used this potential to develop a cheap and effective technique for reclaiming toxic metalliferous wastes in Britain. They have shown that the use of metal-tolerant populations, in conjunction with fertiliser applications, can result in a good vegetation cover on heavily contaminated sites and a number of species are now commercially available for this purpose. These include bent grass (*Agrostis tenuis*) cv. Goginan which will colonise acid lead and zinc wastes, *A. tenuis* cv. Parys for use on copper wastes and red fescue (*Festuca rubra*) cv. Merlin for calcareous lead and zinc wastes. Furthermore, Gemmel and Goodman (1980) have shown that with the addition of nitrogen, phosphorus and potassium fertilisers on smelter wastes in the Lower Swansea Valley UK, these species give rise to a long-term stable sward suitable for grazing. Similarly, reclamation of china clay wastes can be enhanced by the use of fertilisers and the introduction of legumes (especially clovers) which accelerate nitrogen accumulation in a developing sward. With good management and maintenance of lime and phosphorus levels, swards can be developed which can support at least 20 sheep per hectare throughout the summer period (Dancer *et al.* 1977).

31.4 Biogeochemical cycling of pollutants and radionuclides

No description of biogeochemical cycling would be complete without reference to the cycling of a variety of materials which have developed almost entirely as a result of an industrial and technological society requiring large food resources. In many ways, the cycling of such substances (e.g. *heavy metals, pesticides* and *radionuclides*) may be seen as a manifestation of the dominance of the fuel-powered agricultural and urban ecosystems (section 28.5) and their effects on the solar-powered and naturally subsidised ecosystems.

(a) Heavy metals

These are naturally occurring elements which have always been present in ecosystems but human activities have greatly increased their concentrations, to the extent that many occur at toxic levels (e.g. lead, mercury and cadmium). Some 98% of lead in the biosphere is a direct result of the addition of antiknock additives to petrol. Goldberg (1971) has estimated that 2.5×10^5 t yr^{-1} enters the oceans from this source and, as a result, the mean concentration of lead in sea water has increased sevenfold in the past 50 years. In addition, both British Pennine peats and polar ice show lead enrichment dating from the beginning of large-scale industrial development and the use of motor vehicles. Atmospheric lead is thought to be a major health hazard (especially to children) and has resulted in the banning of lead additives in petrol in the USA and a projected programme of lead eradication in petrol in the UK by 1990. Other sources of lead pollution occur as a result of mining activity, metal smelting and, in agricultural areas, where lead arsenate is used as an orchard spray in pest control. Such sources of lead may contaminate the soil, from where the metal is taken up by plants and may eventually enter food webs.

Mercury has long been recognised as a toxic substance but only recently has it attained concentrations which have proved lethal. Such high levels have occurred due to concentration as it passes along food chains and webs and, for example, Berg *et al.* (1966) and Borg *et al* (1969) recorded high concentrations of mercury in wild birds and their predators in Sweden, associated with the application of alkyl mercury as a fungicide to seeds. Smith and Smith (1975) refer to the contamination of fish in Japan, around Minamata Bay and along the Agaro River, which caused the death and the considerable disablement of local peoples. This resulted from local industrial plants discharging mercuric wastes into the waters which were then concentrated in food chains and finally entered the fish.

Much less is known about cadmium but it is geochemically associated with zinc and consequently cadmium pollution occurs as a result of zinc mining and its release by metal-plating industries. It is mobile in both soils and water, is easily assimilated by plants and passes along food webs where it may achieve toxic levels. In addition to the heavy metals mentioned above, other toxic elements of environmental concern are arsenic, beryllium, nickel, chromium, selenium, vanadium, molybdenum, copper and zinc.

(b) Pesticides

Until recently, most *pesticides* (including insecticides, herbicides and fungicides) were artificial synthetic compounds. However, the use of *chlorinated hydrocarbons* (especially DDT) focused much attention on the entire process of nutrient cycling since such compounds are stable, are not easily biodegradable (if at all) and are readily concentrated in fatty tissues. As a consequence, DDT-related compounds and their metabolites are persistent in the biosphere and become concentrated as they pass from one trophic level to the next, in a process known as *biological magnification*. For example, Woodwell *et al.* (1967) have shown that in an estuarine ecosystem on the east coast of the USA, the DDT concentration in water is 0.00005 ppm. However, by the time it passes through a food chain of phytoplankton, shrimp, predatory fish and finally fish-eating birds, the concentration increases to 23–26 ppm which represents a concentration factor of over half a million times. The attainment of such high concentrations of DDT may result in death or impaired reproductive ability. Consequently, the decline in populations of wildlife (especially birds of prey) in the USA and Europe has resulted in legislation prohibiting the use of DDT and similar compounds, such as aldrin and dieldrin.

Polychlorinated biphenyls are produced for use as dielectric fluids in capacitors and transformers and are used in the manufacture of printing inks, solvents and plastics. They are another widespread source of pollution and enter the environment via industrial waste and sewage. They are transported in water or the atmosphere to water bodies, from which they are absorbed by phytoplankton and become concentrated as they are passed along food chains in a similar process and with similar consequences to DDT. Of the many other synthetic compounds which circulate freely in the biosphere, *dioxin* is one which has recently been brought to public attention. It occurs as an impurity in the production of 2,4,5-T (a constituent of some herbicides) and, even in very low concentrations, it is extremely toxic which leads to cancer, abortions and genetic deformities in animals and humans. Agent Orange, a defoliant used by the US Army in South Vietnam, contained dioxin and since it was applied at much higher concentrations than is recommended for domestic use, it has had devastating affects on natural and agricultural systems, as well as on the native peoples and US troops exposed to it.

However, some lessons have been learnt from these examples and in both Europe and the USA, there are strict registration controls on compounds produced for use as pesticides. As a result, large chemical firms have to put a great deal of money and expertise into determining the toxicity and biological effects of such compounds and gear their products to the safety levels required for registration, which is essential for marketing. Perhaps one of the most significant developments in the last two decades has been the modelling of naturally occurring compounds which, while they behave as pesticides, are biodegradable and can be effectively applied at concentrations non-toxic to animals and humans. The *pyrethroids* are a good example of this and are a group of pesticides, based on the pyrethrin molecule, which occurs in pyrethrum (*Chrysanthemum cinerariaefolium*) flowers.

(c) Radionuclides

A third group of compounds giving rise to environmental concern are the *radionuclides*, produced as a result of nuclear weapons-testing and nuclear reactors, when uranium atoms undergo fission or when neutrons (produced in the reaction) combine with high-energy particles. These fission and fusion products may be released into the atmosphere, to be later deposited on the earth's surface as atomic fall-out in dust and rain and, eventually, to be absorbed in food chains. The radionuclides which occur include radioactive zinc-65 (^{65}Zn), iodine-131 (^{131}I), phosphorus-32 (^{32}P), carbon-14 (^{14}C), strontium-90 (^{90}Sr) and caesium-137 (^{137}Cs). However, the latter two are the most important radionuclides entering biogeochemical cycles and because caesium behaves chemically like potassium and strontium like calcium, they can both enter the trophic structure of ecosystems most easily via the grazing food chain.

It is apparent that this passage is more likely to occur in areas such as the Arctic tundra, where the surface storage of precipitation is quite considerable, facilitating the effective transfer of fall-out. At the same time, low weathering rates (often associated with acidic bedrock) give rise to low levels of nutrients. For example, Hanson (1971) has shown that in the tundra area of Anaktuvak Pass in Alaska, USA, lichens (the dominant autotrophs) absorb almost all of the radioactive particles they receive from the atmosphere, particularly ^{90}Sr and ^{137}Cs, which are then passed along the food chain from caribou, reindeer and a variety of carnivores to humans. However, the distribution of ^{137}Cs in caribou shows distinctive seasonal fluctuations as these animals feed all winter on lichens, resulting in a three to six times greater concentration in spring than in autumn. Similarly, because of the springtime hunting by Eskimos, there is often a 50% increase in human ^{137}Cs levels at this time which decreases as the Eskimos change to a diet of fish in the summer. In addition, biological magnification can occur and Hanson (1967) has estimated that a doubling of concentration occurs at each link in the food chain. Consequently, by the time the human diet changes to fish, the Eskimos may already have accumulated one-third to one-half of their permissible amounts. In general, however, little is known about radionuclide biogeochemical cycling and its long-term effects on either plant, animal or human populations.

31.5 Retrospect

Along with energy flows (Ch. 29), biogeochemical cycles are among the most important processes operative in the biosphere. Both influence, and are influenced by, the abundance and type of organisms present on the earth's surface and are therefore responsible for the complexity and operation of all ecosystems, from the global to the local level. In addition, nutrient cycling studies have illustrated how responsive biogeochemical cycles are to anthropogenic activity, such as deforestation and pollution. Adverse, uncontrolled or inadvertent modification of these nutrient cycles may ultimately lead to ecosystem degradation and lower productivity, by reducing the effectiveness of (or even destroying) negative feedback controls. The introduction of high levels of toxic metals, manufactured compounds (such as pesticides) and radionuclides into ecosystems (as a result of industry, high-technology agriculture and atomic fall-out) has created new biogeochemical cycles, many of which have proved lethal to plants, animals and humans. The extant knowledge of these is sparse but points to the need for improved control over the disposal of domestic, industrial and nuclear wastes.

More constructively, however, applications of biogeochemical cycling studies have provided opportunities for increased exploitation of resources, and biogeochemical prospecting, for example, has resulted in the effective location of mineral deposits. Furthermore, the rehabilitation of certain types of derelict land has culminated in the provision of recreation areas and a general aesthetic improvement in landscape. Also, where grazing land has been established in such areas, productivity and economic value have been increased. It is also clear from these two chapters that biogeochemical cycles constitute a major link between the atmosphere, hydrosphere, lithosphere and biosphere and, consequently, are closely allied with all the environmental systems discussed in this book. However, much further research is needed on the relationship between these processes and biogeochemical cycles before it is possible effectively to manage, conserve and improve the often misused environment.

Key topics

1. Contrasting biogeochemical cycles in deciduous, coniferous and tropical forests at the ecosystem level.
2. The interaction between soil, biomass and litter in biogeochemical cycles at the biome level.
3. The application of biogeochemical cycling at the community and species levels to biogeochemical prospecting and the reclamation of derelict land.
4. The effects of biogeochemical cycling on pollutants.

Further reading

Bormann, F. H. and Likens, G. E. (1979) *Pattern and Process in a Forested Ecosystem*. Springer-Verlag; New York. (This is a particularly good review of the Hubbard Brook study.)

Bradshaw, A. D. and Chadwick, M. J. (1980) *The Restoration of Land*. Blackwell; Oxford.

Brooks, R. R. (1983) *Biological Methods of Prospecting for Minerals*. John Wiley and Sons; New York and London.

Fortescue, J. A. C. (1980) *Environmental Geochemistry – a Holistic Approach*. Springer-Verlag; New York.

Hay, A. (1983) *The Chemical Scythe*. Plenum Press; New York.

Jordan, C. F. (1982) The nutrient balance of an Amazonian rain forest, *Ecology*, **63**, 647–54.

Moore, N. W. (1983) Ecological effects of pesticides, in Warren, A. and Goldsmith, F. B (eds), *Conservation in Perspective*. John Wiley and Sons; New York and London, pp. 159–75.

Moriarty, E. F. (1983) *Ecotoxicology*. Academic Press; London and New York.

Chapter 32 Vegetation succession and climax

Vegetation cannot simply be seen as a static, spatial entity but must be considered as a *dynamic system* because of its susceptibility to change. What is seen on the ground, in terms of *structure* (physiognomy) and *species composition* (floristics), reflects an interaction between species as well as the effects of the environment. Furthermore, vegetation responds to changes of both over time and is the result of a reciprocal relationship between plant communities and environment (sections 28.2 and 28.3). Environmental changes, which are responsible for changes in the associated plant community, may be brought about by *allogenic factors*, such as macroclimatic change, or *autogenic factors* instigated by the plants themselves as they determine energy flow (Ch. 29), biogeochemical cycling (Chs 30 and 31) and microclimate. One plant assemblage will be replaced or succeeded by another of different physiognomy and floristics and this process is known as *vegetation succession*. A mature community, that realises the full potential of any given habitat for organic production and its persistence over time, is the ultimate result and this is known as the *climax community*.

According to Cowles (1911), the French biologist Dureau de la Malle (in 1825) was the first to use the term succession in an ecological sense. However, many basic descriptive studies of plant communities provided the inspirational source for Clements to formulate his treatise on *Plant Succession and Indicators* in 1928. Since then, controversy has been rife in the literature regarding the nature of plant communities with alternative views being extolled by Gleason (1917 and 1926) and Whittaker (1953). Even today, there is no resolution to what some workers such as Moore (1983) consider to be essentially a semantic problem. Whether this is true or not is debatable and what follows is a brief review of ideas and philosophy in scientific and geographical circles in the eighteenth century and after. This will provide a historical perspective in order to establish parallels within and between different disciplines and (especially) to relate theories in the plant sciences with those in geography. Established terms will be introduced and examples provided to illustrate the way plant communities develop.

32.1 The parallels between scientific, geographical and plant science philosophies

Clements' theories on the nature of vegetation communities, and the later views of Gleason and Whittaker, can best be understood in the context of the philosophy of their predecessors. Prevalent views of the eighteenth century are a sensible starting-point since this period was one of change and saw the emergence of separate disciplines, such as those now known as geography, botany, geology and zoology (section 28.1). However, a word of caution must be included to remind those who currently study and research in these fields, that the foundation of present-day knowledge goes right back to the early civilisations of the Egyptians, Romans and Greeks. The book, however, does not provide sufficient space to record the ideas of the precursors of eighteenth-century work but can only acknowledge them as having paved the way for what is to be detailed.

In the eighteenth century, scientific thinking was dominated by the *mechanistic approach* of people such as Newton (1642–1727). He advocated the view that all attributes of nature could, given time, be empirically explained in simple, rational and logically quantitative terms. Linnaeus (1707–78) was among the protagonists of this view, as is exemplified by his arrangement of plants and animals into a rationally ordered system. At the same time, Humboldt (1769–1859) and Ritter (1779–1859) developed a more *inductive approach* (section 1.2) to explaining natural phenomena, believing that science should be founded on the objective description of observed facts rather than on logical deductive propositions, such as those of Newton.

Nevertheless, the deductive and mechanistic philosophy was continued in the work of Charles Darwin (1809–82) who emphasised the role of natural laws and causality in his *Origin of Species* (1859). It is from this, and related works, that the philosophy of *determinism* was born, whereby the achievements of human communities were to be explained as a result of natural conditions. Such a view was advanced by a British philosopher. Herbert Spencer (1820–1903) who developed these ideas into *social Darwinism*. Spencer drew analogies between human communities and organisms, such as plants and animals, all of which must struggle to survive in particular environments. Many noted geographers of the time adopted this philosophy, such as Ratzel (1844–1904) in Germany and Semple (1863–1932) and Davis (1850–1934) in the USA. The latter's contribution to scientific and (especially) geographical thinking revolves around his theory of the 'geographical cycle' (Ch. 1) which invokes an idealised landscape beginning with mountain uplift and culminating in lowland plains. This theory warrants especial mention here because there are many parallels with Clements' ideas on vegetation succession and climax (section 32.3).

However, an alternative philosophy to that of

determinism was simultaneously being developed by geographers in France, such as Vidal de la Blache (1845–1918), Bruhnes (1869–1903) and Febvre (1878–1956). The latter advocated 'there are no necessities, only possibilities' and thus *possibilism* became an alternative to determinism. Elements of determinism were, however, retained in so far as the concept of natural limits to human activity was accepted but the emphasis was placed on the choice of activity from a range of possibilities. In many respects, this philosophy reflects a moulding together of the views of Humboldt and Ritter with the determinist approach and was the foundation upon which Vidal de la Blache and others established the French school of regional geography. The study of regions was to dominate geographical research until the 1950s and the more systematic approach of Humboldt and Ritter became the provenance of natural history studies, with the emergence of ecology as an accepted scientific discipline in the early twentieth century (see section 28.1). It was not until the 1960s and 1970s that the area of biogeography began to re-emerge as a major aspect of geographical teaching and research with the move away from regionalism and the adoption of a systematic approach (section 1.3).

Inevitably, biogeographers draw upon the concepts and methodologies of related disciplines, especially those of the biological sciences. Not least are the theories of vegetation succession and climax which will be considered below.

32.2 The concepts of succession and climax

Since the early part of the twentieth century, numerous theories have been put forward to explain the mechanisms of vegetation succession. These have been categorised into two groups (Finegan 1984): the holistic approach, such as that of Clements, which emphasises unity and integration of plants and environment and the reductionist approach, such as that of Gleason, in which elements of Darwinian philosophy and chance are very significant. Neither approach is entirely satisfactory although both seek to explain some aspects of dynamism in vegetation communities.

(a) Clements, Moss and Tansley

Given philosophical developments in scientific circles in the eighteenth and nineteenth centuries, it is not surprising that their influence is apparent in the early twentieth century theories of vegetation development. Clements' concept of vegetation as a functional organism is similar to that of Spencer's (1899) view of human communities. In 1916, Clements wrote: 'As an organism the climax formation arises, grows, matures and dies. Its response to the habitat is shown in processes or functions and in structures which are the record as well as the result of these functions'. His views were also deterministic as he considered the components of a community to have a complex relationship and to develop in unison towards the climax under the control of environmental conditions, which are themselves dictated by macroclimate. He saw the process of vegetation development as progressive, directional and continuous from pioneer via intermediate to climax stages. This development resulted from biological factors (such as plant initiation, selection, community maintenance and eventual termination) and physical factors, such as denudation and habitat modification. Such a model is not unlike the geographical cycle proposed by Davis in 1909, just prior to Clements' own publications.

Although Clements envisaged succession as a continuous process, he believed that there were periods of stabilisation manifested in obvious well-defined communities. To the sequence of these units, representing the passage from initiation to climax, he gave the word *sere* and each step in the sere is known as the *seral stage*. Each sere culminates in the climax and Clements believed that whether a sere occurred on dry land or in water such as a pond or lake, the characteristics of the resulting climaxes would be identical if climatic factors were uniform. Similarly, in any given region, even one with a variety of bedrock types, a monoclimactic vegetation type would develop if climatic factors were uniform. This is the essence of what has become known as the *monoclimax theory*. Clements, however, also recognised that conditions could occur which would prevent complete development of climax communities. Edaphic or anthropogenic factors, for example, may result in *subordinate climaxes* or *subclimaxes* (Fig. 32.1) or in a series of zones, where the normal potential climax is juxtaposed with *pre- and post-climaxes*.

Today, the manifestation of Clements' monoclimax theory is seen in Fig. 28.3 which depicts the climax vegetation communities for the main geographical divisions of the earth. Descriptions of these can be found in most biogeography texts (e.g. Simmons 1979) but are excluded here on the grounds of space and unnecessary repetition. However, the real world is far from being so simplistic and Fig. 28.3 represents a large-scale generalisation. In many regions, the climax types are restricted in their distribution and are often absent

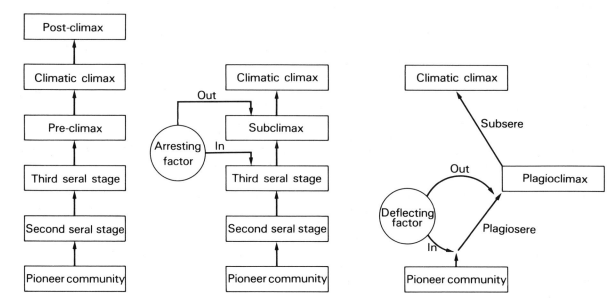

Fig. 32.1 Tansley's (1939) three main types of seral development.

due to local conditions such as unstable substrates, natural hazards or anthropogenic activity. Where physiographic and edaphic factors restrict climax development and maintain it below the climax, Clements applied the term subclimax which usually has characteristics quite different from any of the normal seral stages which would have occurred in the absence of constraints. Alternatively, anthropogenic activity (such as the maintenance of continuous grazing practices and burning) will deflect the normal succession on to a new course known as a *plagiosere* (Fig. 32.1) which culminates in a *plagioclimax*. If the determining factors are removed, the vegetation will resume its development towards the climatic climax along a course known as *subsere*. This will also occur if climax vegetation is partially destroyed and can develop from an earlier seral stage, at which progress has been curtailed. These views of Clements, on the dynamic nature of vegetation communities, became the starting-point for the work of other American and British ecologists such as Tansley. In addition, a plethora of new terms were introduced, many of which are still in use today, as will be seen in later sections of this chapter.

Just as Clements was to advocate the monoclimax theory, Tansley (1939) was responsible for developing the *polyclimax theory* by bringing together the Clementsian view with that of his contemporary, Moss. The latter (Moss 1910) advocated that British vegetation could best be classified on the basis of soil units, rather than on climatic characteristics, and in his description of edaphic climaxes, he recognised both progressive and retrogressive succession. The idea of retrogressive succession was a moot point to Clements, who saw succession entirely as a progressive process. Moss believed that retrogressive succession occurred where the edaphic climax was disturbed by natural events (such as erosion) or by anthropogenic activities (such as coppicing or grazing), while Clements believed that such factors produced new areas on which a different type of succession could proceed. Moss' ideas clearly have their limitations but his work did promote questioning of Clements' overwhelmingly deterministic monoclimax theory: it also led Tansley (1939) to formulate the polyclimax theory whereby he accepted that local variations (such as soil differences) would give rise to different types of climax within any given area, but would still be controlled overall by climate. These two schools of thought, however (viz. the monoclimactic and polyclimactic

theories), can be reconciled by considering the temporal aspect incorporated by Selleck (1960). He believed that on a long-term basis, climate is the overriding control on vegetation development as edaphic factors are reduced (by the climate) to an ineffectual level. Alternatively, in the space of a human lifetime, edaphic factors may override climate.

(b) *Gleason, Whittaker and others*

Moss was not the only worker to question Clements' views. In 1917, Gleason recorded his dissent from Clements' organismal concept of plant communities and advanced the idea that plant communities developed from the random spread of individual plants and were maintained by the sum of the dynamics of its component species. Thus, in 1926 he developed the *individualistic concept of the plant association* and introduced species composition as a major factor in vegetation development, in contrast to the reliance of Clements on physiognomy and structure. Implicit in Gleason's theory was the idea of intergradation of communities along environmental gradients, again in contrast to the more discrete vegetation communities of Clements.

However, Gleason's views were largely ignored at the time but have been reviewed in the work of Whittaker (1975) who also advocated this much more possibilistic approach to vegetation development by taking into account attributes such as floristics, population dynamics and productivity, in addition to physiognomy and structure. As an alternative to the monoclimax and polyclimax theories, Whittaker formulated a *climax pattern hypothesis* in 1953. Contrary to Clements, he believed that climax composition is determined by all factors which are inherent within or act upon the population on a continuous basis. These include edaphic factors, topography, the genetic and morphologic characteristics of the plants themselves, dispersal abilities and anthropogenic practices, as well as climate. He also introduced the idea of *climax relativity* whereby some species may be both seral and climax. Furthermore, types of seral stage may be seral under one set of circumstances but climax under a different set of circumstances and the distinction between seral stages and climax is associated with the degree of stability. His ideas of climax recognition were also very different from those of Clements. They were based more on observations of changes along environmental gradients (and the impossibility of an absolute climax) than on the theoretical concept of the monoclimax and its characterisation of geographical regions.

Whatever the interpretation of the climax concept, a degree of stability is implied which exists between the organisms, and between them and their environment. A situation of dynamic equilibrium exists and the climax is no less dynamic than the succession which led to its formation. The community will be continually renewing itself as individuals die and are replaced. Also, as cyclic changes occur (as in many tropical forests, where the seedlings or saplings of the dominant trees are often absent), the climax species are initially replaced by shade-tolerant species, which then facilitate the growth of further climax species (Richards 1979). In this case, the climax will consist of alternating and often juxtaposed stages.

Furthermore (and returning to the temporal considerations of Selleck), community equilibrium can only be relative. Given what is known about variations in past climate over the short term (such as the postglacial period of the last 12 000 years) or the longer-term glacial/interglacial sequences of the Pleistocene, any idea of long-term stability of vegetation must be in serious doubt. In addition, the occurrence of natural hazards (such as drought, fire, hurricanes, etc.) similarly militates against long-term stability, as does the fact that anthropogenic activities continue to have an ever-increasing effect on the world's vegetation. Such factors render the idea of a 'natural' climax untenable and as Miles (1979) has indicated, it may be more useful to consider the idea of a 'terminal' vegetation type. This would represent a stable community at any site under a given range of environmental and management conditions, in relation to the human life span. In addition, Finegan (1984) proposes that vegetation succession may be more adequately explained by adapting a synthetical approach, incorporating elements of both holistic and reductionist theories. Aspects of these theories are not always mutually exclusive and any explanations of succession based on the operation of a single controlling mechanism are inflexible and likely to be unrealistic. Bearing in mind these controversies and the inadequacy of the terms of succession and climax (and the lack of good alternatives), the continued reference to them in the following sections should not be seen as a blind acceptance but simply convenience to provide continuity with established literature.

32.3 Characteristics of successions

As well as his view of the overriding control of climate on vegetation development, Clements advocated that each component of a seral stage drastically altered the habitat, so that new niches were formed which could be occupied by further invaders. Many recent workers have shown that while site modification is a feature of succession, it is not clear that these changes result in autogenic succession by restricting regeneration or by providing a more appropriate site for newcomers. It may be, for example, that many early invaders are opportunists with efficient seed dispersal, quick growth and a short life-cycle. This latter characteristic would prevent the development of a permanently closed canopy and thus restrict the degree of shade. It would then allow other species to invade, which may be slower-growing with a longer life and greater capacity for biomass and nutrient storage. Eventually, such species would provide too much competition for the early invaders.

However, while change in species composition (for whatever reason) is one of the most obvious features of succession, trophic–dynamic aspects and interspecific relations also change. As a result of ecosystem modelling, developed from the systems approach (Chs 1 and 28), general theories of community development have been derived for comparative and predictive purposes, on the basis of their structural and functional characteristics. Furthermore, such an approach was used by Margalef (1968) and Odum (1969) in an attempt to develop a similar model of succession. For example, Table 32.1 (which is self-explanatory) summarises the trends which Odum believed could be expected to occur during ecosystem development. Current research, however, has shown that some of these premises are debatable, and especially that not all are true of all successions. The work has been heavily criticised by Drury and Nisbet (1973) who cite so many exceptions to these expected trends that they advocate its rejection. In addition, it should be noted that the ecosystem attributes are essentially functions of the species present and could be satisfactorily explained in terms of species characteristics and competition. Drury and Nisbet (1973) further argue that: 'The structural and functional changes associated with successional change result primarily from the known correlations in plants between size, longevity and slow growth. A comprehensive theory of succession should be sought at the organismic or

Table 32.1 A tabular model of ecological succession trends to be expected in the development of ecosystems. (After Odum 1969)

Ecosystem attributes	Developmental stages	Mature stages
Community energetics		
1. Gross production/community respiration ratio	> or <	Approaches unity
2. Gross production/biomass ratio	High	Low
3. Biomass-supported unit energy flow ratio	Low	High
4. Net community production	High	Low
5. Food chains	Simple, linear	Complex, web-like
Community structure		
6. Total organic matter	Small	Large
7. Inorganic nutrients	Mainly in soil minerals	Mainly in organic matter
8. Species diversity-variety component	Low	High
9. Species diversity-evenness component	Low	High
10. Biochemical diversity	Low	High
11. Stratifications and pattern diversity	Poorly organised	Well organised
Life-history		
12. Niche specialisation	Broad	Narrow
13. Size of organism	Small	Large
14. Life-cycles	Short, simple	Long, complex
Nutrient cycling		
15. Mineral cycles	Open	Closed
16. Nutrient exchange rate between organisms and environment	Rapid	Slow
17. Role of detritus in nutrient regeneration	Unimportant	Important
Selection pressure		
18. Growth form	For rapid growth (r-selection)	For feedback control (A-selection)
19. Production	Quantity	Quality
Overall homeostasis		
20. Internal symbiosis	Undeveloped	Developed
21. Nutrient conservation	Poor	Good
22. Stability (resistance to external perturbations)	Poor	Good
23. Entropy	High	Low
24. Information	Low	High

cellular level, and not in emergent properties of communities.'

It is also interesting to note that Miles (1979) sees Odum's model as neo-Clementsian since it is based on ideas of directional (and therefore predictable) community development and control over change by the community, which proceeds to a stable end-point. Thus, the debate continues between the individualistic concept, involving irregularity, diversity and complexity and the holistic concept, with emphasis on orderly development. However, a balance and a degree of reconciliation between the two approaches can be achieved by considering their objectives. The holistic systems-based approach (Ch. 28) allows generalisation, enabling the complexity of nature to be assessed in a simplified logically ordered way. Its main disadvantage must be in the fact that, in reality, many of the generalisations are not substantiated because the approach is too theoretical. It is based on too few observations of the real world which the individualists consider to be too complex for such a simplistic approach to be viable. With this debate in mind, the following examples are given to illustrate a variety of different types of succession and approaches to their study.

32.4 Primary successions

Primary successions usually occur on newly emergent areas (such as bare rock, glacial deposits or sand) and since the environment is often excessively dry due to poor water retention, they are called *xeroseres*. Conversely, those successions which occur on excessively wet, submerged or waterlogged sites are termed *hydroseres*. There have been few direct observations of primary successions because rates of vegetation development are often slow and well beyond the length of a single human lifetime. As a result, most types of succession have been documented either by studies substituting space for time (where recent and older substrates are juxtaposed) or by examining (where possible) the plant micro- and macrofossil record of preserved stratigraphic sequences.

(a) Xeroseral successions

Shure and Ragsdale (1977) have documented the patterns of primary succession occurring in soil-island communities of granite outcrops in the Panola Mountains in Georgia, USA. They examined the flora, macro and micro-arthropod populations and a variety of environmental characteristics such as soil depth, soil temperature variations, soil moisture, pH and cation exchange capacity of 10 communities in each of 3 seral stages. A summary of their results is given in Fig. 32.2, which illustrates some of the trends occurring during succession. Shure and Ragsdale conclude that primary succession on the outcrops occurs as a result of mutual interaction between the substrate and the biota. The initial colonisers (the *Diamorpha*) have large nutrient and material fluxes and they, together with subsequent increases in plant cover, are essential for the accumulation of organic matter which in turn favours further vegetation development. In addition, interspecific competition is influenced by changing environmental conditions. Increased soil moisture, for example, which occurs as soils become deeper, favours the survival of larger and more competitive plant species. Changes in consumer populations tend to respond to vegetation and substrate changes, especially increasing primary productivity which is paralleled by increasing food

web complexity. Furthermore, it is clear that the few species which occupy the earliest seral stages are those best adapted to survive under stress and, as succession proceeds in these soil-islands, they ultimately converge in their characteristics of density, biomass and species diversity. In this case, the Clementsian and neo-Clementsian views of succession are not without some substantiation.

In addition, evidence of primary succession has been documented from areas where glacier recession has provided fresh substrates for vegetation development. A classic example is the work of Cooper (1926, 1931) from the Glacier Bay area of Alaska, USA, consisting of comparative observations of sites exposed at different times and the monitoring of changes over 19 years in permanent quadrats. These observations showed a vegetation development in which three principal stages could be recognised. The initial or pioneer stage was dominated by herbaceous and mat-forming species including *Rhacomitrum* mosses, the broad-leaved willow herb (*Epilobium latifolium*), the horsetail (*Equisetum variegatum*) and a low-growing evergreen shrub (*Dryas drummondii*). The second stage consisted of thickets of *Dryas* and dwarf willows (*Salix* spp.) with the establishment of tree willows (*Salix* spp.), alder (*Alnus crispa*) and the cottonwood (*Populus trichocarpa*). A forest stage is the final development, dominated initially by sitka spruce (*Picea sitchensis*) and later by a mixture of spruce and hemlock species (*Tsuga* spp.)

The result of Cooper's permanent quadrats also illustrate some of the features characteristic of primary successions in such areas. For example, there was a steady increase in the density of herbs and in the size and cover of mat-forming species. However, despite the presence of apparent stages, the individuals of all stages (including the forest trees) appeared throughout. Such changes in the vegetation can be explained in terms of seed banks, dispersal rates and varying growth and reproductive rates (section 32.3). Thus, species with especially mobile propagules (such as moss spores or the plumed seeds of *Epilobium*) arrived in the greatest numbers and consequently formed the majority of the initial vegetation. The more tardy shrub invaders progressively outshaded this earlier vegetation and were themselves later outshaded by the conifers.

As well as in natural habitats described above, primary successions can occur on artificially created habitats, such as old spoil heaps. Roberts *et al.* (1981) have documented succession on china clay wastes in Cornwall where there are a number

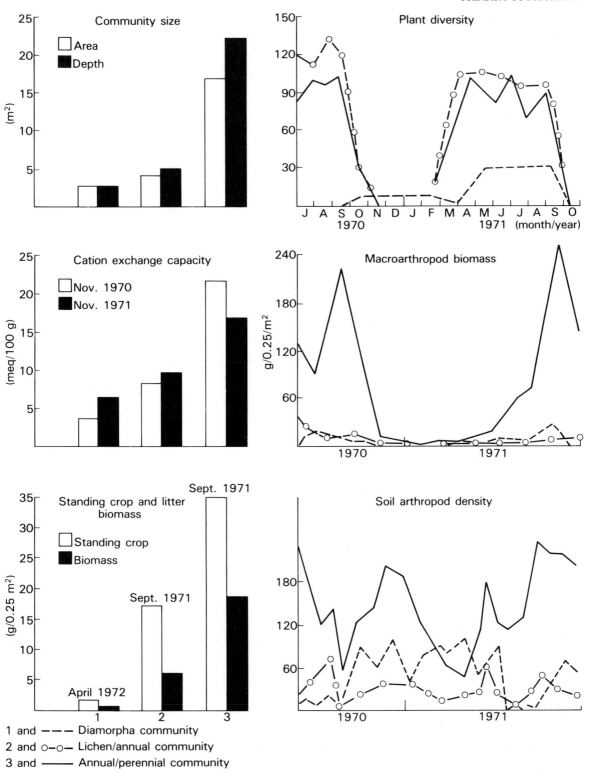

Fig. 32.2   A summary of some of the changes occurring in successional stages on granite outcrop surfaces. (Adapted from Shure and Ragsdale 1977)

Table 32.2 Details of the sampling sites used to examine the colonisation of china clay wastes in Cornwall, UK. (After Roberts *et al.* 1981)

Site and National Grid reference		Date of last tipping	Age in 1976 (yr)	Number of quadrats	Major species
Carpalla (SW 963538)		1958	18	2	*Lupinus arboreus**
Goonvean (SW 951550)		1960	16	2	*Lupinus arboreus*
Lower Lansalson (SX 005548)	I	1878	98	2	*Rhododendron ponticum*
	II	1934	42	2	*Rhododendron ponticum, Ulex europaeus*
Ruddle (SX 012548)	I	1925	51	4	*Sarothamnus scoparius, Ulex europaeus*
	II	1937	39	2	*Salix atrocinerea, Sarothamnus scoparius, Ulex europaeus*
South Caudle Downs (SX 006578)	I	1919	57	4	*Calluna vulgaris, Ulex gallii*
	II	1928	48	2	*Calluna vulgaris, Ulex gallii*
	III	1933	43	2	*Salix atrocinerea, Sarothamnus scoparius*
Wheal Henry (SX 022590)	I	1900	76	2	*Calluna vulgaris, Sarothamnus scoparius, Ulex europaeus*
	II	1925	51	2	*Salix atrocinerea, Ulex europaeus*
	III	1936	40	6	*Salix atrocinerea, Ulex europaeus*
Wheal Rashleigh (SX 060552)		1860	116	4	*Betula pendula, Quercus robur, Ulex europaeus*
Winnipeg (SW 967560)		1950	26	2	*Sarothamnus scoparius, Ulex europaeus*

* Species nomenclature follows Clapham *et al.* (1962).

Group	Species	Group	Species
A	*Digitalis purpurea*	B	*Ulex europaeus*
	Holcus lanatus		*Ulex gallii*
	Lupinus arboreus		*Vaccinium myrtillus*
	Rumex acetosa		
B	*Agrostis tenuis*	C	*Rumex acetosella*
	Calluna vulgaris		*Salix atrocinerea*
	Cotoneaster simonsii	D	*Betula pendula*
	Crataegus monogyna		*Corylus avellana*
	Deschampsia flexuosa		*Hedera helix*
	Erica cinerea		*Lonicera periclymenum*
	Glechoma hederacea		*Quercus robur*
	Knautia arvensis		*Rhododendron ponticum*
	Lamium purpureum		*Sorbus aucuparia*
	Osmunda regalis		
	Pteridium aquilinum		
	Rubus fruticosus agg.		
	Sarothamnus scoparius		
	Taraxacum officinale		
	Teucrium scorodonia		

A and B = primary colonising groups; C = intermediate groups; D = mature group.

of abandoned waste tips, ranging in age since last tipping from 16 to 116 years (Table 32.2). Four main species groups were recognised, of which two were considered as primary colonising groups, one an intermediate group and a final (to 1976) mature group (Table 32.2). Of the pioneer groups, species colonise waste tips between 16 and 55 years after tipping ceases. The common sallow (*Salix atrocinerea*), the chief species of the intermediate group, was found on sites 40–75 years old while Rhododendron (*Rhododendron ponticum*) of the mature group colonised some sites after as little as 40 years. However, many of the tree species were confined to the oldest site of 116 years. In this particular study, nitrogen cycling appears to have been central to vegetation development. The two pioneer groups included leguminous shrubs such as gorse (*Ulex europaeus*) and the tree lupin (*Lupinus arboreus*), whose contribution to the nitrogen budget of the wastes was very significant, especially in the early stages. This facilitated a build-up of nitrogen within the ecosystem, highlighted by the fact that sites 50–60 years old had accumulated as much nitrogen as was found in older sites. Thus, after approximately 50 years, other species can invade, such as the common sallow, which do not fix nitrogen. In this particular environment, nitrogen fixation and its accumulation in organic matter appears essential for vegetation development.

(b) Hydroseral succession

Until recently, the main approach to studies of hydroseral succession was based on Clements' (1928) statement 'zonation is the epitome of succession'. This implies that the zonation of vegetation presently occurring around lakes is a spatial representation of hydroseral development. Such an approach was adopted by Tansley (1939) who devised a generalised scheme for hydroseral development in relation to climate. Figure 32.3 illustrates the sequential development of reedswamp through fen and carr to forest in suboceanic climates and bog in oceanic climates. Numerous workers in the 1940s to 1960s, however, produced evidence to show that this sequence did not always occur and that many factors other than climate were effective in determining hydroseral character.

This led Walker (1970) to synthesise published records, based on stratigraphic descriptions and sub-fossil plant remains, of British postglacial hydroseres. Since his aim was to test the autogenic and unidirectional hydroseral theories of Clements and Tansley, his synthesis is restricted to hydroseres in which allogenic influences such as silting are minimal. Despite the fact that vegetation types are rarely distinct spatially or temporally, Walker (for convenience of analysis) recognised 12 vegetation stages or ecological units. These are given

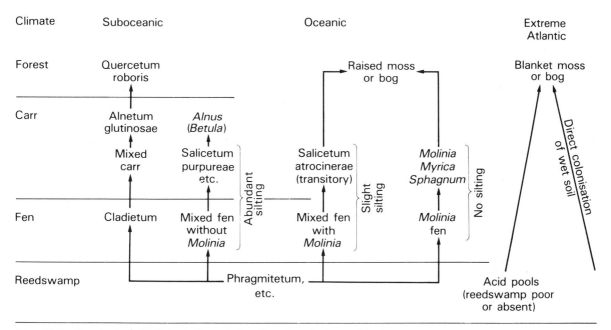

Fig. 32.3 Tansley's (1939) depiction of the climatic relations of hydroseres.

Table 32.3 Frequencies of transitions between vegetation 'stages' free from obvious allogenic influences derived from 20 pollen diagrams. (After Walker 1970)

	*'Stages'	Succeeding vegetation												
		1	2	3	4	5	6	7	8	9	10	11	12	Total
	1	.	.	.	.	.	.	.	.	.	.	.	.	0
	2	.	.	2	4	1	.	3	.	.	.	1	.	11
	3	.	4	.	3	3	.	.	.	.	.	.	.	10
Antecedent vegetation	4	.	4	1	.	8	.	1	.	.	.	1	.	15
	5	.	.	.	.	.	.	6	3	2	.	4	.	15
	6	.	.	.	.	.	.	.	.	.	.	.	.	0
	7	.	1	.	1	.	.	.	.	6	.	3	.	11
	8	.	.	.	.	1	.	.	.	.	.	2	.	3
	9	.	.	.	.	.	.	.	.	.	.	6	.	6
	10	.	.	.	.	.	.	.	.	.	.	.	.	0
	11	.	.	.	.	.	.	.	.	.	.	.	.	0
	12	.	.	.	.	.	.	.	.	.	.	.	.	0
	Total	0	9	3	8	13	0	10	3	8	0	17	0	71

* (1) biologically unproductive open water; (2) open water with micro-organisms; (3) open water with submerged macrophytes; (4) open water with floating-leaved macrophytes; (5) reedswamp; (6) tussock sedge swamp; (7) fen (herbs on organic soil); (8) swamp carr (trees on unstable peat); (9) fen carr (trees on stable peat); (10) aquatic *Sphagnum* spp.; (11) *Sphagnum* bog; (12) marsh (herbs on mineral soil).

in Table 32.3 which also shows the number and variety of vegetation transitions which were identified and classified according to the pair of vegetation stages to which they related. This data set shows that 46% of the recorded transitions are accounted for by the preferred course illustrated in Fig. 32.4, with the final stage being bog rather than woodland. In addition, the remaining 54% showed considerable variation in succession, especially from reedswamp onwards. Of the total, 40% of transitions from reedswamp are to fen, 27% to bog and 33% to carr. Of these, 17% are in a direction reverse to that indicated by the majority of the records. Walker concludes 'the course of a particular hydrosere in the past or in the future can only be hypothecated in a probabilistic manner from a consideration of all possible transitions to and from existing vegetation types weighted for environmental site conditions and species availability'. While there is no doubt that succession is a characteristic of lake ecosystem development, it is in no way as clear cut as the Clements–Tansley tradition would suggest.

32.5 Secondary successions

Succession which occurs on a previously vegetated site is designated as a secondary succession. Such developments may occur on formerly cultivated or

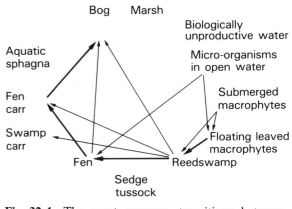

Fig. 32.4 The most common transitions between vegetation stages in the hydrosere, drawn from Table 32.3 (Adapted from Walker 1970). Thickened lines show dominant courses.

grazed farmland or in climax vegetation such as forests, where fire or felling has occurred (as in the Hubbard Brook forest ecosystem described in section 30.1). Secondary succession is usually a much more rapid process than primary succession, because it occurs in areas which have already been biologically modified by a previous vegetation cover and attendant energy flow and nutrient cycling. In general terms, a large part of the world's vegetation has undergone disturbance. Much of what is described as natural forest cover, for example, is the result of secondary growth on naturally burnt or flooded areas and many of the world's grasslands have been subject to periodic catastrophic disturbance by fire and drought. Thus, secondary succession is a widespread occurrence in a great variety of situations, both natural and artificially created, and a comprehensive treatment here is impossible. The examples which follow illustrate some of the characteristics of secondary succession.

Figure 32.5 summarises the results of a study undertaken between 1952 and 1959 on abandoned cropland in South Carolina, USA (Odum 1960). Of particular note is the pattern of gradual and continuous change in species composition and species diversity throughout the study period. In addition, there is a rapid trend from an erratic to a more even seasonal distribution of productivity together with the early stabilisation of average litter and total annual production. If, as Odum proposes, the structural attributes of the ecosystem are represented by species composition and diversity and the functional attributes are represented by productivity, then structurally the community developed continuously and gradually. However, in terms of function, a steady state was achieved at an early stage. The first characteristic is a feature commonly observed in secondary succession but the second characteristic shows that productivity does not automatically increase as succession proceeds, as has been commonly assumed.

The causes of rapid change in species composition and diversity during the early stages of succession are unclear but factors such as availability of seed banks, competition and life-cycle length, as discussed in section 32.4, must all be contributory factors. In terms of productivity, Odum proposes the advent of further surges as the succession proceeds, corresponding to the replacement of the initial annual and biennial growth forms by perennial species, such as broomsedges (*Andropogon* spp.), as they themselves are replaced by trees. The mechanism for this is not obvious but may be due to the exploitation of water and nutrients not easily available to the previous life-form. After the initial surge, a new steady-state pattern of energy flow becomes established. Thus, Odum's work echoes, to some extent, the Clementsian view of succession but is more probabilistic in its consideration of factors such as competition, rather than emphasis on site modification. Furthermore, it introduces the productivity factor (often disregarded in successional studies) but which, as a feature common to all vegetation

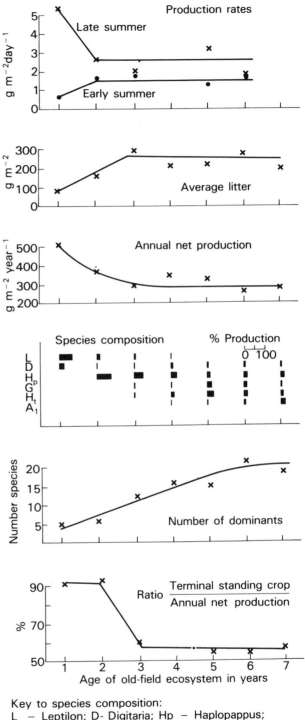

Key to species composition:
L – Leptilon; D- Digitaria; Hp – Haplopappus;
G – *Gnaphalium obtusifolium*; Ht – Heterotheca;
A – Andropogon.

Fig. 32.5 Features of succession in an old-field ecosystem. (Adapted from Odum, 1971)

types, provides a universally applicable and comparative measure.

A further example of secondary succession is that of Swaine and Hall (1983) who monitored succession in an area of upland evergreen forest in Ghana, during the first five years following clearance for mining. Some of their results are illustrated in Fig. 32.6, which shows that a density of 2.5 trees m^{-2} was achieved within one year, of which 95% were secondary species. Then, density declined exponentially to 1 tree m^{-2} after five years, due chiefly to the death of secondary species. In addition, more than 90% of the secondary and 60% of the primary species encountered throughout the study periods, colonised during the

first year. However, by the end of the period, the number of secondary species was declining and new primary species were appearing.

Thus, succession here does not follow the Clementsian theory in so far as a large primary flora appeared during the initial stages. Also, since soil characteristics beneath nearby primary forest and the cleared area were similar, site modification is not a prerequisite for primary forest development. As an alternative, Swaine and Hall invoke Horn's (1976) idea of a comprehensive hierarchy, whereby many species colonise simultaneously. However, only the long-lived persist, a view substantiated in this case because many of the secondary species are short-lived and, even in a five-year period, competition has resulted in their progressive elimination. Furthermore, the inspread of new primary species is still occurring and since species diversity is 25% higher than that of mature forest, this elimination is likely to continue until an equilibrium is achieved.

32.6 Retrospect

This chapter is not presented as a comprehensive survey of approaches to the development of vegetation communities. Neither is it intended to be partisan, but rather it is written to provide a background against which the controversies surrounding the holistic and individualistic approaches can be more clearly appreciated. In the long term, there must be reconciliation between the two approaches since both have much to contribute. The holistic approach provides a useful framework, in common with the approach adopted for the other types of environmental systems discussed in this book. On the other hand, many ecologists perceive it as too general, while the individualistic approach is often considered too detailed for sensible study.

No apology is made for the absence of descriptions of the so-called climax communities of the geographical regions of the earth. This is well covered in a wide variety of biogeography and ecology texts, and given limited space here, a consideration of approaches to the study of vegetation systems must take precedence. For the same reasons, other attributes of vegetation dynamics have been excluded (such as cyclic changes, regeneration and pattern) but their absence does not signify inconsequence. Nevertheless, the dynamic character of vegetation com-

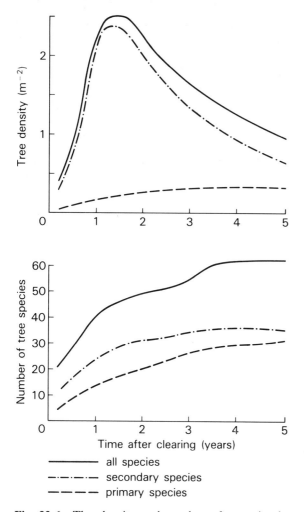

Fig. 32.6 The density and number of trees in the first five years after clearance in the Atewa Range Forest Reserve, Ghana. (Adapted from Swaine and Hall 1983)

munities is apparent from the examples given. This is a feature shared with all other environmental systems within which vegetation plays an important role and which, themselves, have direct effects on vegetation communities. As a consequence, processes in vegetation systems are just as important as those in atmospheric and sediment systems, etc. in shaping the landscape and, as such, are particularly important to the physical geographer.

Finally, a word should be said in praise of Clements and Tansley since much criticism has been directed at their views. Their works remain landmarks in ecological and biogeographical thinking, not least for incorporating the idea of dynamism into an otherwise static concept of ecology.

Key topics

1. The limitations of theories on vegetation characteristics.
2. The controversy of climax theory and the significance of stability to modern approaches to vegetation studies.

3. The characteristics of primary successions on natural and 'man'-made substrates.
4. The nature of secondary successions.

Further reading

Connell, J. H. and Slatyer, R. O. (1977) Mechanisms of succession in natural communities and their rôle in community stability and organisation, *Amer. Nat.*, **3**, 1119–44.

Golley, F. B. (ed) (1977) *Ecological Succession*. Benchmark Papers in Ecology, vol. 5, Dowden, Hutchinson and Ross; Stroudsburgh, Pennsylvania (contains reprints of Clements, Gleason, Whittaker, Odum and Cooper papers).

Holt-Jenson, A. (1980) *Geography: Its History and Concepts*. Harper and Row; London and New York.

Miles, J. (1979) *Vegetation Dynamics*. Chapman and Hall; London.

Newman, E. I. (ed) (1982) *The Plant Community as a Working Mechanism*. Spec. Pub. Ser. Brit. Ecol. Soc. No. 1, Blackwell Sci; Oxford.

West, D. C., Shugart, H. H. and Botkin, D. B. (eds) (1981) *Forest Succession: Concepts and Applications*. Springer Verlag, Berlin and New York.

Part VI
Oceanic systems
Chapters 33–35

The following three chapters introduce the ocean systems which dominate the total area of the earth's surface. The formation and morphology of the ocean floor are discussed, including the complex continental margins (with seismically active or passive features) and the ocean deeps, with contrasting flat plains and conspicuous volcanic ridges. Sea water is analysed in terms of its physical and chemical composition and its movement by waves, tides and ocean currents.

Marine deposition is considered including the source and type of deposits and the processes involved in sedimentation both in the ocean basins and coastal zones. Finally, the discussion of marine ecosystems emphasises the diversity of organisms and communities which inhabit the various marine habitats, which range from the ocean deeps to the littoral zone.

Breaking waves, Perranporth, Cornwall, UK.

Chapter 33 Formation and morphology of the ocean floor

Oceanic systems are evident over some 72% of the earth's surface and areas of concern range from the beaches and coastal lagoons on the edge of the continent, across the peripheral shelves and slopes and down to the deepest parts of the ocean, around 11 km deep. It is apparent that even though the mid-ocean areas are relatively isolated from the land systems discussed so far, the oceanic–continental margins act as repositories for the sediment transfer systems discussed in Part III. Indeed, the link between oceanic and terrestrial systems is literally cemented when this sediment accumulation in active geosynclines eventually leads to orogenesis (section 15.1) and the initiation of new terrestrial systems. Ocean systems also influence atmospheric systems in many regions of the earth, mainly through the absorption, transfer and release of substantial amounts of heat. For example, section 2.3 emphasises that the oceans use about 90% of the net radiation to evaporate water and the resultant storage of latent heat in the oceanic boundary layer represents a significant energy source for the genesis of oceanic storms, like wave depressions and tropical cy-

clones (Ch. 13). Furthermore, Chapters 3 and 11 reveal that the advective transfer of sensible heat, associated with the movement of mass and energy in ocean currents, represents an important negative feedback mechanism which helps to offset the excessive heat accumulation in the tropics.

In addition to their importance in sediment transfer and atmospheric systems, the oceans are a great, largely untapped, source of minerals and food which are distributed over the globe by distinctive currents and these features will be discussed in the following two chapters. This chapter examines the formation and morphology of the ocean foor, particularly in terms of the vast increase of knowledge over the last 30 years since, until 1950, the geology of the oceans was mainly based on intuition, despite the introduction of echo-sounding and gravity-measuring techniques in the 1930s. However, by the 1950s, echo-sounding had been greatly refined to operate accurately and cheaply at considerable depths where, for example, modern *precision depth recorders* can detect changes as small as 1 m at depths of 5000 m. Coupled with an increased oceanographic fleet, this detection has expanded the *bathymetric mapping* of the oceans and, at the same time, seismic reflection techniques became routine procedure. Furthermore, deep-sea drilling has been continuing since 1968 and more recently, deep-diving submersibles have been used to observe the features of the ocean floor, especially on the *mid-oceanic ridges*.

This chapter examines the characteristics and morphology of the major oceans and emphasises the three major topographical provinces that make up the oceanic basin. Firstly, the ocean floor features marginal to the land masses which are part of the continental crust, viz. *continental shelves*, *continental slopes* and *continental rises*. Secondly, *deep ocean floors* which are dominated by flat *abyssal plains* and sharply defined *abyssal hills*, *seamounts* and *fracture zones*. Finally, the oceanic ridges which, together with the deep *trenches* in the Pacific continental margins, are the most conspicuous and spectacular topographic feature in all ocean basins.

33.1 The major oceans: characteristics and general morphology

The ocean surface covers about 72% of the surface of the earth whereas the ocean basins together amount to 65% of the earth's surface. Consequently, the volume of water slightly exceeds the amount required to fill the basins and the surplus water spills over and covers the lowest slopes of the continents around their margins. Therefore, the ocean floor can be divided simply

into two major physiographic regions. Firstly, the spill-over zone or *continental margins*, which are represented by the continental shelf and slope, and form only a small part of the ocean system (21%). Secondly, the main ocean basins which have an average depth of 3.8 km but can reach depths of 11.04 km, as in the Mariana Trench which is located in the Pacific Ocean southeast of Japan (Fig. 33.1). These two regions are linked by the continental rise, a transition zone which is known to occur over a distance of 200 km or less (Kennett 1982). It should be emphasised that the ocean basin region is characterised by a complex pattern of topographic features, dominated by ridges or rises (mostly in mid-ocean) with intervening deep-sea basins between the ridges and continents (Fig. 33.1). In fact, the deepest trenches in the oceans occur close to the land, facing the high mountain ranges on the continental edges, so that both areas exhibit greatest vertical change over a narrow crustal zone.

Table 33.1 indicates that there are three major oceans, viz. the Pacific, Atlantic and Indian. The

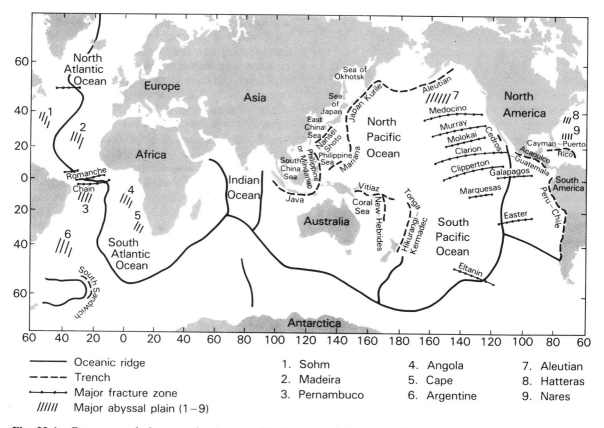

—— Oceanic ridge	1. Sohm	4. Angola	7. Aleutian
- - - Trench	2. Madeira	5. Cape	8. Hatteras
•—•— Major fracture zone	3. Pernambuco	6. Argentine	9. Nares
////// Major abyssal plain (1–9)			

Fig. 33.1 Ocean morphology: major topographic features of the ocean floor.

Table 33.1 Area, volume and mean depths of the ocean. (After Kennett 1982)

Ocean and adjacent seas	Area (10^6 km²)	Volume (10^6 km³)	Mean depth (m)
Pacific	181	714	3940
Atlantic	94	337	3575
Indian	74	284	3840
Arctic	12	14	1117
Total	361	1349	3729
Area of earth's surface	510		

Pacific is the largest ocean, occupying about 30% of the earth's surface and about 50% of the total area of ocean systems. It is also the deepest ocean, with a mean depth some 200 m deeper than the total oceanic average, associated with a preponderance of surrounding deep trenches which isolate the deep-sea basins from continental sedimentation. Figure 33.1 illustrates that these trenches extend almost continuously from the east of the North Island, New Zealand, through the South Pacific islands to the Philippines, Japan, the Aleutians and Alaska. However, the trench is absent (and indeed forms a conspicuous 'gap') along most of the Canadian and American seaboard but it reappears off the southern California peninsula and dominates the Central/South American coast down to southern Chile. Two other distinctive morphological features in the Pacific are the large number of volcanic arcs, particularly in western and central areas, and the presence of extensive *marginal basins* up to 2000 km wide. The latter feature dominates the western Pacific between the offshore deep trenches/volcanic islands and the land masses of Asia and Australia (e.g. the Seas of Okhotsk and Japan, and the China, Philippine and Coral Seas; Fig. 33.1).

The Atlantic Ocean is the second largest (about half the area of the Pacific) extending as a relatively narrow, elongated and winding basin between the arctic and subantarctic regions. The Atlantic is about 150 m shallower than the total oceanic average and this is associated with a preponderance of continental shelves and the mid-oceanic ridge, which extends almost continuously from north of Iceland to southeast of Tristan da Cunha (Fig. 33.1). In fact, the deep trenches which characterise the Pacific are generally absent in the Atlantic Ocean apart from the Puerto Rico–Cayman Trench in the Caribbean and the South Sandwich Trench in subantarctic waters, both of which are about 8.4 km deep.

The Indian Ocean is the third largest and its mean depth is very close to the total oceanic average, resulting from a small area (9%) of continental shelves. The deepest locality is the Java Trench off Indonesia (Fig. 33.1), where the recorded depth is 7.5 km, which is the only trench known in the Indian Ocean and appears to be a westward continuation of the group of arcuate trenches of the western Pacific (Shepard 1963). The Arctic Ocean completes the listing in Table 33.1 but it is a minor, shallow, land-locked ocean with a large part of its area (68%) composed of continental shelves and slopes. There are also a large number of much smaller, mainly land-locked seas lying between continental blocks which were not listed in Table 33.1, including the Mediterranean Sea, Black Sea, Gulf of Mexico and North Sea. Each one has a distinctive morphology related to the physiography of the adjacent land masses, but they are generally shallow water bodies composed mainly of large basins separated by *sills* or rises and seamounts.

33.2 The continental margins

This topographical province lies between the land masses and the ocean basins and although it varies considerably from region to region, there are two main types. The first is termed the *Atlantic-type margin* and is characterised by a gradual transition and increasing depth over 800 km or so from the continental shelf to the continental slope and continental rise (Fig. 33.2). This type of margin is often termed an *aseismic* or *passive margin* because it is seismically inactive, lacking earthquakes and widespread volcanism. It occurs on the edge of stable, subsiding continental blocks, with little deformation since Palaeozoic times, and surrounds most of the Atlantic and Indian Oceans. Furthermore, the Atlantic-type margin is a *divergent* or *constructive zone* associated with rifting of the continental crust and *sea-floor spreading* (section 33.4). Eventually, it becomes a site of massive subsidence with a thick accumulation of sediments, which obscure the precise location of the boundary between the continental and ocean crust.

The second type of margin, which dominates

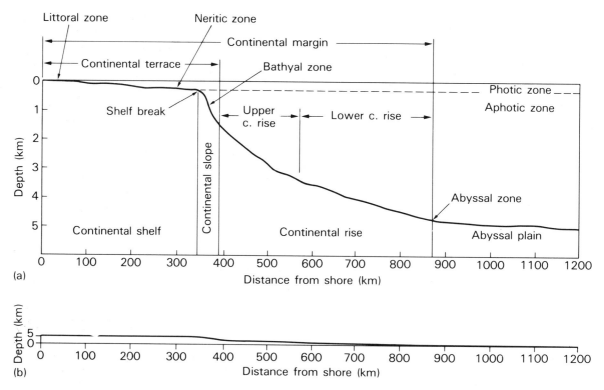

Fig. 33.2 Principal features of the continental margin: (a) vertical exaggeration 1/50 (b) no vertical exaggeration. (Adapted from Kennett 1982)

the Pacific Ocean, is termed the *Pacific-type margin* and has been further subdivided into the *Chilean-type* and *island arc* or *Mariana-type*. The former type is characterised by a narrow shelf with a trench immediately below the continental slope whereas the Mariana-type exhibits a shallow marginal basin up to 2000 km broad and landward of the island arc and trench system, which typifies the Philippine plate (Fig. 33.3). Both these Pacific-type margins are often called *seismic* or *active margins* because of the proximity to active earthquakes (which can extend to a depth of as much as 700 km) and volcanic activity. They represent the destructive boundary of two converging plates (Fig. 33.3) of contrasting densities which results in the sinking or *subduction* of the more dense plate to the *Benioff zone* (Fig. 33.4(a)). The crust is deformed into a deep curved trench by plate subduction (Fig. 33.4(a)) and volcanic island arcs can appear (as in the Mariana-type) from the upwelling of the slab-melted lithosphere in the form of *magma* with earthquakes developing from the stresses along the descending plate.

A good example is found in New Zealand where the Pacific plate from the east subducts below the Indo-Australian plate to the west (Fig. 33.3). The resultant deformation forms the Hikurangi trough or trench off the east coast of the North Island (Fig. 33.1) and the associated *slab melting* in the Benioff zone produces stresses and vulcanicity which are responsible for the volcanic centre of the North Island, between Tongariro and Rotorua, and earthquake epicentres along the east coast. In fact, *convergent* or *destructive margins* are found around most Pacific coasts, apart from the British Columbia to northern California 'gap' (discussed in the last section) which can be classified as a *conservative margin*, and does not gain or lose material. Indeed, the Pacific and North American plates slide past each other here at the *San Andreas Fault* (Fig. 33.3), which represents a massive type of *tear fault* known as a *transform fault*. However, convergent margins are much more restricted in the Atlantic and Indian Oceans where the margins are more commonly divergent. The exceptions are located at the Puerto Rico–Cayman, South Sandwich and Java trenches described earlier, where convergence and subduction occur.

It is apparent that the continental shelf is the

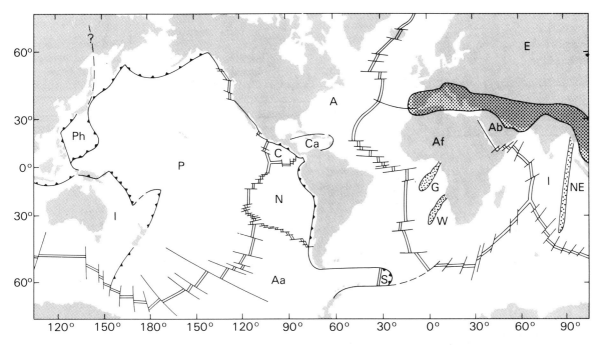

Fig. 33.3 The present system of plates. Ridges: double lines; trenches: toothed lines; transform and strike – slip faults: single lines. (A) American plate; (Aa) Antarctic plate; (Ab) Arabian plate; (Af) African plate; (C) Cocos plate; (Ca) Caribbean plate; (E) Eurasian plate; (I) Indo-Australian plate; (N) Nazca plate; (P) Pacific plate; (Ph) Philippine plate. The Alpine-Himalayan Tertiary collision zone is stippled. Aseismic ridges: open stipple (W) Walvis ridge; (G) Cameroon line; (NE): Ninety East ridge; (S) Scotia arc. (After Oxburgh 1974)

most conspicuous part of the topography of the continental margins, especially along the divergent coasts of the Atlantic, Indian and Arctic Oceans. Furthermore, the average width of the shelf is some 78 km (although the widest shelves, about 400 km, are found in the Arctic Ocean) and everywhere, the shelf slope is very gentle (less than 1 : 1000 or 1 m km^{-1}) with very subdued relief forms less than 20 m high. The classical hypothesis associated with the origin of continental shelves emphasised the combination of wave-cut and wave-built terraces forming under present conditions, with exposed rock terraces on the landward side and sediment deposition to the seaward, sloping down gradually to the ocean basins. It is now evident that shelves have a multiple origin and that the present morphology is the cumulative effect of erosion and sedimentation (composed of a mixture of terrestrially derived debris and calcareous organic remains) associated with a number of large-scale sea-level changes during the last 1 million years or so (Kennett 1982). There is no doubt that lowering sea-levels, resulting from continental glaciation, must be paramount in shelf development, especially since low levels had a duration of many thousands of years which allowed wave erosion and deltaic deposition to proceed. The present seaward edge of the shelf or *shelf break* (Fig. 33.2) is believed to have been formed about 18 000 years ago, when sea-level stood at that point due to the removal of water from the oceans as large ice sheets. The position of the shoreline at that time coincided with, or was located slightly inland of, the present shelf break (Kennett 1982).

At the shelf break, the continental slope (probably of *diastrophic* origin, i.e. crustal dislocation) deepens rapidly from between 100–200 m to 1500–3500 m. The sea-bottom gradient is steep and can exceed 1 : 40 although the average slope is 4° over a narrow zone of some 200 km. However, the slopes are more precipitous in the Chilean-Pacific type discussed earlier, where a deep trench is located at the foot of the slope. It is more common (especially in the Atlantic and Indian Oceans) for the slope to merge gently into the continental rise (Fig. 33.2), which represents the physiographical division between the slope and the

323

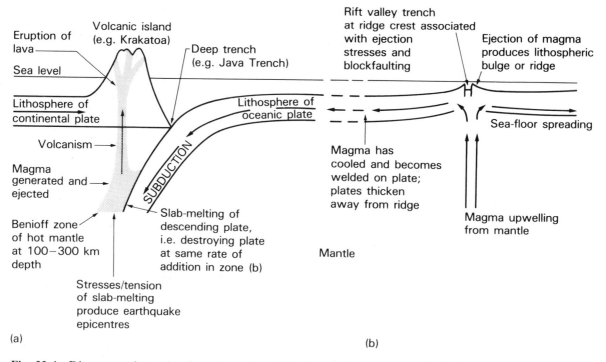

Fig. 33.4 Diagrammatic section (not drawn to scale) of ocean-floor topography in: (a) convergent; and (b) divergent zones.

ocean basins. The continental rise is a zone up to 200 km wide, characterised by a gentle seaward gradient (1 : 100 to 1 : 700) terminating abruptly in the flat *abyssal floor* discussed in the next section. Furthermore, the amplitude of local relief is less than 40 m, since the rise is formed of sediment (several kilometres thick) which has been transported from the land masses and deposited at the base of the continental slope.

However, when this transport of sediment is combined with high-velocity *turbidity currents* (section 34.8) it produces a swirling submarine avalanche and the resultant abrasive action is responsible for the erosion of deep *submarine canyons* or *deep-sea channels* in the continental rise. Furthermore, these channels may continue upwards and downwards incising the continental slope and shelf above and the abyssal plain below. These canyons are winding V-shaped channels with *deltas or deep-sea fans* of sediment at the downslope end, associated with a decrease in slope and a reduced current speed. The best-known examples have been studied off the coast of California, USA and include the La Jolla, Scripps, Monterey, Coronado and San Lucas Can-

yons. Other examples include the Tokyo Canyon (Japan), the Trincomalee Canyon (Sri Lanka), the Hudson Canyon (northeast USA) and the Mediterranean Canyons (west of Corsica and off the French/Italian Riviera).

It should be noted from Fig. 33.2 that other important zones characterise the continental shelf, which broadly correspond with the physiographic divisions discussed so far. For example, the *littoral zone* represents the shoreline area between high and low water spring-tides and merges with the *neritic zone*, which forms the major part of the shelf and terminates at the shelf break. Finally, the *bathyal zone* corresponds with the continental slope and the upper continental rise and merges with the *abyssal zone* (below 2 km on the lower part of the rise), which extends down to the abyssal floor at depths exceeding 4 km.

33.3 The deep ocean floor

The ocean basin floor lies between the continental margins discussed in the above section and the mid-oceanic ridges (section 33.4), and is characterised by three major physiographic subdivisions. Firstly, the abyssal floor which is further divided into plains and hill provinces. Abyssal plains were first discovered in the late 1940s in the Atlantic and Indian Oceans and are now known to be widespread features in both these oceans. They also occur in smaller seas, like the western Mediterranean, the Gulf of Mexico and the Caribbean, but are not common in the Pacific where the deep marginal trenches act as sediment traps. The only exception here is the Aleutian or Alaskan Plain (Fig. 33.1) which is situated to the west of the British Columbia to northern California 'gap' described in earlier sections (where deep trenches are not found). The best examples have been observed in the Atlantic Ocean (Fig. 33.1) with the Hatteras, Nares and Sohn Plains to the east and south of the USA, the Madeira and Cape Verde Plains (off northwest Africa), the Pernambuco Plain (off Brazil), the Angola and Cape Plains (off southwest Africa) and the Argentine Plain (off Uruguay and Argentina).

Abyssal plains occur at a depth of between 3000 and 6000 m and are extremely flat ocean-floor areas, with a slope of less than 1 : 1000 and an horizontal extent of up to 2000 km. They are characterised by considerable thicknesses of deep-sea sedimentary deposits of *oozes* and *red clays* (section 34.7), some 2500 m thick on the Argentine Plain, which have obliterated the original topographic irregularities to produce flat, featureless submarine surfaces. The largest and flattest plains are found in oceans, like the Atlantic, where large rivers drain off the continent depositing vast amounts of sediment over the ocean floor. Conversely, in the Pacific, with fewer large river systems and a preponderance of trenches which act as sediment traps on the ocean margins, the ocean basin receives considerably less sediment and fewer plains form (e.g. only the Aleutian Plain described above). On the landward edge of the abyssal plains, the slope increases rapidly as the continental rise takes over and, in some cases, the abyssal plains merge with the deep-sea fans on the seaward edge of the rise. Alternatively, the plains are incised by abrading turbidity currents when the submarine canyons on the rise (discussed earlier) are extended into the abyssal deposits.

Abyssal hills are small, sharply defined hills which rise from the abyssal plains to altitudes of less than 1000 m (averaging 200 m) with slopes ranging from 1 to 15°. The hills usually occur in groups (up to 10 km long) which are located between the abyssal plain and the slopes of the mid-ocean ridges (section 33.4) where the sediment deposition is thinnest and does not obliterate original basement irregularities, like ocean-floor ridges. However, even though most of these features have a structural origin, some abyssal hills are composed of thin sedimentary deposits which have buried small volcanoes. Abyssal hills are very common in the three major oceans but they dominate the Pacific where they cover about 80% of the ocean floor, compared with a 50% dominance in the Atlantic (Gross 1972).

The second physiographic subdivision of the ocean floor is associated with volcanoes and volcanic ridges, which rise about 1000 m from the sea floor as seamounts (e.g. Gilbert Seamount, Gulf of Alaska, 52° N, 150° W). They normally occur as isolated conical hills rising abruptly out of the abyssal plain but they are sometimes clustered as *seamount chains*, which can rise above sea-level to form island chains like Hawaii and Madeira. Seamount features are common in all major oceans (particularly the Pacific Ocean) and, over time, the volcanic cones are truncated by subaerial or shallow-water wave erosion prior to subsidence of the ocean floor or submergence following a rise in sea-level. The resultant drowning produces submerged, flat-topped seamounts which are called *guyots* or *tablemounts*, which now occur about 2 km below the sea surface (e.g. Pratt guyot, south of Greenland, 56° N, 42° W).

The third and final ocean-floor subdivision is an extensive area of *transverse faults*, represented by linear fracture zones of irregular topography marked by troughs and escarpments. In fact, the ocean floor is cut by innumerable transverse faults (10–100 km wide) which displace the mid-oceanic ridges (section 33.4) and form a pattern of semi-parallel fractures on an ocean-floor map. Some major, individual fracture zones exceed 2000 km in length and consist of a number of linear ridges and valleys, with cliff-like sides exceeding 3 km. These occur in all the major oceans but they are most conspicuous in the eastern North Pacific (Fig. 33.1), extending westwards from the continental slope of California and Mexico as five major fracture zones some 2000–3000 km long which, from north to south, are named the Medocino, Murray, Molokai, Clarion and Clipperton fracture zones. Other major zones exist in the Pacific off Central/South America (the Galapagos and Easter fracture zones) and in equatorial parts of the Atlantic (Romanche and Chain zones) separating the Cape Verde and Pernambuco abyssal plains. Most of the major fracture zones appear to have some connection with the active seismicity of the adjacent land since, for example, the Medocino and Murray zones appear to be related to bends in the San Andreas Fault, whereas the Clarion fracture appears to connect with the east–west volcanic belt of southern Mexico (Shepard 1963).

33.4 Mid-oceanic ridges

Figure 33.1 reveals that the most outstanding topographic feature of all ocean basins is the mountainous ridge which extends some 80 000 km along the middle part of the ocean floors (except in the North Pacific, where it merges with the west coast of the USA and Canada). This mid-oceanic ridge represents a broad submarine mountain range some 1000 km wide, and rising to a crest between 1000 and 3000 m above the adjacent ocean floor. On average, the submerged crest has a depth of about 2 500 m but the highest parts can rise above the sea surface to form scattered islands. These islands are particularly conspicuous in the Atlantic where, from north to south, they are represented by the Azores, St Paul Rocks, Ascension, Tristan da Cunha and Diego Alvarez islands.

The actual topography of the ridge varies along its length, ranging from broad (up to 4000 km wide), smooth non-faulted features in the southeast Pacific to a much more rugged faulted relief in the Atlantic and Indian Oceans with regular lateral displacement by fracture zones (section 33.3). Here, a *central rift valley*, up to 2 km deep and about 30 km wide, lies along the ridge axis with rugged inward-facing scarps produced by block-faulting. These submarine rift valleys correspond with the *graben structures* on the high East African Plateau and, indeed, the mid-ocean rift valleys merge with the African rift valleys in the Gulf of Aden. The ridges are composed of basalt, gabbro and serpentine and have been formed by the upwelling of magma from the mantle to the surface, with associated sea-floor spreading and plate movement away from the ridge axis (Fig. 33.4(b)). Volcanoes, both at the surface and underwater, are still active at a few locations along the ridge, particularly in the Atlantic where they form the basaltic islands which extend from the Azores to Tristan da Cunha, as described above.

33.5 Retrospect

The ocean floor has a complex structure and morphology although *plate tectonic theory* has provided a framework for interpreting the evolution of all topographic features. The continental margins represent the most complicated province in the ocean system which is subdivided into Atlantic and Pacific types, representing respective divergent and convergent zones with markedly different topography (i.e. a predominance of broad accumulation zones in the Atlantic/Indian Oceans, and subducted trenches, with widespread volcanism and earthquakes, in the Pacific). The topography of the continental margins is characterised by a transition over 800 km from gently sloping shelves to steep slopes and gentle rises (incised with submarine canyons), terminating abruptly in the deep ocean floor. This latter province again has widespread relief, ranging from flat abyssal sedimentation plains to sharply defined abyssal hills and scattered volcanic seamounts and eroded, submerged guyots. The floor has also been fractured into major linear zones of irregular topography, which are most conspicuous in the North Pacific. Finally, the most outstanding topographic feature of all ocean basins is the mid-ocean ridge which is an extensive submarine mountain range (with a central rift valley in the Atlantic and Indian Oceans) formed by the upwelling of magma from the mantle, and associated sea-floor spreading. Furthermore, in the Atlantic Ocean, the ridge has been displaced laterally by innumerable transverse faults, with major fractures located particularly in the vicinity of the Romanche–Chain zones (Fig. 33.1)

Key topics

1. Problems in the classification of continental margins.
2. The physical characteristics of ocean margins.
3. The physiography of the deep ocean floor.
4. The influence of tectonic processes on mid-oceanic ridges.

Further reading

Anikouchine, W. A. and Sternberg, R. W. (1973) *The World Ocean, An Introduction to Oceanography*, Prentice-Hall; London and New Jersey.

Couper, A. (ed.) (1983) *The Times Atlas of the Oceans*, Times Books Ltd; London.

Emery, K. O. (1980) Continental margins – classification and petroleum prospects, *Amer. Assoc. Petrol. Geol. Bull.*, **64**, 297–315.

Heezen, B. C. and Hollister, C. D. (1971) *The Face of the Deep*. Oxford University Press; London and New York.

Holcome, T. L. (1977) Ocean bottom features – terminology and nomenclature, *Geojournal*, **6**, 25–48.

Uyeda, S. (1978) *The New View of the Earth, Moving Continents and Moving Oceans*. W. H. Freeman and Co.; San Francisco.

Chapter 34 Sea water movement and marine deposition

Movements in sea water occur on various scales of depth, from those that affect the topmost few metres (wind waves) to large horizontal flows in the top few 100 m (tides and currents), to circulations extending the entire ocean depth (*thermohaline* circulations). Apart from its mobility, which renders it responsive to external forces, the properties of sea water which are implicated in its movement are basically temperature, salinity and density. In general, temperature decreases with depth but not linearly. Typically, there is a warm surface layer some 100–200 m thick, well mixed by turbulence, and below this is a layer of rapid temperature decrease with depth. This represents the so-called *thermocline* (Fig. 34.1) which separates the surface layer from cold deep water, approaching 0 °C, and which is especially well marked in strongly heated subtropical waters.

The salinity of sea water averages 35‰ (parts per thousand) but ranges from 33 to 38‰ in open water and exceeds 40‰ in the Red Sea. It should be noted that 86% of the salt content is sodium chloride, but salts of magnesium, calcium and potassium are also present. Salinity is increased by evaporation and ice formation and decreased by precipitation, entry of stream runoff and ice melt. In high latitudes in particular, with abundant precipitation and reduced evaporation, surface waters tend to be less saline than average. However, there is a subsurface layer of marked salinity increase which characterises the *halocline*.

The density of sea water increases as temperature decreases and salinity increases. While fresh water has maximum density at 4 °C, water of 36‰ salinity reaches its maximum density at −2 °C, just before freezing. Density is lowest in the warm surface layers and increases sharply with depth (the *pycnocline*), due to a temperature decrease or a salinity increase or both. In the ocean basins, water tends to find its own density level and the resulting stable density stratification dictates that, unlike the atmosphere, strong vertical movements are discouraged, except in certain limited areas. There are, however, points of similarity between atmospheric and oceanic circulations, among them the recognition of *water masses*, characterised by distinctive temperature–salinity relationships (King 1975).

34.1 Waves

Apart from the shock waves or *tsunamis* (that result occasionally from earthquakes, volcanic eruptions or landslides on the ocean floor), the familiar sea waves are disturbances of the water surface due to the action of the constantly eddying wind

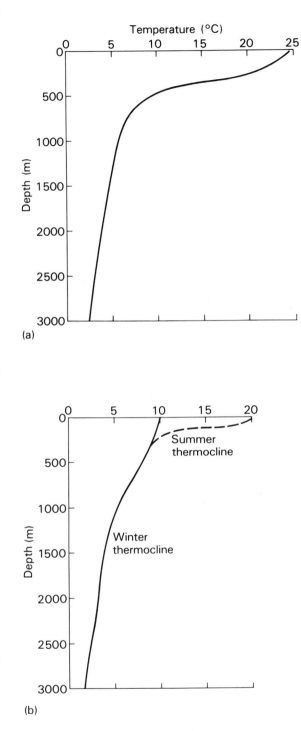

Fig. 34.1 Typical temperature profiles in the open ocean. The thermocline is permanent in: (a) tropical waters; and (b) changes seasonally in middle-latitude waters: it is non-existent in polar waters.

above it. Although waves may move at considerable speeds, it is clear from laboratory wave tank studies that, in deep water, molecules have little net horizontal motion. Indeed, they move in circular orbits in the vertical plane but in shallowing water, as shore is approached, the circles become ellipses which eventually collapse. The wave is then said to 'break' and the motion is translated into the horizontal transport familiar as *surf*. Waves may also break in deep water, when the wind speed increases, and the first 'whitecaps' or 'white horses' appear at Beaufort force 3 (3.4–5.4 m s⁻¹). Wave stability breaks down when the ratio of height (trough to crest) to wavelength (crest to crest) exceeds 1 : 7. Wave heights rarely exceed about 6 m, although a height of 34 m has been reliably recorded in the North Pacific (Gross 1972).

Waves generated by strong winds will spread out initially in all directions but will persist and even undergo amplification only in the downwind direction. In deep water, the speed of wave progression depends only on the wavelength. The long waves, raised under a storm, will travel faster and further than short waves and may be experienced as *swell* far from the parent disturbance, especially with considerable *fetch* (i.e. the longest distance the wind can act in a constant direction). For example, the south coast of England has known swell generated 10 000 km away in the South Atlantic. The actual wave structure at a given location may be extremely complex, combining swell from distant storms and waves generated by winds on the way or *in situ*.

Wave formation and behaviour involve mass and energy transfers between air and water. The kinetic energy of turbulent wind creates potential energy in the head of water raised, which is translated into the kinetic energy of water movement. At coasts, enormous amounts of energy become available for erosive action, though much of it is wasted as frictionally produced heat. Consequently, research into harnessing wave energy is said to hold promise for the future. When wave crests break, air bubbles are introduced into the oceans, replenishing their store of dissolved gases (oxygen, nitrogen and carbon dioxide) and water droplets are left suspended for a short while in the air as spray. The evaporation of these droplets represents a latent heat transfer (section 4.2) and also supplies the bulk of the atmosphere's condensation nuclei in the form of salt particles (section 5.1).

34.2 Tides and tidal currents

Tides are periodic rises and falls of sea-level due to astronomical forces. It is convenient to consider initially earth–moon relationships and to assume that the earth is completely water-bound and non-rotating. Earth and moon are bound together in a circulation about a common centre, which is actually located within the earth (because of its far greater mass) about 4700 km from its own centre. In this revolving system, mutual gravitational attraction (which pulls earth and moon together) is balanced by an outwardly directed centrifugal force. As far as the water is concerned, directly under the moon (the sub-lunar point), the attractive force exceeds the centrifugal, so that the water surface is pulled towards the moon. On the diametrically opposite side of the earth, the centrifugal force (which is the same all over the earth's surface) exceeds the attraction (which is weakened by the greater distance), so that the water surface is again pulled away from the earth. At these two points only, the tidal force acts vertically: everywhere else, there is a horizontal traction force which moves water horizontally towards the sublunar point and its opposite. In this way is created a somewhat egg-shaped skin of water covering the almost spherical earth (Fig. 34.2).

As the earth rotates about its axis, the distended water envelope remains fixed by the location of the moon, so that each longitude experiences two high tides and two low tides every 24 hours 50 minutes (i.e. the period between successive tracks of the moon over a given earth point). Moreover, in a lunar month, the moon moves from a position $28\frac{1}{2}°$ N of the equator to $28\frac{1}{2}°$ S of the equator, which introduces inequalities in the two daily tidal bulges. Tidal behaviour is further complicated by the solar influence. Although the sun is so much bigger than the moon, it is also much further from the earth (Table 34.1) so that its tidal influence is only about 46% of the

Table 34.1 Some comparisons – earth, sun and moon. (*Source*: Uvarov *et al.* 1979)

	Equatorial diameter (km)	Mass (× earth mass)	Mean distance from earth (million km)
Earth	12 756	1*	–
Sun	1 392 000	332 958	149.60
Moon	3 476	0.0123	0.3844

* Mass of the earth 5.976×10^{24} kg.

moon's. Nevertheless, the sun excites a once-daily tide in a varying relationship with the lunar tide. When both influences act together (i.e. the sun, moon and earth are in line), the highest tides and greatest tidal ranges occur (*spring tides*). When sun and moon are at 90° to the earth, the lowest tidal ranges (*neap tides*) are experienced. The 50-minute difference between the lunar and solar tidal days, as well as the variable distance of both sun and moon from the earth, further contribute to the complexities of many tidal patterns.

All these complications would occur on an entirely water-bound globe. As it is, only in the Southern Ocean does water completely encircle the earth, allowing an unimpeded tide (largely lunar in origin) to be generated. Such a tide was in fact formerly envisaged as the earth's basic tide, being regarded as a progressive wave with a wavelength half the earth's circumference at that latitude (the *progressive wave theory*). The tides of the Pacific, Indian and Atlantic Oceans were thought to be northward-progressing offshoots of the basic Southern Ocean wave. However, it is now accepted that the three ocean basins, or even parts of them, are capable of generating their own tides of a *standing wave* form. The analogy is with a rectangular tank of water, which can be rocked about a central axis. This would simulate a high tide at one end and a low tide at the other, with a *nodal line* in the middle where no rise and fall occur but the maximum horizontal movement or *tidal current* is found. In the actual oceans, 'tanks' can be identified of such dimensions as to have a natural period of oscillation of about 12 hours. These can resonate to the lunar pull and develop strong standing waves.

A further development of the standing wave theory recognises the potency of the Coriolis effect (section 8.2). This imparts a rotatory twist to the tidal currents and the nodal line is replaced by an amphidromic point of no rise and fall. The ac-

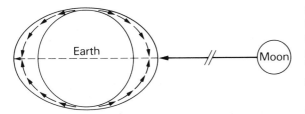

Fig. 34.2 The horizontal flow of water on a water-bound earth, in response to the moon's tidal pull.

tual tidal pattern at a given location may be a mixture of progressive and standing waves and some areas are at or near an amphidromic point for the lunar oscillation and experience only solar tides. Standing waves are particularly well developed in bays of suitable dimensions (e.g. the Bay of Fundy in Canada). Furthermore, narrowing and sharply shallowing estuaries can concentrate the tidal rise so markedly as to produce an advancing wall of water (the tidal *bore*), as in the estuary of the River Severn in the UK or the mouth of the Colorado River in the Gulf of California, USA.

34.3 Ocean currents

Similarities between the global pattern of prevailing winds and that of the main ocean currents (Fig. 34.3) suggests that the ocean circulation is basically wind-driven. However, wind stress is only one of several forces influencing current formation, although it is responsible for the so-called *drift currents*. Observations of winds and ice drift in the Arctic Ocean during the 1890s showed that the wind-driven surface currents deviated to the

right of the wind direction, an effect readily attributed to the Coriolis deflection. Ekman (1905) established theoretically that the angle of deflection should be 45° at the surface, increasing with depth to 180°, though the speed of movement here is frictionally reduced to very little (the original Ekman spiral, section 8.4). The net movement throughout the layer affected (the *Ekman transport*) is at 90° to the surface wind. The relationships of, for example, the northeast trades to the north equatorial current, or of the southwesterly winds to the North Atlantic Drift are now more clearly seen. The significance of wind is also well illustrated in the northern Indian Ocean, where the seasonal wind reversals are accompanied by current reversals (west–east in summer and east–west in winter).

The drift currents necessarily flow broadly east–west or west–east. When, as in the Pacific, Indian and Atlantic Oceans, they meet the adjacent continental coasts, a head of water is piled up, which is relieved by currents flowing north or south. These are the so-called *boundary currents*. Since earth rotation tends to displace water bodies westwards, this reinforces the effect of the trade winds, so that western boundary currents are stronger than those at eastern boundaries. Western boundary currents, like the Gulf Stream or the Brazil current, flow polewards as warm currents. Conversely, eastern boundary currents, like the Peru current or the Californian current, return equatorwards as cold currents. The contribution of these water movements to the global heat balance has already been mentioned (Chs 3 and 11).

Along some western continental margins, the prevailing winds blow broadly parallel to the coast. The Ekman transport now removes surface water offshore, which causes an *upwelling* of cold *benthos* water from below the thermocline. Off the coast of Peru, this upwelling water forms the colder Peru coastal current which is sometimes distinguished from the less cold Peru oceanic current (a return boundary current). The upwelled water is usually nutrient-rich and very productive in terms of plankton and fish populations. However, the periodic replacement of the cold current by abnormally warm water (*El Niño*), when the prevailing along-shore winds fail (Johnson 1976), temporarily wrecks this ecological balance, causing high mortality among fish and birds. El Niño events (which occur roughly one year in seven) also bring dramatic changes in the weather of the region, replacing almost complete aridity with tropical downpour, and also seem to have more

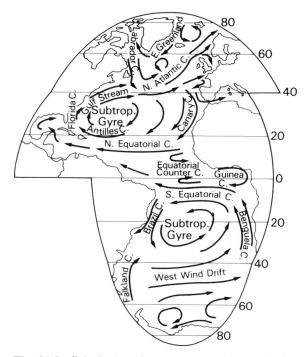

Fig. 34.3 Principal surface currents of the Atlantic Ocean. The Pacific and the southern Indian Oceans have broadly similar patterns.

widespread climatic repercussions (Perry and Walker 1977).

The combined effect of drift and boundary currents produces surface water circulations or *gyres* in subtropical latitudes (Fig. 34.3) in an anticyclonic sense (i.e. clockwise in the northern hemisphere and anticlockwise in the southern). The Sargasso Sea, for example, is at the centre of the North Atlantic subtropical gyre. Also, somewhat smaller cyclonic subpolar gyres exist, mainly in the North Atlantic and North Pacific. In the anticyclonic gyres, water accumulates due to the Ekman transport, thickening the surface layer to produce a minor 'hill' in the ocean surface topography. On the other hand, there are also minor 'hollows' in the surface, for example, off west coasts where upwelling occurs. The maximum difference of water-level is normally of the order of 2 m, though the extremes are vastly separated in space. These gradients in ocean surface topography (analogous to pressure gradients in the atmosphere) induce water movement, which is deflected by the Coriolis force to flow along the water contours. Such geostrophic currents around the anticyclonic gyres operate at subsurface levels, but certain surface currents (e.g. boundary currents) are also geostrophic in the accepted sense. The Equatorial counter-current, however, flows directly downslope since the Coriolis force here is nil or negligible.

34.4 The thermohaline circulation

The depth or thermohaline circulation of the oceans is essentially density-driven. Vertical movements result from instability (just as in the atmosphere) but in the oceans, this occurs in rather limited areas where surface water of high density can form. Such water sinks as a distinctive water mass and spreads horizontally, where it reaches its own density level. The densest water mass is created in the Weddell Sea, Antarctica, due to ice formation and cooling of surface water to about 1.9° C. This sinks as *Antarctic bottom water* (AABW) to the ocean floor and creeps northwards at speeds of only 1–2 cm s^{-1} to reach as far as 45° N in the Atlantic (Fig. 34.4). A comparable Arctic bottom water is largely confined to that ocean basin by submarine ridges, extending from Greenland to Scotland, and the overflow mixes with other Atlantic waters. Of greater importance is the *North Atlantic deep water*

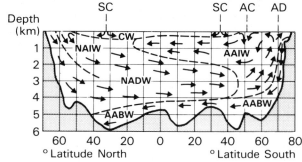

Fig. 34.4 Main features of the depth circulation of the Atlantic Ocean: SC – subtropical convergences; AC – Antarctic convergence; AD – Antarctic divergence; CW – central water; AAIW, NAIW – Antarctic and North Atlantic intermediate water; AABW – Antarctic bottom water; NADW – North Atlantic deep water.

(NADW), which forms in the Greenland and Norwegian Seas where warm saline water, deriving from the Gulf Stream, is chilled to about −1.4 °C by mixing with the Arctic overflow. NADW sinks to form the bottom water in the North Atlantic but, being less dense than AABW, flows above it south of 45° N (Fig. 34.4).

The shallow central water masses (CW) form by downwelling at the *convergences*, which coincide with the subtropical gyres. *Intermediate water* is formed by slow mixing of adjacent water masses or by downwelling at subpolar convergences. Upwelling (such as off the Peru coast) represents an oceanic *divergence*. An important divergence is found around Antarctica, where prevailing westerly winds (reinforced at times by strong katabatics blowing off the ice plateau) create an offshore Ekman transport. Upwelling here brings a distinctive water mass (Antarctic circumpolar water) to the surface. At about 50° S, northward-blown Antarctic surface water meets less cold water at the *Antarctic convergence* (sometimes called the Antarctic polar front) and sinking occurs here to give Antarctic intermediate water (AAIW). A similar effect in the North Atlantic (south of the Labrador Sea) is on a much smaller scale.

34.5 Marine deposition

The ocean floors are largely covered with sediments of a depth averaging around 600 m, though it may exceed 2 km in places. There are also some areas bare of deposits, either because they are newly formed parts of the ocean floor or because bottom currents are here strong enough to sweep them clean, but these are relatively few. The character and thickness of ocean-floor sediments vary according to the available sources (which means in effect the nature of adjacent land) and the processes of transport (including, among others, those considered earlier in this chapter) and of deposition. This branch of oceanography owes much to the pioneer studies carried out by John Murray and his associates on the famous cruise of HMS *Challenger*, during the years 1872–76 (Murray and Hjort 1912).

34.6 Sources of marine deposits

Most marine deposits originate over land. It has already been pointed out (Ch. 33) that the oceans act as repositories for all manner of materials resulting from weathering and subaerial erosion on neighbouring continental areas. For this reason, sediments are usually thickest near the continental margins. The materials are transported into the oceans by a number of sediment transport systems discussed in Part III, particularly by rivers and, to a lesser extent, by glaciers and winds. Some of these materials are used by marine organisms (Ch. 35), lowly one-celled animals or plants, to build their skeletons or tests and these eventually add a *biogenic* content to the accumulation of deposits on the ocean floor. The inorganic or *lithogenic* deposits include water-transported material, which is widely distributed but is found especially in low and middle latitudes, where large rivers are located. Inorganic deposits also include glacial material (mainly in the fiords of high latitudes and sea areas where icebergs commonly melt), volcanic material (which may originate from marine or subaerial vulcanicity) and wind-borne material (near the northern hemisphere deserts). Small amounts of ocean-floor material derive from meteoric dust and from anthropogenic wastes of various kinds, which are found mainly on continental shelves near urban and industrial sites.

34.7 Types of deposit

The original classification of *pelagic* or deep-sea deposits, due to John Murray (Murray and Hjort 1912), has survived with little basic change, except in nomenclature. Murray differentiated inorganic (lithogenic) sediments, which he grouped together as 'red clays', from largely organic (biogenic) deposits which he characterised as 'oozes', if they contained more than 30% by volume of biogenic material. He further subdivided the latter into *calcareous* or *siliceous* oozes, according to the chemical nature of the skeletal parts they contained. The so-called 'red clays' consist of very fine mineral grains of terrigenous origin, mainly in the clay grade (i.e. less than 4 μm, and often around 1 μm, in diameter). The colour is due to conversion of the iron content to ferric oxide (rusting) by the oxygen dissolved in sea water. However, by no means all such deposits have undergone so much oxidation and other terms such as 'brown clay' or 'lutite' or 'deep-sea mud' are also in use. The fine size of these particles allows easy transport and wide distribution.

The calcareous oozes (sometimes now referred to as 'muds') comprise material rich in the skeletal remains of the pelagic foraminifera *Globigerina* (widely spread) and the less common mollusc *pteropod*, as well as the *coccoliths* or platelets formed by the unicellular algal plant coccolithophoridae. The siliceous oozes result from the accumulation of the hard frustules of the siliceous algae called *diatoms* and the skeletons of the planktonic animals *Radiolaria*, both one-celled. The organic oozes reflect rich concentrations of calcium carbonate or silica, according to the dominant type, in solution in the overlying waters and this in turn may represent biologically highly productive surface waters, often in areas of upwelling (sections 35.1 and 35.2).

Another type of deposit is metalliferous. *Ferromanganese nodules* (which may also contain copper, cobalt and nickel) are widespread, sometimes buried in other sediments, sometimes lying on them. They consist of concentric coatings around nuclei which may be fragments of volcanic material or of sharks' teeth or coccoliths. Nodules of up to 850 kg weight have been found. Although their origin is by no means clear, it may be related to a solution phase of manganese and iron, perhaps in hot water associated with submarine vulcanicity. Volcanic material on the ocean floor may be in the form of rock fragments, volcanic glass and minerals (especially montmorillonite) formed as a result of reaction between newly extruded volcanic material and hot water. Volcanic fragments and glass are mainly basaltic (as are most volcanic islands) and these elements may be relatively large (greater than 50 μm). On the other hand, particles deriving from volcanic eruptions over land and settling after considerable transportation by winds may be very fine (less than 10 μm).

Detrital materials, obviously related in origin to adjacent land areas, are found at the oceanic margins often with a thickness exceeding 1 km and displaying a range of particle size. The finest clays may be difficult to distinguish from the pelagic red or brown clays and indeed, as is now thought, contribute to them through transport by currents. However, the process that introduces these varied lithogenic materials on to the deep ocean floor is very different from those responsible for the accumulation of the typical pelagic deposits.

34.8 Processes of marine deposition

The bulk of the pelagic deposits form far from land by slow down-settling through the overlying depth of ocean water. This particle-by-particle accumulation proceeds at a rate which is governed theoretically by Stokes' Law for the size of particle concerned. This states that for particle diameter up to 200 μm, assuming spherical particles of similar density, the fall speed of each particle through a fluid is proportional to the square of its radius. Thus, larger particles settle faster than smaller ones. Table 34.2 indicates settling times for some typical grain sizes through an average 4 km depth of water to the underlying floor. Particles of diameter exceeding 200 μm, or of irregular shape, fall more slowly than is predicted by Stokes' Law because of increased resistance due to turbulence set up around them. Accumulation of sediments by this means is at rates less than 1 cm in 1000 years.

Table 34.2 presupposes still water, but the suspended material in ocean water is also subject to horizontal transport in the complex oceanic circulation described in section 34.3. Thus, it would appear that while sands and larger-sized material

would settle very near the point of discharge, silts could be transported some distance and clays vast distances before finally settling out. A relationship must exist between particle size, settling velocity and current velocity, which determines whether suspended material will settle or be transported. Some estimates are embodied in Fig. 34.5 which also indicates the possibilities or erosion with larger particles and higher current velocities. It is clear that the fine size, slow settling speed and amenability to horizontal transport of the red clays must result in a very wide distribution throughout the oceans. Furthermore, since this material is little altered in transit, high concentration occurs at greater depths where more vulnerable sediments cannot survive. Red clays have been found at depths between about 4000 m and deeper than 8000 m, with a mean depth for 126 samples collected of about 5400 m (King 1975).

When the biogenic elements are considered, other processes become important. For example, the abundance and depths of their deposits depend on the productivity of surface waters (always high at divergences), the proneness of skeletal remains to destruction by fragmentation and solution, and the presence of lithogenic sediments. Siliceous and phosphatic constituents apparently dissolve with equal readiness at all depths and survival rates depend largely on the strength of the skeletal material. Some radiolaria shells are strongly silicified and these deposits have much the same depth range as the red clays. On the

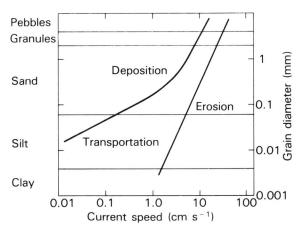

Fig. 34.5 The relation between particle size, current speed and the possibilities of deposition, transportation and erosion. (Simplified after Turekian 1976 and others)

Table 34.2 Some typical settling times.

Radius (μm)	Settling velocity (cm s^{-1})	Time to fall through 4 km
1 (clay)	0.00025	51 years
10 (silt)	0.025	185 days
100 (sand)	2.5	1.8 days

other hand, diatom remains are more destructible and these are most often found at depths between 100 and 7000 m, with a mean depth of 3900 m.

Calcareous elements dissolve little at shallow depths (where the water is usually saturated with calcium carbonate), but this process occurs much more readily at greater depths, aided by lower water temperature and more dissolved CO_2. Below about 4500 m, solution is especially easy, so that calcareous sediment is rarely found deeper than 5000 m. *Globigerina* has a mean deposition depth of about 3600 m, but pterapod shells, which are particularly fragile, are characteristic of shallower depths (mean around 2000 m). Near the continents, a greater lithogenic content often dilutes the biogenic proportion to less than the 30% required to be classified as an organic ooze.

It has been estimated that some 40% (by volume) of total ocean sediment is concentrated in the continental rise. This zone (see Fig. 33.2) represents the thick apron of material comprising the base of the continental slopes (section 33.2) and consists of coalesced deposition fans at the mouths of submarine canyons (section 33.2). Furthermore, it appears from recent studies (Allen 1970) that both canyons and sediments have been formed by the action of turbidity currents (section 33.2), a process unrecognised by Murray and his colleagues on the *Challenger*. The source of this material is to be found in the highly variable shelf and near-shore sediments. These may be muds, sands, gravels, shells, coral reefs, etc. depending on the character of the adjacent land, the climate, the type of weathering, the presence or absence of river and the activity of waves and currents.

The turbidity currents begin with slumping on the upper continental slope, which may be triggered off by shocks due to earthquakes or nearby hurricanes, or simply by unusually heavy sediment loads carried downslope as a dense slurry in existing canyons and deposited as a fan where the break of slope causes loss of momentum and carrying power. As with subaerial alluvial fans, the material increases in fineness both outwards and upwards. A typical *turbidite* sequence has a lowest and innermost layer of coarse unlayered pebbles and sands. This is followed above and beyond by laminated sands, then by cross-bedded fine sands and silts, finally grading into fine silts and clays. Much of the finer material is capable of further transport by quite slow-moving currents and, in this way, the turbidity flows contribute to the lithogenic sediment cover of the deep ocean floor in general.

34.9 Processes at coastlines

Coastal scenery can be highly varied responding as it does to the structure and material of the land margin, the dominant subaerial processes acting thereon and the legacies of isostatic or eustatic movements of sea-level (section 15.2). All coastlines, however, experience the impact of oceanic energy, which manifests itself in processes of erosion due to destructive waves, of transport effected by wave action and longshore drift and of deposition, due to constructive waves sometimes aided by tidal currents. The sediments involved in these processes are predominantly land-derived through delivery by rivers or by mass movements from cliffs, such as landslides and mudflows (Ch. 18).

Erosion proceeds by a combination of pounding by the considerable weight of water hurled against the land margin by breaking waves, plus abrasion from the sediment load carried by the water, together with the hydraulic pressure of air compressed into cracks. With a much jointed or faulted rock material, this powerful direct attack results in familiar features such as caves, arches and stacks. There may also be a contribution from chemical corrosion, though this is most important on limestone coasts. The most destructive waves are steep (i.e. their height is large in relation to their length) and are wind-generated in the locality: on the other hand, the long low wave forms due to swell are less energetic and more associated with constructional processes.

Wave action depends to a large extent on the depth of water and the configuration of the coastline. As the shore is approached and the depth decreases, advancing waves suffer refraction and the crest lines tend to become parallel to the bottom contours. This necessarily concentrates wave energy on exposed headlands where the erosional features of cliff and wave-cut platform are best developed. Bays, on the other hand, are low-energy zones where constructional forms may dominate. Here, erosion can affect the *backshore*, i.e. the zone above normal high tide (Fig. 34.6), only with very destructive storm waves and more usually, the presence of beach material acts as a protective buffer zone in which energy is dissipated in moving material and in friction and percolation. Erosion rates on cliff coasts depend on rock hardness as well as exposure, with cliffs in 'soft' drift material being especially vulnerable to rapid recession. For example, during the North Sea storm surge of 1953, a 2 m high cliff in East

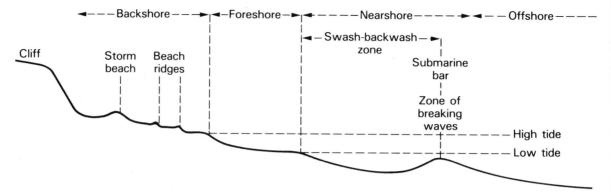

Fig. 34.6 Some terms used in connection with coastal processes. Beach ridges are also called berms. Submarine bars may grow to become barrier islands. (Adapted from King 1980)

Anglia was 'eaten back' more than 30 m in a single night.

The material thus eroded, together with that made available through subaerial processes, is now subject to transport by the action of waves and (less so) of tidal movement. Apart from changing the level at which wave action operates, tides are significant mainly in certain situations where one direction of tidal streaming is dominant. Transport by wave action varies with the nature of the coastline. Within sheltered bays, sediment movement is largely normal to the beach. The steeper (storm) waves remove material from the upper part of the beach and this sediment is carried down by the backwash to be deposited somewhere in the *nearshore* zone (Fig. 34.6). At other times, the strong *swash* or surf associated with long swells moves this material back to the upper beach. This to-and-fro movement thus varies in direction with weather conditions but is of limited extent and by and large the total amount of sediment material in the bay may show little change. With a crenulate coastline, such as is found in southwest England or southwest Ireland, each bay forms a separate system and the rate of transport of material between them is low.

In the case of more regular coastlines, particularly where the dominant winds blow over a considerable fetch, refraction is incomplete and the waves then approach the shore obliquely. As the waves break, the swash is impelled on to the beach at the same angle, but the backwash responds only to gravity and runs down the maximum slope (i.e. normal to the beach). The sediment load thus shifts along a zigzag path and transport can take place over considerable distances. Longshore drift is most effective when the angle of wave incidence is approximately 30–45° to the beach line. Longshore movement can change direction if that of the prevalent waves changes, but usually one particular direction is dominant. For example, on the Channel coast of southern England, maximum fetch and prevailing winds from the southwest ensure a predominantly eastwards transport. Thus, longshore drift can operate over long periods with the general effect of moving sediment from erosion areas to deposition areas.

Coastal accretion occurs in several situations, each representing an excess of sediment income over loss. To take an obvious example, in seas with small tidal range, the sediment brought down by large rivers cannot easily be removed and consequently deltas build out, like those of the Rhône, Nile and the Mississippi. Less obviously, large storm waves (which are usually destructive agents) can throw up shingle beaches through their ability to shift the larger, heavier particles. The coarseness of this material ensures that most of the advancing swash sinks in and the backwash is much reduced, so that little material moves back downbeach. The result of this positive balance of material may be a small temporary ridge destined to be demolished by a later destructive phase, but some high-level storm beaches persist. Chesil Beach in Dorset, England, is 30 km long and up to 13 m high: strictly, this is a *tombolo* linking the island of Portland to the mainland.

While Chesil Beach, with its northwest–southeast alignment, is attributed to storm waves arriving from the southwest, more complex constructional forms can result when there are two or more directions of dominant wave approach. The so-called *cuspate foreland* of Dungeness, further east on the coast of southern England, consists of marshland behind beach ridges built up by storm waves both from the southwest along the English Channel and from the east, which is open to the southern North Sea. Ridges that build out to continue the line of the coast are commonly called *spits*. These may also change alignment to give characteristic angular forms at points where a different direction of wave action becomes operative. Hurst Castle Spit, in the Solent off the south coast of England, is an example: here, a series of such laterals reflects a succession of ridge-building episodes by northeasterly waves.

Sandy spits are built by long swells rather than by storm waves and are again aligned normal to their direction of approach. Wave refraction around the points of growing spits produces rounded, hook forms. Various combinations of coastal configuration, wave incidence, availability of material through longshore drift and sometimes of tidal streaming can result in complex spit formations, such as double spits growing from both headlands of a bay. Some coasts are lined with *barrier islands* which (leaving aside partly submerged ridges of solid rock) may represent remains of breached spits or the build-up of sand bars in the nearshore zone (Fig. 34.6). Landward of these barriers, fine material readily accumulates and conditions become suitable for the development of salt-marshes, with their characteristic vegetation (section 35.5). Yet another barrier form that depends on a different mode of transport can be seen on many flat constructional beaches with a wide foreshore. At low tide, the sands dry out and onshore winds blow fine material inland to give lines of dunes, which are again often backed by coastal marshes.

34.10 Retrospect

The use, in sections 34.3 and 34.4, of terms familiar from Chapters 10, 11 and 12 invites comparison between oceanic and atmospheric circulations. There are obvious similarities, yet also important differences. Sea water movement is very much slower than air flow: for example, even the drift current has a speed only 2–3% of that of the wind causing it. It has been estimated (Turekian 1976) that the average residence time of water molecules in the oceans must be of the order of 40 000 years, such is the slow creep and mixing that comprise most of the oceanic circulation system. The oceans are by far the major water reservoir in the global hydrological cycle (see Fig. 2.1) and the residence time of water molecules in the atmosphere (Ch. 5) and on (or in) land is highly variable but very much shorter. Within the oceans, while surface circulation (which closely 'apes' wind patterns) is largely zonal, deep-water movement is predominantly meridional and transports vast quantities of water between northern and southern hemispheres.

The gravitational accumulation of deep ocean-floor sediments provides a depth sequence which must also represent a time sequence. Various coring techniques, facilitating careful examination of samples with regard to colour, texture, chemical content, fossil remains and other properties (together with techniques of radioactive dating), have encouraged the establishment of a chronology. This has thrown light on problems of global climatic change, as well as the history of the ocean floors (King 1975).

Key topics

1. The control of the density stratification of sea water.
2. The conditions under which sea water moves at different scales
3. Ocean floors as repositories for material produced by terrestrial sediment transfer systems.
4. The importance of erosion, transportation and accretion along coastlines.

Further reading

Dietrich, G. (1963) (trans. by Ostapoff, F.), *General Oceanography: An Introduction*. Wiley-Interscience; New York and London.

Pickard, G. L. and Emery, W. J. (1982) *Descriptive Physical Oceanography: An Introduction*. Pergamon; Oxford and New York.

Weyl, P. K. (1970) *Oceanography: An Introduction to the Marine Environment*. John Wiley and Sons; New York and London.

Chapter 35 Marine ecosystems

As stated in Chapter 33, the oceans cover approximately 72% of the earth's surface which is equivalent to almost two and a half times the area of the land masses and freshwater bodies together. All the oceans are interconnected and their physical characteristics such as currents, depth, temperature and salinity (Ch. 34) restrict the abundance, movement and productivity of marine organisms.

In common with terrestrial ecosystems, NPP (section 29.3 and Table 29.1) values vary greatly in the range 2–3000 g m^{-2} yr^{-1}, with the highest values recorded from littoral ecosystems (those adjoining the land masses) such as estuaries and *coral reefs* and the lowest values from the open oceans. In contrast, however, the average NPP of 152 g m^{-2} yr^{-1} for marine ecosystems is much lower than the 773 g m^{-2} average for terrestrial ecosystems. The range of marine NPP values is a reflection of the variety of marine habitats which are briefly described in this chapter. Nevertheless, it should be borne in mind that although the oceans occupy the largest portion of the earth's surface, marine ecosystems are among those least

documented and understood. This is despite the fact that the oceans contain extensive, but largely untapped, inorganic and organic resources.

35.1 Marine habitats: their global distribution and factors affecting them

The structure of the ocean floor largely determines the nature of marine habitats (see Fig. 33.2). The chief morphological subdivisions of the oceans (which correspond to different types of marine ecosystems) are the littoral zone, which occurs adjacent to the coast, the neritic zone, which occupies the continental shelf, and the oceanic zone which covers the continental slope and rise, the abyssal plain and mid-oceanic ridges. Ecosystems relating to these environments will be detailed in sections 35.3–35.5 but, in general, these habitats are a function and a reflection of three important environmental gradients to which marine organisms respond. These are the depth gradient from the ocean surface to the ocean floor, the coastal to open water gradient (which reflects in large part the depth gradient) and finally, the latitudinal gradient, which determines the amount and seasonality of solar radiation.

One of the most important features of marine habitats is the depth of the *photic zone* (see Fig. 33.2) which represents the depth to which light can penetrate. Consequently, it is a determinant of primary productivity since photosynthesis can only occur in illuminated waters. Its depth varies from approximately 250 m in clear waters of the open ocean to 50 m in clear coastal waters but may be less than 1 m in highly turbid waters. Immediately below the photic zone is the *compensation zone* which is the depth where photosynthesis balances respiration. Below that lies the *aphotic zone* which is not illuminated and therefore no primary production occurs. The depth gradient applies to heat exchange as well as light intensity since insolation heats surface waters which become less dense and continue to float on the surface where they receive more heat. Thus, a body of relatively hot, less dense water floats on top of a much larger mass of cold dense water. The junction between the two is the thermocline (Ch. 34), the position of which (like the compensation zone) is variable (see Fig. 34.1) and is a function of the non-uniform distribution of solar radiation around the globe (see Fig. 2.2).

In terms of ecology, however, the presence and position of the thermocline are important because the temperature difference above and below it inhibits mixing and this has implications for nutrient supply. Nutrients absorbed by organisms in the photic zone, and incorporated into living tissue, may sink into the aphotic zone due to gravity and currents and cannot be replaced by mixing. Thus, photic waters may become depleted of nutrients while the aphotic zone and the benthos hold unavailable nutrient reserves. Wind-induced turbulence and waves can cause mixing to a maximum depth of 200 m. If mixing extends below the photic zone, primary producers may have to occupy the aphotic zone for long periods where they survive on their own reserves but do not add to food availability. Alternatively, where mixing does not extend below the photic zone, primary production may still be limited to low levels due to nutrient exhaustion. Consequently, marine productivity is determined by both mixing and the position of the thermocline.

In tropical waters, for example, there is a low continuous level of productivity, because the depth of the wind-induced mixing is within the photic zone above a permanent thermocline. Thus, while light intensities are sufficient to facilitate productivity, nutrient availability is limited. In waters of the temperate zone, however, seasonality occurs since winter-light intensities are low, so the photic zone and thermocline are well above the. depth to which wind-induced mixing occurs. Thus, nutrients are plentiful but productivity is inhibited by light availability. This increases with the onset of spring when the extent of the photic zone increases at a time when nutrients are still abundant, resulting in a burst of productivity. Simultaneously, a seasonal thermocline begins to become established which restricts productivity by reducing mixing. This in turn limits nutrient supply, which has also been depleted by the spring and early summer consumption. This seasonal thermocline disappears towards the end of summer when light intensities decline and further mixing increases nutrients. Before the onset of winter, another burst of productivity may occur which utilises the new stock of nutrients until light intensities fall to such low levels that productivity ceases. A different situation occurs in high latitudes where nutrients are always abundant but where sufficient light to allow photosynthesis is available for only five months of the year. Even so, the relationship between the availability of nutrients and light intensity is only favourable for three months of the year when a burst of productivity occurs.

The gradient from coast to open water is shown in Fig. 33.2. The littoral zone is distinctive because its waters are shallow and light may penetrate to the ocean floor. Where tidal fluctuations occur (section 34.2), parts of this zone may be exposed daily to the atmosphere and thus directly receive solar radiation. As a result, productivity is higher than in the neritic or oceanic habitats and this transitional zone between the marine and terrestrial environments supports a variety of ecosystems (section 35.5) despite the fact that it occupies less than 1% of the marine environment. Similarly, the neritic zone is small in extent and occupies only 3% of the marine environment. Although light rarely reaches the ocean floor, wind action results in extensive mixing and facilitates nutrient availability, which is also enhanced by the direct receipt of drainage from the land masses. Average productivity is 360 g m^{-2} yr^{-1} in comparison to 125 g m^{-2} yr^{-1} for the open ocean (see Table 29.1) where productivity is limited by nutrient-poor surface waters.

Other factors which influence marine habitats and organisms include upwelling and downwelling (section 34.3). These are ecologically important because they effect mixing and thus influence nutrient availability. The former is particularly significant because it brings about movement of nutrients from the aphotic to the photic zone and is especially important in coastal regions. Here, upwelling renders nutrients available from the continental shelf, which results in the relatively high productivity of neritic waters (Fig. 35.1). Both drift and density-driven currents (section 34.4) also play an important role in marine ecology. The former affect the distribution of surface-

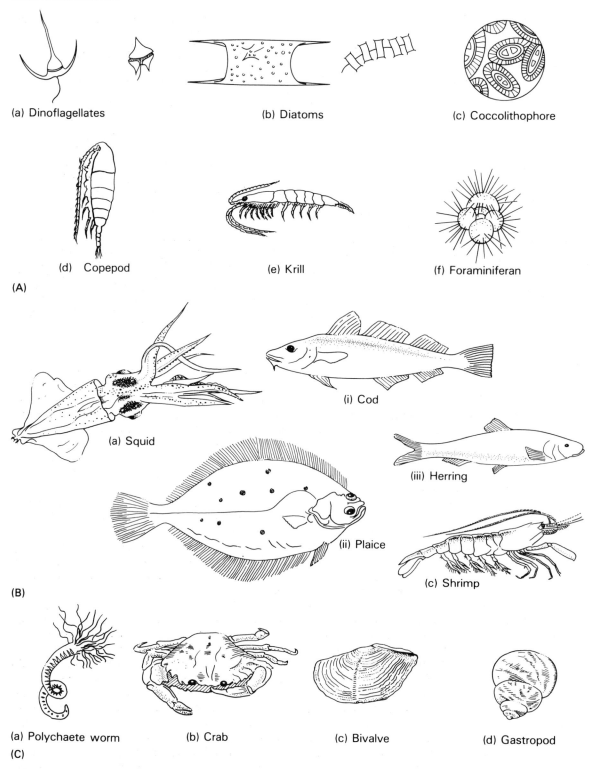

(a) Dinoflagellates   (b) Diatoms   (c) Coccolithophore

(d) Copepod   (e) Krill   (f) Foraminiferan

(A)

(a) Squid

(i) Cod

(ii) Plaice

(iii) Herring

(c) Shrimp

(B)

(a) Polychaete worm   (b) Crab   (c) Bivalve   (d) Gastropod

(C)

Fig. 35.1 Examples of marine organisms. **(A)** Planktonic organisms; **(B)** Nectonic organisms; **(C)** Benthic organisms.

living organisms while the latter are responsible for ensuring that oxygen, dissolved in surface waters, is transported to the ocean depths. Thus, habitat characteristics, especially light intensity and mixing, are important determinants of marine productivity, an indication of which is given in Fig. 29.2. The organisms responsible for this global pattern will be discussed below.

35.2 Marine organisms

Three groups of organisms inhabit the marine environment. The first group are the *plankton* which consist of free-floating or free-swimming organisms, most of which are microscopic and occur with greatest abundance in the photic zone. Secondly, there are the *neckton* which include larger swimming animals such as fish, which may be found in all marine habitats. Together, these may be described as pelagic since they inhabit the water mass, while the third group, the *benthos*, comprises species which live in or on the ocean floor.

(a) The plankton

These organisms tend to be carried along by currents or drifts. They range in size from bacteria of less than 1 μm in diameter to relatively large jellyfish in excess of 0.5 m in diameter and are generally classified on the basis of size (Table 35.1). Apart from in the littoral zone where macrophytic plants are present (section 35.5), *phytoplankton* are the sole source of food energy. They are the primary producers (sections 29.3 and 29.6) and thus form the first trophic level in marine food chains and webs. Since these organisms require light for photosynthesis, they are almost entirely confined to the photic zone (see Fig. 33.2) although sometimes wind-induced turbulence may carry them downwards into the aphotic zone. Their small size is advantageous in so far as it ensures that available energy is used in reproduction rather than in individual growth and it facilitates buoyancy. Thus, large phytoplankton populations are maintained in the photic zone.

Each ocean, or region within an ocean, has particular groups of plankton that are dominant. In regions of downwelling, for example, dinoflagellates (Fig. 35.1) are abundant especially in warmer waters. These comprise a group of micro-algae distinguishable by whip-like flagella that allow mobility and help to maintain dinoflagellate popu-

Table 35.1 The classification of plankton according to size

Diameter size (μm)	Plankton group	Characteristics
< 2	Ultraplankton	Monera,* chiefly bacteria which may exist as individuals or in colonies
2–20	Nanoplankton	Protista* and phytoplankton – primary producers
20–200	Microplankton	
200–2000	Macroplankton	Zooplankton – consumers
>2000	Megaplankton	

* Protista are single or acellular eukaryotes and their immediate relatives. Eukaryotes are organisms containing a true nucleus with DNA arranged in chromosomes and have organelles such as plastids and mitochondria. Monera, however, are prokaryotes and do not contain nuclei or organelles. Both Monera and Protista may be either autotrophs or heterotrophs.

lations in the photic zone. In regions of upwelling, *diatoms* (Fig. 35.1) are the dominant forms of phytoplankton. These are also unicellular algae but consist of an outer siliceous shell or frustule enclosing the cell itself, which contains brownish-green chloroplasts that enable diatoms to photosynthesise. The design of the frustule and the presence of oil droplets inside the cell facilitate buoyancy so that diatoms can float in the photic zone. Some diatoms are benthic, rather than planktonic, and these occupy the ocean floor to which light can penetrate in shallow littoral regions. Diatoms are the most important primary producers in cold polar waters, where their frustules may accumulate after death as diatomaceous ooze (section 34.7).

Coccolithophores (Fig. 35.1) are a further (though minor) component of the phytoplankton. They are *nanoplankton* and their cells are encased by calcareous plates (coccoliths) embedded in a gelatinous sheath and oil droplets within the cells facilitate buoyancy. They are abundant in warm tropical waters (where coccoliths may form part of the sediment on the ocean floor) but occur everywhere except in cold polar seas. Other phytoplanktonic organisms include the blue-green algae (Cyanophyceae) and the green algae (Chlorophyceae). Also important in the plankton are the *ultraplankton* which are mainly bacteria, occurring at all depths and latitudes. These play a vital role in the detrital food chain just as they do in terrestrial environments (section 29.6). They are especially abundant where concentrations of organic matter are high, as in the photic zone and littoral waters. They obtain their energy by breaking down organic material released by the phytoplankton, the faeces of *zooplankton* and dead organisms. The bacteria themselves are grazed by components of the protist zooplankton (Table 35.2) which, together with detrital substrate, are fed upon by omnivorous and detrital-feeding zooplankton so that the energy re-enters the grazing food chain.

Zooplankton are also part of the grazing food chain and are a diverse assemblage of mainly animal life (section 29.4) containing herbivorous, omnivorous and carnivorous species. The herbivorous zooplankton are especially important since they graze directly on the phytoplankton, converting primary productivity into secondary productivity. The zooplankton are a complex group and include

Table 35.2 The main groups of planktonic organisms excluding parasites. (After Barnes and Hughes 1982)

Monerans	Bacteria
	Cyanophytes (blue-green algae)
Protists	Cryptophytes
	Dinophytes (dinoflagellates)
	Euglenophytes
	Chrysophytes
	Bacillariophytes (diatoms)
	Haptophytes (e.g. coccolithophores)
	Prasinophytes
	Xanthophytes
	Chlorophytes (green algae)
	Several groups of amoeboid and flagellate protists
	Foraminiferans
	Radiolarians
	Acantharians
	Ciliophorans (ciliates)
Animals	Medusozan coelenterates (medusae, jellyfish, siphonophores, etc.)
	Ctenophores (sea gooseberries)
	Gastropod and cephalopod molluscs
	Polychaetes
	Ostracod, copepod and malacostracan crustaceans
	Chaetognaths (arrow-worms)
	Thaliacean and appendicularian tunicates
	Fish

(Together with the larvae of otherwise benthic groups)

holoplankton and mesoplankton. The former are species which are permanent members of the zooplankton while the latter are members only in their larval stages (e.g. species of necktonic fish and benthic invertebrates). Copepods (Fig. 35.1) are among the most important of the zooplankton, along with the shrimp-like euphausiids, commonly known as *krill* (Fig. 35.1). The copepods are small crustaceans which are the food source for many species of fish such as herring, sprat and pilchard and thus form the basis of many fishing industries. Krill are also crustaceans and are especially common in Antarctic waters where they provide a food source for large baleen whales. As a result of the Falkland Islands conflict in 1982, a British government review body, established to investigate possible ways of economically developing the South Atlantic area, has suggested the exploitation of krill reserves as a source of protein for human food, an activity already being undertaken by several countries (Beddington and May 1982).

Other components of zooplankton include arrow-worms (of the genus *Sagitta*) which are distributed in the shallow waters of northwest Europe. These are torpedo-shaped carnivores which are economically important as they are an important prey of herring. In addition, protozoans such as Foraminifera (Fig. 35.1) and Radiolaria often occur in the plankton and are abundant in warmer tropical and subtropical waters where their shells form part of the benthic sediments, for example *Globigerina* ooze. Two further groups of phytoplankton may also be recognised, the neuston and pleuston. The former are species such as the Portuguese man-of-war (*Physalia caravella*), which maintain themselves partly submerged in the water and partly in the air. This species is carnivorous, feeding on animals captured by its trailing tentacles. The neuston are species of bacteria and protozoa which occupy the shallow (1 mm depth) layer of sea water immediately adjacent to the atmosphere.

In common with the phytoplankton, many zooplankton have evolved anatomical features to enhance their survival in ocean waters. These include gelatinous tissue and the presence of oil droplets to improve buoyancy, and structures (such as rakes and bristles) to increase their surface area for more efficient food and nutrient uptake. In particular, many have the ability to move vertically because of the presence of cilia and flagella. Such appendages enable them to move downwards where they may become entrained in a current which carries them to a new area. Upward movement then enables them to reach the photic zone again and a new food source. Such movements are often diurnal, with zooplankton rising to the surface to feed on phytoplankton at night and descending during the day to darker waters where they avoid heavy predation. Similar seasonal trends occur in polar waters where zooplankton overwinter in a dormant state in deeper waters and ascend to graze diatom blooms in the summer.

(b) The neckton

The chief feature differentiating the neckton (Fig. 35.1(b)) from the plankton is their swimming ability, enabling them to counteract currents and drifts. Their anatomical features, such as muscular systems and streamlined shapes, facilitate horizontal and vertical movement. Some fish and cephalopods (e.g. squids) have evolved gas-filled structures to aid buoyancy, other species sustain themselves in the water by constant locomotion and a third group rest on the ocean floor while feeding or between feeding excursions.

Most of the neckton are vertebrates (e.g. fish) but include cephalopod molluscs (e.g. squid) and crustaceans (e.g. shrimps) and are carnivores, occurring at the highest trophic levels (section 29.5). The most important constituents of the neckton are the fish, which can be categorised into two groups, namely the pelagic species (such as herring and anchovy) which live mainly in the upper levels of the ocean and the demersal species (such as plaice and cod) which live mainly on or near the sea-bed. Species from both groups are important economically as a source of food for humans. Other members of the neckton include mammals such as seals, which feed in the sea but reproduce on land, the toothed and whalebone whales, which predate fish or krill, and seabirds.

Due to their dependence on swimming for both preying and as a defence against predation, the neckton are pelagic organisms with high energy requirements. Food availability is a particular problem for deep-sea fish whose habitat has diffuse and low-levels of food reserves. Thus, many species have gelatinous body tissue which minimises weight in relation to size, reduces energy expenditure in relation to metabolism and reduces gravitational attraction. Many species have also evolved means of luring prey such as bioluminescence, mimicry of prey, cavernous mouths and expandable abdomens allowing them to consume prey twice their own size. Some species, such as the baleen whales, have specialised structures allowing them to strain krill from the water and others, such as the seals, can dive to great depths to secure food.

(c) The benthos

Benthic species (Fig. 35.1(c)) are those which live on the ocean floor and the community is known as the benthos. Such communities are extremely varied, ranging from those of the littoral zone to those of the abyssal plain (see Fig. 33.2). Like the plankton, benthic organisms are classified according to size (Table 35.3), and substratum type is the chief factor controlling their distribution, although the amount of light, temperature and water quality are also important. As a broad distinction, benthic species can be classified as epiflora and epifauna (which occupy the sediment surface) and inflora and infauna, which live totally or partially buried within the sediment. These groups roughly correspond to hard substrates, where epifloral and epifaunal species are dominant and soft substrates, where inflora and infauna predominate.

Benthic organisms (Table 35.4) have evolved three mechanisms for feeding. They may be filter-feeders, filtering suspended organic material from the water, deposit-feeders collecting organic particles which have settled on to the substrate surface or they may be predators. The first two groups are most common over the continental shelf and abyssal plain (see Fig. 33.2) where bacteria also play a vital role in energy flow (section 35.2(b)). Food which is not used by pelagic consumers and faecal pellets eventually settle on the ocean floor, providing a nutrient source for bacteria, which can also synthesise protein from dissolved nutrients. These bacteria are often consumed by either filter- or deposit-feeders. Examples of filter-feeders include molluscs such as

Table 35.3 The classification of benthic organisms according to size

Size range (μm)	Benthic group	Characteristics
< 100	Microbenthos	Mostly Monera and Protista
100–500	Meiobenthos	Mostly animals with some protists such as foraminifera
> 500	Macrobenthos	Mostly animals except in the littoral zone (see text)

Table 35.4 The main groups of benthic organisms, excluding parasites. (After Barnes and Hughes 1982)

Monerans	Bacteria
	Cyanophytes* (blue-green algae)
Protists	Bacillariophytes* (diatoms)
	Chlorophytes* (green algae)
	Euglenophytes*
	Phaeophytes* (brown algae)
	Rhodophytes* (red algae)
	Several groups of amoeboid protists
	Foraminiferans
	Ciliophorans (ciliates)
Plants	Tracheophytes*
Animals	Poriferans (sponges)
	Medusozoan and anthozoan coelenterates (hydroids, corals, etc.)
	Turbellarians (flatworms)
	Gnathostomulids
	Nemertines
	Gastrotrichs
	Kinorhynchs
	Nematodes
	Priapulids
	Entoprocts
	Molluscs (all groups)
	Sipunculans
	Echiurans
	Polychaete and oligochaete annelids
	Pogonophorans
	Tardigrades
	Pycnogonid and merostomatan chelicerates (sea spiders, king crabs, etc.)
	Crustaceans (all groups)
	Phoronids
	Bryozoans
	Brachiopods
	Echinoderms
	Hemichordates
	Ascidian tunicates
	Cephalochordates
	Fish

* Groups confined to the littoral zone.

oysters and cockles, sponges and the Pacific anemone (*Anthopleura xanthogrammica*). Deposit-feeders include molluscs, for example *Macoma* spp. and *Yoldia* spp. and the polychaete *Hobsonia* spp. Carnivores include the drilling gastropod *Thais* and the decapod crab *Callinectes sapidus*.

The benthos of the littoral zone is distinct and more diverse than that of the deeper parts of the ocean. Here, photosynthesis is undertaken by a variety of organisms including the protists which are associated with soft benthic sediments or which may be symbiotic within littoral animal communities such as coral reefs (section 35.5). Other photosynthetic organisms are the multicellular (e.g. kelps and seaweeds) and filamentous algae and species of higher plants, which have become adapted to marine coastal ecosystems. These include eel grass (*Zostera* spp.), turtle grass (*Thalassia* spp.), salt-marsh and mangrove communities. All of these directly provide a food supply for herbivores or indirectly, via the detrital food chain (section 29.6), provide a food source for filter- and deposit-feeders which in turn may be consumed by fish.

35.3 Communities of the oceanic zone

Organisms of the oceanic zone have already been described (section 35.2) although some of their relationships require further explanation. In surface waters of the photic zone, photosynthesis by phytoplankton results in GPP (section 29.3) some of which is used by phytoplankton itself for metabolic processes. The remainder (NPP) is available as a food source for heterotrophs and detrivores or some of it may leak into the surrounding water. Energy transfer is not uniform or continuous in the entire oceanic zone but shows considerable variation in time and space. As already explained, nutrient availability is important in determining productivity and some species can absorb nutrients where only high concentrations occur, while others are adapted to low nutrient concentrations. In addition, herbivores have a direct effect on phytoplankton populations. In some areas of the oceanic zone, the phytoplankton may be grazed as fast as they reproduce whereas in other areas, there is a lag between phytoplankton productivity and the response of herbivore populations. This is characteristic of high latitudes where summer blooms of phytoplankton occur but the build-up of herbivore populations may have a lag time of six weeks.

In general, secondary productivity mirrors the distribution of primary productivity. At each trophic level, there is energy transfer, of variable efficiency, and Cushing (1971) found that in regions of food scarcity, efficiency was greatest at 24%, dropping to as low as 3% where food was abundant. Although trophic efficiencies in marine ecosystems have been little investigated, they are generally considered to be higher than in terrestrial ecosystems. Likewise, the significance of the detrital pathway varies and estimates of its importance differ from almost nothing to 90%. Similar to phytoplankton, the distribution of zooplankton varies considerably but the latter also migrate vertically, occupying deeper water in daylight and ascending in darkness. Since the depth of migration relates to size, deep scattering layers (Dietz 1962) are produced with each layer containing different organisms. In addition, seasonal migrations may occur, as in polar regions, which represents a response to food availability. Bacteria also provide a food source and a mechanism for nutrient cycling. They live suspended in the water column from which they extract dissolved organic substances and in some tropical waters, Sorokin (1971) has suggested that bacterial productivity may exceed that of the phytoplankton.

In terms of resources, the relationships between planktonic organisms and commercially significant fish are important (Hardy 1956). In the North Atlantic, for example, herring (*Clupea harengus*) feed mainly on copepod crustaceans and arrow-worms and the copepod *Calanus* has been used as an indicator species to locate herring shoals. Beddington and May (1982) have also described the trophic relationships between phytoplankton, zooplankton and whales in the South Atlantic (Fig. 35.2), where the food web is relatively simple.

In the abyssal and bathyal regions (see Fig. 33.2) of the ocean, phytoplankton are absent and marine life is sustained by detrital particles and excretion products. These regions are thus dominated by carnivores, omnivores and detrivores. Trawls at 1650 m depth (Marshall 1979), for example, yielded an abundance of copepod crustaceans, molluscs, worms and comb jellies and Vinogradova (1962) has shown that between 2000 and 4000 m depth, the *macrobenthos* is diverse and contains 400–500 species. Below 6000 m, however, the number of species declines rapidly to less than 10 at 9000 m, most of which are ooze consumers. Furthermore, the *meiobenthos* (Table 35.3) shows variations with deposit-feeders (section 35.2(c)) dominant where sediments are organically rich and suspension-feeders (section 35.2(c)) dominant on compacted sediments. Mobile scavengers and carnivores (e.g. sharks and octopuses) also tend to congregate where food is available, as shown by experiments where food bait is lowered to the ocean floor (Isaacs and Schwartzlose 1975).

The final trophic level within the marine

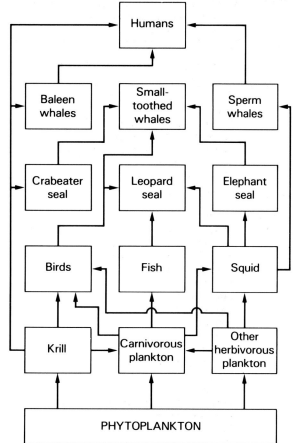

Fig. 35.2 An example of a Southern Ocean ecosystem (Adapted from Beddington and May 1982). Arrows indicate energy flow.

ecosystem is dominated by the neckton (section 35.2(b) and Fig. 35.1), many species of which are harvested to provide food for humans. Many necktonic species have varying stages in their life-history and, especially where fish are concerned, this has important implications for commercial exploitation. Figure 35.3 shows the three most important foci on the migration routes of most fish. For individual species, data from the fishing industry show that the phases of migration and the routes used are persistent over long time periods, presumably reflecting reproductive success and survival.

35.4 Communities of the neritic zone

This zone occurs over the continental shelf (see Fig. 33.2) between the littoral and oceanic zones.

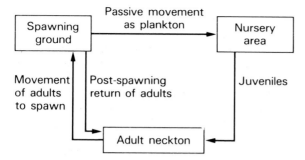

Fig. 35.3 A diagrammatic representation of routes taken by migratory neckton.

Planktonic relationships in the neritic zone are similar to those already described (sections 35.2(a) and 35.3), although productivity is generally higher than in the oceanic zone because shelf areas directly receive nutrients from nearby land masses. In addition, detrital material is received from the producers of the littoral zone, which also enhances secondary productivity. The benthic community is controlled by the nature of marine sediments which are oxygenated only in their surface layers and organisms can only exist at deeper anaerobic levels if they can oxygenate their immediate environment. This is the case for some species of macrofauna (e.g. worms, clams and crabs) whose burrowing habit results in the downward movement of oxygenated water and a supply of food. Most benthic species are, however, confined to the surface aerobic sedimentary layers, below which chemosynthetic and anaerobic bacteria constitute the largest populations.

Benthic organisms have two sources of food. Detritus and living phytoplankton occur in the water and living and dead organic matter is contained in the sediment itself. The direct assimilation of phytoplankton is part of the grazing food chain but the detrital pathway of energy transfer is also important and much more complicated (section 35.2(a)). Organic constitutents of detritus (e.g. faeces) may be lost by leaching into the water or absorbed as it passes through organisms. If it reaches the ocean floor, it is colonised by bacteria and may then be utilised by protists and mesofauna. These groups provide a food source for suspension- and deposit-feeders (section 35.2(c)) which in turn provide a food source for benthic carnivores. The supply of food particles from the pelagic zone determines the productivity of both suspension- and deposit-feeders (section 32.2). As explained above, bacterial action on some of this food may be necessary before it can be utilised by either suspension- or deposit-feeders and consequently there is a direct relationship between these species and bacteria populations. On the one hand, the latter control food availability to some extent but are themselves often limited by nutrient deficiencies, especially nitrates and phosphates. On the other hand, detrital feeders release these substances and other nutrients facilitating nutrient circulation, which helps to maintain bacterial populations.

In general, biomass and productivity for all benthic organisms decrease with increasing water depth. Furthermore, there is a continual interaction between the pelagic and benthic systems, especially as many macrobenthic species produce larvae which swim and feed in the water column. The proportion of species with this type of life-cycle varies geographically with latitude. In tropical waters, for example, it is typical of 80% of benthic species but is absent in species of polar waters. The diversity of benthic species, especially the epifauna (section 35.2(c)), also varies with latitude, being highest in tropical waters and declining polewards. This seems to be the result of speciation over long periods of time in tropical waters thus favoured by the alignment of the continents along a north–south axis creating geographical barriers. Diversity is maintained as a result of favourable interaction between food availability and mortality rates. Infauna do not show this variation in diversity because conditions within marine sediments are more uniform, thus reducing latitudinal variation.

35.5 The littoral zone

The position of this zone adjacent to the land masses is shown in Fig. 33.2. Its distinguishing feature is the shallowness of its waters through which light can penetrate to the ocean floor. Pelagic and benthic communities, and interactions, are similar to those described for the oceanic (section 35.3) and the neritic (section 35.4) zones, but considerable variations occur within the primary producers which, apart from the phytoplankton, do not occur in other marine habitats. Communities of the littoral zone include sea grass meadows, *salt-marshes*, *mangrove swamps*, the communities of rocky and sandy shores, estuarine communities and coral reefs. Each one of these communities has been the subject of much research but only salt-marshes, mangrove swamps and coral reefs will be considered here due to lack of space. Further reading texts given at the end of this chapter provide more detailed accounts of this diverse array of ecosystems which occupy a unique position at the junction between the marine and terrestrial environments.

(a) Salt-marshes

The global distribution of salf-marshes is shown in Fig. 35.4. Long and Mason (1983) define them as areas of alluvial or peat deposits, colonised by herbaceous and small shrubby terrestrial vascular plants, almost permanently wet and frequently inundated with saline waters'. However, salt-marshes over the globe are by no means uniform and Beeftink (1977) has distinguished six types. These are lagoonal marshes, beach plains, barrier island marshes, estuarine marshes, semi-natural marshes and artificial marshes. Conditions for their formation are nevertheless similar. They tend to form where sediment accumulates just above the mean high-water mark of spring tides or where mixing of salt and fresh water in estuaries effects mud deposition.

Plants colonising these substrates tend to accelerate deposition by obstructing water currents from which particles sediment out of the water column and their roots and rhizomes consolidate the substrate. Such processes are also facilitated by filamentous and mat-forming components of the benthic microflora, especially algae (e.g. *Enteromorpha intestinalis*). This accretionary process thus raises shore levels and reduces tidal submergence. *Macrophytes* colonising this environment are halophytes which can tolerate high salinity levels as well as periodic inundation. They include marsh samphire (*Salicornia* spp.) and cordgrass (*Spartina alterniflora*) which, together with eelgrass (*Zostera* spp.) are the primary colonisers dominating the lower marsh. At higher levels, submergence and salinities are reduced so that a wider diversity of species can colonise, such as the sea aster (*Aster tripolium*) and sea thrift (*Armeria maritima*). These are perennial herbs which may form a close sward and provide competition for the initial colonisers which may persist only in isolated patches in the middle and upper marsh where physiographical conditions allow.

The upper marsh is a transition zone between marsh communities proper and dry-land communities.

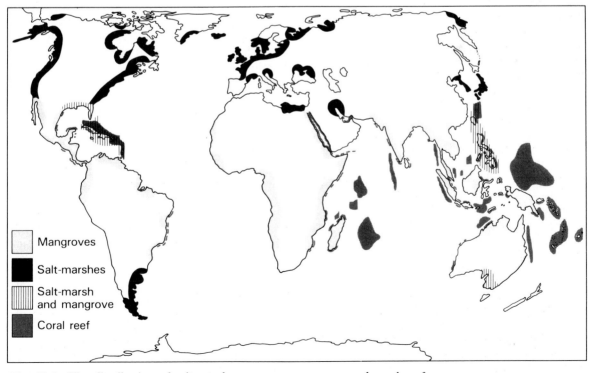

Mangroves

Salt-marshes

Salt-marsh and mangrove

Coral reef

Fig. 35.4 The distribution of salt-marshes, mangrove swamps and coral reefs.

Its floristics will depend on the nature of adjacent vegetation from which species can colonise. In British salt-marshes, this zone is often dominated by the red fescue grass (*Festuca rubra*) which can provide a good pasture and will form a dense turf under grazing. Just as species composition between marshes varies geographically, so does productivity which is generally high, especially in comparison with the oceanic zone. Wiegert (1979) has shown that in Georgian and South Carolinan salt-marshes of the USA, net production (of both *Spartina* and algae) is 1653 g cm^{-2} yr^{-1}, of which 880 g cm^{-2} is exported via tidal action mostly to the neritic zone. Export of organic matter (and nutrients), however, is often much reduced from sheltered marshes and some may even be importers as Nixon (1980) showed for the Flax Pond salt-marsh of New York, USA.

(b) Mangrove swamps

As shown in Fig. 35.4, salt-marshes and mangrove swamps have a geographical relationship. In general, the latter replace the former in the warm coastal waters of the tropics and subtropics, with some overlap in areas adjacent to the Gulf of Mexico and southeast Australia. Variation in species composition also occurs between the mangrove swamps of the USA and West Africa and those of East Africa, Southeast Asia and Australasia. Substrate deposition is similar to that described for salt-marshes and mangrove swamps may develop in estuaries, lagoons and adjacent to coral reefs. Mangroves are evergreen shrubs and trees well adapted to colonising mobile saline mud by means of prop roots (or rhizophores) which inhibit tidal currents and effect deposition. Seeds germinate while still on the parent plant, from which they drop to float in the water until a suitable habitat is encountered where the already well-developed roots may become attached. Once established, the roots project above the mud surface and pneumatophores (breathing pores) allow oxygen to enter the root system. This extensive root system provides surfaces for other organisms, both plants and animals, which also enhance the accretionary process.

As with salt-marshes, mangrove communities exhibit a distinct zonation. In Florida, USA, for example, the red mangrove (*Rhizophora mangle*) dominates the outermost zone with the black mangrove (*Avicennia nitida*) in shallower water, often in association with cordgrass (*Spartina alterni-flora*). The next zone is characterised by the white mangrove (*Laguncularia racemosa*) which is transitional to dry-land communities. Figure 29.4 illustrates the important role that mangroves play in the food chains of many marine ecosystems. From mangrove swamps of the Florida coast, Odum (1971) has shown that 9 t ha^{-1} of mangrove leaves are deposited annually. This represents 95% of NPP and fuels the detritus food chains of this and adjacent ecosystems. Detritus consumers ingest leaf particles, or organically enriched substrate, and then provide a food source for small carnivores which in turn are fed upon by game fish and birds. Thus, mangroves are economically important, providing a large proportion of the energy input into food webs which support large fish populations harvested for human consumption.

(c) Coral reefs

Coral reefs are distributed (Fig. 35.4) in shallow waters of the ocean, usually where sea temperatures are 20 °C or above. Charles Darwin compiled the earliest distribution maps of coral reefs and identified three categories which are still recognised today. These are *barrier reefs, fringing reefs* (Fig. 35.5) and *atolls*. Apart from temperature, other factors which influence coral reef distribution include light availability, water clarity, salinity and geological structure (sections 33.2 and 33.3). Since light is essential for reef-building communities, they are confined to the photic zone of coastal waters where rock platforms are present, as in the Pacific where basaltic rock platforms have been implaced near the ocean surface by volcanic activity. As Fig. 35.4 shows, the Red Sea has extensive reef development at its margins, promoted by the presence of an offshore platform over which lies shallow, warm, non-turbid water. Light can penetrate down to the platform surface and the enclosed nature of the basin precludes the upwelling (section 35.1) of cold ocean currents.

Reef building is effected by the deposition of calcium carbonate where the coastline is gradually sinking or where there are eustatic rises in sea-level (Ch. 15). Thus, during glacial stages, when sea-level was much reduced, many reefs would have been subjected to erosion and as a corollary, greatest reef growth occurs during interglacial stages (Marshall and Davies 1984). Coral reefs will emerge where their rate of upward growth exceeds the rate of submergence and eventually a sequence of landforms will develop as illustrated in

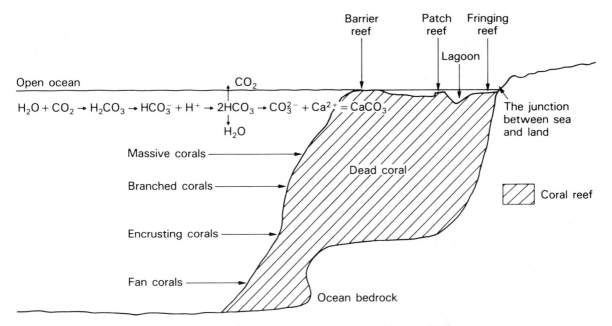

Fig. 35.5 Characteristics of coral reefs. (Adapted from Furley and Newey 1982)

Fig. 35.5, which shows the position of fringing and barrier reefs in relation to the coastline. Alternatively, atolls are formed as a result of reef development around emergent volcanic peaks which have subsequently subsided, for example the Marshall Islands. In addition, some reefs like the Great Barrier Reef, which occupies some 207 000 km² off the northeast coast of Australia, are an accumulation of a variety of different types of reef.

The organisms responsible for coral reef building are coelenterate animals or polyps (similar to sea anemones) which are colonial. Each individual coral has a calcareous skeleton which it secretes by assimilating calcium from sea water, although the exact mechanism of this is not fully understood. In this way, layers of calcium carbonate are built up eventually forming a limestone reef. However, despite the fact that the chief structural components of coral reefs are animals, the community is not strictly heterotrophic. This is because it is a discrete ecosystem containing large populations of green plants, especially dinoflagellates (section 35.2(a)) and other algae which live in the tissue of the coral polyps or within their calcareous skeletons. This is a symbiotic relationship in which the algae utilise carbon dioxide and nutrients provided by the corals which, in turn, benefit from oxygen and carbohydrates produced as a result of photosynthesis by the algae. This interdependence also results in very efficient intra-system nutrient cycling.

A diverse range of animal life is also associated with coral reefs and many species are adapted for survival in the numerous microhabitats provided within the reef itself. Large populations of filter-feeders (such as barnacles and sponges) also occur on coral reefs along with large fish populations some of which like the parrot fish (*Sparisoma* spp.) feed directly on the coral. As shown in Table 29.1, coral reefs are among the most productive ecosystems in the world with an annual NPP in the range 1500–3500 g m⁻² which compares favourably with an average of only 125 g m⁻² yr⁻¹ for the open ocean. The former range is not dissimilar to that of tropical rainforests and exceeds that of many agricultural systems. Unfortunately, this highly organised and productive ecosystem can be impaired or destroyed by a variety of natural and human agencies, such as high winds, storms and pollution, including agricultural runoff from the land and sewage effluent. Furthermore, many reefs are under threat of destruction and even extinction from a population explosion of a specific predator known as the 'crown of thorns' starfish (*Acanthaster planci*). Rapid population growth of this species in the last two decades due to unknown factors has caused considerable degradation on Red Sea reefs and the Great Barrier Reef.

35.6 Retrospect

In summary, this chapter illustrates the diversity of marine ecosystems which exists as a function of the relationship between the geological structure of the oceans, surface and deep-water movements and latitudinal variations in solar energy receipt. Energy flow occurs via both the grazing and detrital pathways but the intricacies of community structure and function are considerably less well documented and understood than those of terrestrial ecosystems. Primary production is confined to the photic zone which, in relation to the ocean deeps, is small in area and since overall productivity mirrors primary productivity, it declines away from the coast as water deepens.

At present, the most significant marine biological resources are surface-water pelagic fish populations, some of which are commercially harvested for human consumption. Looking to the future, it may not be too much longer before zooplankton, and perhaps even phytoplankton, are similarly exploited. Finally, this chapter is intended to provide a basic introduction to marine organisms, their interactions and community development. Such an approach seems to be absent from most physical geography and ecological texts. This exposition is by no means comprehensive but it serves to illustrate the importance of the marine environment and reinforces the importance of energy flow and biogeochemical cycling as fundamental components of ecosystem functions as described in Chapters 29–31.

Key topics

1. The relationship between geology, ocean water characteristics and marine habitats.
2. Trophic relationships in marine ecosystems.
3. The biotic characteristics of the oceanic and neritic zones of marine ecosystems.
4. The diversity of ecosystems in the littoral zone where marine and terrestrial environments merge.

Further reading

Chapman, V. J. (ed.) (1977) *Wet Coastal Ecosystems*. Elsevier; Amsterdam, London and New York.

Couper, A. (ed.) (1983) *Times Atlas of the Oceans*. Times Books Ltd; London.

Gray, J. S. (1981) *The Ecology of Marine Sediments: An Introduction to the Structure and Function of Benthic Communities*. Cambridge University Press; Cambridge.

Kinne, O. (ed.) (1970–82) *Marine Ecology*, vols 1–5. John Wiley and Sons; New York and London.

Levinton, J. S. (1982) *Marine Ecology*. Prentice-Hall; London and New Jersey.

Long, S. P. and Mason, C. F. (1983) *Saltmarsh Ecology*. Blackie; London and Glasgow.

Saenger, E. J., Hegerl, E. J. and Davie, J. D. S. (1983) *Global Studies of Mangrove Ecosystems*. IUCN Commission on Ecology Papers No. 3, IUCN; Gland, Switzerland.

Overview and Prospect

Chapter 36 The integration and modification of physical processes

This concluding chapter emphasises the significant interrelationship between earth surface processes which are evident at every level from the global scale, through the regional scale (tundra zone) to a more local scale (moorland environment). It also examines the effect of human activities on the operation and interaction of these processes. As mentioned in Chapters 1 and 28, these activities have a vital role to play in the functioning of all systems, through the deliberate and inadvertent modification of the natural physical environment.

36.1 Interrelationships at the global scale

Intricate linkages between the various processes operating at or near the earth's surface on the global scale are implicit in the following pages. They are part of a complex earth–atmosphere system which itself comprises countless subsystems. The whole picture is difficult to comprehend but some guidance may be obtained from Fig. 36.1,

which is simplified from a diagram by Lockwood (1974). The general circulation of the atmosphere is here regarded as the fundamental set of processes, from which all others flow. However, it must be recognised that the configuration of the earth's surface represents another fundamental given disposition. If the arrangement of continent and ocean and the relief of the land surface were other than they are at present, the same processes would operate but the resulting distributions would be different, as indeed they must have been in earlier geological times.

The general circulation of the atmosphere (box 1 in Fig. 36.1) depends on net radiation receipts and earth rotation, which has been described as Nature's way of correcting the imbalances of radiation distribution. This is incompletely achieved but results in an energy balance (box 2) and an atmospheric moisture balance (box 3) appropriate to each part of the globe. The energy balance determines such parameters as surface temperature and energy fluxes into the overlying atmosphere, which in turn influences the possibilities of precipitation. The availability of moisture for this process depends on the atmospheric water balance. In this way, boxes 2 and 3 combine to determine not only the water and energy availability for plant life (box 5), but also the water free to move within and above the ground, i.e. the runoff (box 4). The runoff determines to a large extent the water resources and hence the water supply available for use by both the human and the animal populations (box 8). At the same time, runoff is a prime influence on the types and rates of weathering and erosion over much of the globe (box 6). In the absence of moving water, or where water is locked up in the frozen state, different erosional pro-

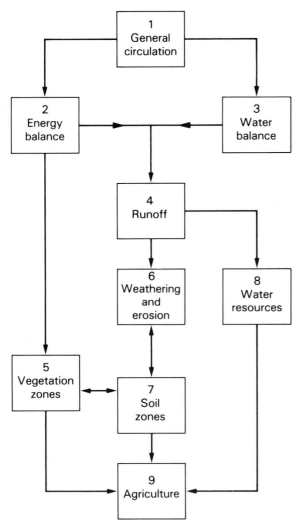

Fig. 36.1 Interactions within the physical environment.

cesses are at work. Weathering processes combine with vegetation type to produce characteristic soil types (box 7).

Water supply, soil zonation and vegetation type are all obvious determinants, among others, of the type and success of agriculture (box 9). To proceed further along this path would be to enter the realms of human geography. However, this should be regarded not as a wall blanking off further thought, but as a reminder that the processes of the physical world cannot be isolated from the activities of the human world. In terms of Fig. 36.1, only box 1 can (at present anyway) be regarded as immune to human interference. As demon-

strated in earlier pages, many so-called ecosystems are in fact control systems, involving human intervention (not always well directed).

Although, in the main, anthropogenic influences have manifested themselves on the mesoscale, the pace and expansion of activities such as agriculture/land management and urbanisation/industrialisation have been such as to encroach on the global scale. The removal of forest increases the albedo and thus decreases absorption of solar energy (it has been estimated that complete deforestation would lower the mean surface temperature of the earth by 0.2–0.3° C) and, it would seem, also increases the CO_2 concentration in the atmosphere. Increased tillage exposes bare soil and substantially increases the dust content of the air. The burning of fossil fuels for industrial or space heating purposes unleashes a variety of unwelcome pollutants, but the ultimate combustion product of carbonaceous material is CO_2 and it is well known that this gas, together with water vapour, is responsible for the 'greenhouse effect' (section 2.2). Industrial processes and domestic fuel consumption also contribute to the dust-loading of the atmosphere. All these processes have implications for changes in global climate.

Interest has been directed towards the CO_2 question since the beginning of this century. Around 1880, the CO_2 concentration was probably 280–290 ppm and firmer measurements since have placed the figure at 310–315 ppm in 1950 and 335–340 ppm in 1980. The concentration is currently increasing at something approaching 1 ppm yr^{-1} and the rate of increase is itself accelerating due to mounting energy requirements. Forecasts are hedged around with uncertainties, but some estimates suggest that by the year 2050, CO_2 levels may be double those of the mid-nineteenth century. Much conjecture, and not a few numerical models, have addressed themselves to the question of what a doubling of CO_2 would mean for global surface temperatures.

Different models have yielded answers to that question varying from a mere 0.1° C to over 4° C, but most opinion favours an increase of 2–3° C for a doubling of CO_2 concentration. However, the warming effect would undoubtedly be greatest in high latitudes, where increasing warmth would imply reduction of ice and snow (thereby lowering albedo), more evaporation leading to more moisture and cloud in the lower troposphere and less tendency for the development of thermally induced high pressure. All this would further increase surface temperatures and one model predicts an increase of more than 7° C north of latitude 70° N for a doubling of CO_2 content. Such an increase would mean an ice-free Arctic and a poleward shift in the climatic zones of the northern hemisphere.

It is not clear whether any incontrovertible signals of the enhanced CO_2 effect have yet been detected and in the early 1980s, much research was being directed to this question (e.g. Beatty 1982). Between about 1880 and 1940, mean global temperatures increased by 0.6°C (and by 0.9°C in high latitudes) and this was accepted as confirmation of the CO_2 impact. However, from 1940 to 1980, the temperature curve has levelled-off or dithered uncertainly (though warming has dominated the 1980s), while CO_2 concentrations continued to grow. These fluctuations do not necessarily negate the significance of CO_2 but point to the difficulty of separating a CO_2 signal, if any, from the general 'noise' induced by the other factors that could influence climatic variation. For example, an alternative school of thought predicts that increased dust-loading of the atmosphere by particulate pollution and surface dust (increased *turbidity*) will reduce surface temperature by increased scattering of incoming solar radiation. The little turbidity monitoring there has been suggests a 50–80% increase this century. Volcanic dust-veiling has the same effect, but there were no significant eruptions during the period 1940–60.

It is clear that increased CO_2 and increased dust-loading have directly opposite effects and it is possible that a rough equilibrium has been established since the 1940s. Fears exist about the release of chlorofluorocarbons (i.e. the notorious CFCs produced by a wide range of domestic products), which has been associated with ozone depletion in the stratosphere (Fig. 3.1) and increased lethal u.v. radiation at the earth's surface. The discovery of the infamous ozone 'hole' over Antarctica in 1987 has led to major cutbacks in CFC production, initiated by the Montreal Protocol and later updates. This reduction will eventually minimise the expected impact of increased u.v. radiation in the form of a dramatic rise in the incidence of skin cancers and eye cataracts. Despite many uncertainties which argue for caution, there is general consensus amongst environmentalists around the world that global warming will become an increasing problem over the next century due to the expected build-up of CO_2. The consequences would be serious sea-level rises and wide ranging environmental disruption as outlined in Thompson (1989), listed in the Further Reading at the end of section 36.3.

36.2 The interaction of physical processes in the tundra zone

The intricate linkages between earth surface processes in the tundra are dominated by its cold, dry climate, with a mean annual temperature between −1 and −15° C and mean annual precipitation usually between 127 and 1400 mm. These climatic characteristics are associated with distinctive atmospheric energy and circulation controls which, in turn, influence the operation of earth surface processes (as discussed below). Due to the obliquity of solar radiation and the inequality in the duration of insolation (with pronounced winter deficits), the energy balance of the tundra is controlled by very modest net radiation values. For example, over the Siberian tundra at 65° N, 120 °E, the net radiation ranges from small negative values in winter (-24 W m^{-2}) to modest positive values in summer ($+ 121$ W m^{-2}).

However, it should be noted that the available net radiation will vary considerably at the earth's surface since it depends on local cloud conditions and surface albedo variations associated with changing vegetation cover. This latter factor is most variable over the tundra since it ranges from 13% for treeless, lichen-covered surfaces, to 10% for muskeg and 7% for spruce bog (Thompson 1974). Furthermore, in general heat transfer terms, the latent heat flux consumes most (70%) of the available net radiation over the tundra but it is apparent that specific energy balances depend entirely on local terrain factors. For example, dry polar desert conditions (characterised by nonvascular plants or glacial outwash gravels, which are not conducive to standing water) are highly resistant to evapotranspiration and have an energy balance dominated by sensible heat transfers into the air and ground, which can consume more than 80% of the net radiation. Conversely, wet tundra conditions are associated with a saturated active layer (especially following the ablation of the winter snow pack) and *Sphagnum* moss environments, where latent heat transfers utilise more than 75% of the net radiation.

The interaction of climatic factors with the operation of sediment transport, soil formation and plant growth in the tundra is well documented, and it is only possible to include a brief summary in this section. Sediment transport systems are influenced by the annual and diurnal freeze-thaw cycles (associated with the oscillations of air temperature around 0° C), although hydration shattering must also be considered as a viable alternative process. However, the efficacy of freeze-thaw processes is constantly under review and Fahey (1973) indicated that the diurnal cycles rarely penetrated below 10 cm. He concluded, therefore, that annual freeze-thaw is a more effective geomorphic agent which is responsible for the dominant physical weathering processes. These produce typical tundra landforms such as patterned ground, pingos and mass wasting in the form of creep and gelifluction. The emphasis on physical weathering is at the expense of chemical weathering, which is inhibited by the low temperatures and meagre amounts of precipitation. Consequently, chemical reactions are halved for every 10° C reduction in air temperature and this feature influences tundra pedogenesis.

Tedrow (1958) indicated that the large amount of silt loam in 'cold' soils is related to this silt size being the stable end product of the mechanical breakdown of parent material by freeze-thaw cycles. Furthermore, this frost action leads to salt crusting at the soil surface, when the salts are lifted by cryogenic disturbances at unvegetated sites, and to involutions of soil horizons. Two soil types dominate cold environments, namely Arctic brown and tundra soils. The former soils develop over well-drained dry sites (like ridges) free from permafrost, where a weakened podsolisation occurs due to the reduced leaching and the fewer ferro-organic complexes present in the dry, cold conditions. Tundra soils dominate wet, ill-drained lower slopes over melting ground ice, where soil saturation promotes the mobilisation of iron. However, the gleying process is reduced by the low temperatures which restrict the number of micro-organisms present in the soil.

Plant growth is also influenced by the windy, cold and dry characteristics of the tundra. The high wind speeds result in plant desiccation and structural damage, which leads to stunted, prostrate, low-lying plants like cranberry and dwarf willows and flag-form trees (i.e. flagged krummolz). The low temperatures restrict all plant processes (i.e. growth, respiration, photosynthesis and reproduction) and dormancy is a common feature of tundra vegetation. Small precipitation amounts impose physiological drought conditions which again limit all plant processes, although the accumulation of the winter snow pack provides useful insulative protection against the severe cold. Consequently, herbaceous plants dominate the tundra (including grasses, flowering herbs, sedges, rushes, dwarf shrubs, lichens and mosses), which are perennials with large ground root and stem storage systems. The plants themselves can adapt to polar conditions by metabolising at low temperatures, by storing carbohydrates over the winter and by initiating a rapid life-cycle immediately following snowmelt. Also, preformed shoots/flower buds can develop within 10 or 20 days following exposure, and up to 38% of tundra species are wind-pollinated. Other plant features, designed to withstand the frigid conditions, include dehydration of cells (to minimise frost damage) and the release of an anti-freeze-type plasma into the cells.

There is no doubt that the most significant feature of the tundra is the permafrost, which covers some 21 million km^2 in the northern hemisphere. The relationship between climate and permafrost is rather tenuous although, in Canada, the southern boundary of the continuous and discontinuous permafrost zones is associated with a mean annual temperature (MAT) of −1.1 and −8.0° C respectively. However, the poor correlation between mean temperature and permafrost formation is evident in the Great Slave Lake and Ungava regions of Canada, where the discontinuous boundary is close to the −4° C MAT. Also, field observations in the discontinuous permafrost zone demonstrate that permanent ice may exist in patches where the MAT is only −0.5° C but can be absent where this value is −6.7° C (Crawford and Johnston 1971). Obviously, the interaction between climate and permafrost is controlled by a variety of terrain factors (including drainage, vegetation/snow cover and soil type) which influence the energy exchange mechanisms at the tundra surface, particularly the soil heat flux.

For example, tundra soil heat fluxes which can consume in excess of 15% of the available net radiation are associated with wet, coarse-grained soils where the thermal conductivity (K) is greatest. Brown (1963) measured low K values in dry peat (about 0.00017 g cal s^{-1} cm^2 °C, *sic*), which hinder heat transfer to the underlying soil compared with saturated peat (K = 0.0011 g cal s^{-1} cm^2 °C), where the active layer is thickest. The low thermal conductivity of snow cover insulates the ground from frost penetration/freezing during autumn and soil warming/thawing in spring, which leads to anomalous permafrost distributions. For example, the absence of continuous permafrost on the east side of Hudson Bay in Canada, compared to the west side, is attributed to an autumn snow pack in excess of 50 cm (Thompson 1974).

Ground drainage systems also affect permafrost formation since deep water penetration facilitates heat transfers, maximum thawing and thick active layers. In terms of vegetation cover, certain plant types (especially *sphagnum* moss) possess a low thermal conductivity which insulates the ground ice from thawing and maintains cold soil conditions. For example, soil temperatures at 0.3 m depth ranged between 4.4° C under grass, 2.5° C under sedge and 0.3° C under moss (Brown 1961). Plants also use considerable amounts of net radiation for evapotranspiration and under these conditions, the latent heat flux dominates at the expense of sensible heat transfers into the air and ground. However, it should be noted that these ground ice/terrain interactions are disturbed by human activities in the tundra, which are mainly associated with the exploitation of Arctic oil and natural gas reserves. Permafrost is particularly sensitive to energy balance changes associated with the clearing of tundra vegetation, which alters the natural albedo and the receipt of net radiation. Furthermore, the removal of the protective 'mat' of vascular plants destroys the natural surface insulative properties and the marked reduction in latent heat flux accelerates ground heat transfers, which are controlled by the basic K properties of the soil.

The best example of this disturbance followed the bulldozing of a seismic line west of the Mackenzie Delta in northwest Canada, which removed the vegetation cover and set into operation the thaw processes discussed above. Within four years, a gully had formed some 7 m wide and 2.4 m deep (Crawford and Johnston 1971) which represented a classic example of 'Man'-induced thermokarst in the sensitive tundra zone. Furthermore, it should be noted that the environmental impact of this thermokarst development extends well beyond the immediate area disturbed by seismic surveys and road construction. For example, Kerfoot (1974) observed near Sitidgi Lake in Canada that the width of the disturbed area ranged between 15 and 20 m, even though the actual road surface averaged 5–7 m across (about one-third of the total area affected). Another worrying fact is the estimation by Lachenbruch (1970) that a 1–2 m pipeline buried 1.8 m in permafrost, and carrying oil at an operating temperature of 80° C, could thaw the ground to a depth of more than

9 m within five years. It is important to monitor these activities closely so that they do not lead to irreparable damage in the so-called fragile tundra. Furthermore, it is imperative that the natural interaction of climatic, biotic and pedogeomorphic processes is allowed to continue undisturbed in order to preserve this unique environment.

36.3 The interaction of physical processes in moorlands

So far in the chapter we have studied the interaction of physical processes and human agencies at global and regional scales. In this section, we examine these relationships at a much more localised scale in terms of moorlands which are environments largely devoid of trees and covered by vegetation, which is less than 1 m in height. Heather (*Calluna vulgaris*), a dwarf shrub of the Ericaceae family, usually predominates and forms associations with species such as bilberry (*Vaccinium myrtillus*), bearberry (*Arctostaphylos uva-ursi*) and purple moor grass (*Molinia caerulea*). Occupying 1.2 million ha in the UK and extensive areas of northwest Europe, moorlands are ecosystems which dramatically reflect the interplay between biotic, abiotic and anthropogenic factors. At first glance, these sparsely inhabited scenic tracts of the upland zone create the impression of a truly natural landscape, conditioned only by the postglacial climatic changes of the last 13 000 years and unaffected by human activity. However, observations of present-day processes, coupled with palaeoecological evidence of past vegetation change, show that most moorlands have developed as a result of both climatic and anthropogenic factors and can be maintained only as a control system (section 1.4).

Climatically, moorlands occur in areas which experience a cool temperate oceanic climate with mild winters and cool summers. Even in the warmest months, temperatures do not rise above 22° C but for at least four months of the year average temperatures are above 10° C. Rainfall is abundant, between 600 and 1100 mm yr^{-1} which is fairly evenly distributed over the year, and humidity is high. In consequence, the potential water deficit is low overall and water availability is not a limiting factor in ecosystem function. Equally, however, such climatic conditions could support forest vegetation, a fact which will be considered below. The parent materials underlying most moorland ecosystems consist of acid rocks, such as granites or schists, or superficial deposits such as fluvio-glacial sands and gravels, glacial till or wind-blown sand. Soils associated with these parent materials, and which occur under moorlands, include several types of podsol, oligotrophic brown earths and rankers, all of which are markedly acid (with pH values in the range 3.4–6.5), are deficient in nitrogen and are poor in exchangeable cations, especially calcium. Additionally, moorland vegetation may develop on peat deposits where the surface has dried out, resulting in improved drainage and aeration.

As stated above, evergreen dwarf shrubs, especially heather (*Calluna vulgaris*), are the dominant vegetation type. The floristic composition of such communities varies as a result of differences in climate, soil type and management and, for example, McVean and Ratcliffe (1962) have recognised three main variations in heather moorland which occur below the tree line in Scotland. Their category of dry heather moor consists of almost pure stands of heather with a well-developed moss layer and patches of locally abundant bell heather (*Erica cinerea*) and bilberry (*Vaccinium myrtillus*) which occur over podsolic soils, especially iron-humus podsols. A variant of this is the *Arctostaphylos* (bearberry)-rich heather moor where heather still predominates but there is a greater variety of both dwarf shrubs and herbs and the soils are stony podsols with only a thin humus layer. McVean and Ratcliffe's (1962) third category is damp heather moor where bilberry may be co-dominant with heather and crowberry (*Empetrum nigrum*) is often present, along with a well-developed moss layer over acid peaty humus, peaty podsols or drying bog peat.

As well as these floristic variations, heather-dominated communities exhibit morphological changes which are a result of the four-phase life-history of heather (Gimingham 1972). Up to age 15 years, heather communities are in the pioneer phase, when new growth occurs from seeds or from new buds at stem bases, and the building phase when the new growth develops a dense canopy that shades out other species. In the mature phase, between 15 and 25 years, growth is less vigorous and the central branches begin to spread out, opening the canopy so that light can penetrate to the ground, where mosses and lichens flourish. In the final degenerate phase the central branches, heavily encrusted with lichens, die. This life-cycle has important implications for management which will be considered below. Other constituents of the biota of moorlands include animals. These may be domestic herbivores such as cattle and sheep, game birds (such as grouse and ptarmigan) rabbits, hares, red deer and insects as well as a transitory population of wild birds. It is in the management of both plant and animal populations that anthropogenic influences are most clearly seen.

Except in areas above the climatic limit for tree growth, moorlands occur where both climatic and usually edaphic conditions can support either coniferous or broad-leaved forests, as is evidenced by the large tracts of planted conifers which thrive in what were once European moorlands. In addition, it is not uncommon to witness the invasion by trees of moorlands where management is poor. However, the origins of most moorlands are debatable. Firstly, it is highly likely that 'natural' moorlands have existed at high altitudes for almost all of the postglacial and additionally, the lower strata of Boreal zone forests are similar floristically to moorlands. Secondly, palaeoecological evidence from a number of moorland areas indicates that forest preceded moorland. In some areas, this shift occurred around 7000 years ago with the onset of cooler and wetter climatic conditions, but a considerable body of evidence points to a much more pronounced shift beginning about 5000 years ago and is attributed to forest destruction by prehistoric agriculturalists. Their burning and felling of trees produced open spaces for cultivation and grazing and when these areas were subsequently abandoned, moorland communities spread, especially in poorer ground. Such a process was probably accelerated by a further shift to cooler and wetter conditions about 2700 years ago.

There is also considerable evidence to show that forest clearance continued throughout historic time as a result of agricultural expansion, the use of wood for building houses and ships and for iron-ore smelting. By the end of the seventeenth century, almost all of the natural woodland of Britain and large parts of northwest Europe had been destroyed and were replaced in upland areas by extensive tracts of moorland, which provided free-range grazing for domestic herbivores. In the 1780s, stocks of hill sheep were greatly increased in Scotland and by 1800, burning had been introduced to augment the effects of grazing which had been only partially successful in preventing the invasion of trees and other undesirable species, such as bracken (*Pteridium aquilinum*). Today, burning is a major management technique in moorland communities and its use is related to the heather life-cycle discussed earlier. The objective is to provide a continuous supply of young heather which is more palatable to animals and with a higher nutrient content than the old woody tissue characteristic of the mature and degenerate phases of the life-cycle.

Controlled rotational burning is practised so that small patches or long narrow strips of vegetation are fired once every 5–25 years to release nutrients in the ash and promote new growth. This results in a mixture of short and tall heather and is particularly suitable for grouse which utilise the former for feeding and the latter for cover. Where sheep or deer are grazed over extensive areas, patch burning is unnecessary and it is not uncommon to find tens of hectares being burnt at once. Burning, however, can have as many drawbacks as advantages and can result in a complete destruction of heather shoots and seeds. This prevents regeneration and encourages fire-resistant (but non-palatable) species and soil impoverishment. Nevertheless, and whatever the original causes of moorland formation, management by fire and grazing plays a dominant role in what must be considered as a control ecosystem. Climatic and edaphic factors, though still significant, are subordinate to the anthropogenic creation of a tight intra-system manipulation of energy and nutrient transfers.

Further reading

Detwyler, T. R. (1971) *Man's Impact on the Environment*. McGraw-Hill; London and New York.

Goudie, A. (1981) *The Human Impact. Man's Role in Environmental change*. Blackwell; Oxford.

Gregory, K. J. and Walling, D. E. (eds) (1979) *Man and Environmental Processes*. W. M. Dawson and Sons Ltd, Folkestone and Westview Press; Boulder.

Thompson, R. D. (1989) Short-term climatic change: evidence, causes, environmental consequences and strategies for action, *Progress in Physical Geography*, 13, 315–347.

References

Abercromby, R. and Marriott, W. (1883) Popular weather prognostics, *Quart. J. Roy. Met. Soc.*, **9**, 27–43.

Afanasiev, J. N. (1927) *The Classification Problem in Russian Soil Science*. Acad. Sci. USSR-Russ, Pedol. 5; Leningrad.

Allen, J. R. L. (1970) *Physical Processes of Sedimentation*. Allen and Unwin; London and Boston.

Anderson, J. M. (1981) *Ecology for Environmental Sciences*. Edward Arnold; London.

Andrews, R. D., Coleman, D. C., Ellis, J. E. and Singh, J. S. (1974) Energy flow relationships in a short grass prairie ecosystem, *Proc. 1st Internat. Congr. Ecol.*, 22–8.

Bacastow, R. and Keeling, C. D. (1973) Atmospheric carbon dioxide and radiocarbon in the natural carbon cycle, in Woodwell, G. M. and Pecan, E. V. (eds) *Carbon in the Biosphere*. AEC Technical Information Centre; Washington, pp. 86–136.

Baes, C. F., Goeller, H. E, Olson, J. S. and Rotty, R. M. (1977) Carbon dioxide and climate: the uncontrolled experiment. *Amer. Sci.*, **65**, 310–20.

Bagnold, R. A. (1941) *The Physics of Blown Sand and Desert Dunes*. Chapman and Hall; London.

Baker, V. R. and Ritter, D. F. (1975) Competence of rivers to transport coarse bedload materials, *Bull. Geol. Soc. America*, **86**, 975–8.

Barnes, F. A. and King, C. A. M. (1953) The Lincolnshire coastline and the 1953 storm flood, *Geography*, **38**, 141–60.

Barnes, R. S. K. and Hughes, R. N. (1982) *An Introduction to Marine Ecology*. Blackwell; Oxford.

Barry, R. G. (1969) The world hydrological cycle, in Chorley, R. J. (ed.) *Water, Earth and Man*. Methuen; London and New York, pp. 11–29.

Barry, R. G. (1970) A framework for climatological research with particular reference to scale concepts, *Trans. Inst. Brit. Geogr.*, **49**, 61–70.

Barry, R. G. (1981) *Mountain Weather and Climate*. Methuen; London and New York.

Barry, R. G. and Chorley, R. J. (1987) *Atmosphere, Weather and Climate*. Methuen; London and New York.

Beatty, N. B. (ed) (1982) Carbon dioxide effects, in *Proceedings of the Workshop on First Detection of Carbon Dioxide Effects*. US Department of Energy.

Beddington, J. R. and May, R. M. (1982) The harvesting of interacting species in a natural ecosystem, *Sci. Amer.*, **247**, 42–9.

Beeftink, W. G. (1977) The coastal salt marshes of western and northern Europe: an ecological and phytosociological approach, in Chapman, V. J. (ed.) *Wet Coastal Ecosystems*. Elsevier; Amsterdam, London and New York, pp. 109–55.

Belasco, J. E. (1952) Characteristics of air masses over the British Isles, *Met. Office, Geophysical Memoirs*, No. 87. HMSO; London.

Berg, H. (1940) Die Kontinentalität Europas und ihre Anderung 1928/37 gegen 1888/97, *Annals Hydrogr. Berlin*, **124**.

Berg, W., Jonnels, A., Sjostrand, B. and Westermark, T. (1966) Mercury content in feathers of Swedish birds from the past 100 years, *Oikos*, **17**, 71–83.

Bjerknes, J. and Solberg, H. (1922) The life cycle of cyclones and the Polar front theory of atmospheric circulation, *Geofys. Publ.*, **3**, No. 1.

Bolin, B, (1970) The carbon cycle, *Sci. Amer*, **223**, 124–32.

Borg, K., Wanntrop, H., Erne, K. and Hanko, E. (1969) Alleyl mercury poisoning in terrestrial Swedish wildlife, *Viltrevy*, **6**, 301–79.

Bormann, F. H. and Likens, G. E. (1979) *Pattern and Process in a Forested Ecosystem.* Springer-Verlag; Berlin and New York.

Boulton, G. S. (1972) The role of the thermal regime in glacial sedimentation, in *Polar Geomorphology.* Institute of British Geographers, Special Publication No 4; London, pp. 1–19.

Boulton, G. S. (1975) Process and patterns of subglacial sedimentation: a theoretical approach, in Wright, A. E. and Moseley, F. (eds) *Ice Ages: Ancient and Modern.* Seel House Press; Liverpool, pp. 7–42.

Boulton, G. S. (1979) Processes of glacier erosion on different substrata, *J. Glaciology*, **23**, 15–37.

Bowen, H. J. M. (1979) *Environmental Chemistry of the Elements.* Academic Press; London and New York.

Brady, N. C. (1990) *The Nature and Properties of Soils.* Macmillan; London and New York.

Brinkmann, W. A. R. (1971) 'What is a foehn?', *Weather*, **26**, 230–9.

Brooks, R. R. (1983) *Biological Methods of Prospecting for Minerals.* Wiley; New York and London.

Brown, R. J. E. (1961) Potential evapotranspiration and evaporation at Norman Wells, N.W.T., *Proc. Hydrol. Symp. No 2, NRC, Canada*, 123–7.

Brown, R. J. E. (1963) The influence of vegetation on permafrost, *Int. Conf. Permafrost*, Purdue Univ., 20–5.

Browning, K. A. and Ludlam, F. H. (1962) Airflow in convective storms, *Quart. J. Roy. Met. Soc.*, **88**, 117–35.

Bruce, J. P. and Clark, R. H. (1969) *Introduction to Hydrometeorology.* Pergamon; London and New York, pp. 33–56, 288–9.

Brunsden, D. and Jones, D. K. C. (1976) The evolution of landslide slopes in Dorset, *Phil. Trans. Royal Soc. of London*, **A283**, 605–31.

Bryan, R. and Yair, A. (eds) (1982) *Badland Geomorphology and Piping.* Geobooks; Norwich.

Budyko, M. I. (1962) The heat balance of the surface of the earth, *Soviet Geog.*, **3** (5), 3–16.

Byers, H. R. and Braham, R. R. (1949) *The Thunderstorm.* United States Weather Bureau; Washington, D.C.

Cannon, H. L. (1960) The development of botanical methods of prospecting for uranium in the Colorado Plateau, *United States Geological Survey Bulletin*, **1085-A**, 1–50.

Carey, S. W. (1976) *The Expanding Earth.* Elsevier; Amsterdam, London and New York.

Carson, M. A. and Kirkby, M. J. (1972) *Hillslope Form and Process.* Cambridge University Press; Cambridge.

Chalmers, A. F. (1978) *What is this Thing called Science: an Assessment of the Nature of Status of Science and its Methods.* Open University Press; Milton Keynes.

Chandler, R. J. (1982) Lias clay slope sections and their implications for the prediction of limiting of threshold slope angles, *Earth Surface Processes and Landforms*, **7**, 427–38.

Chandler, T. J. (1965) *The Climate of London.* Hutchinson; London.

Chandler, T. J. (1969) *The Air Around Us.* Nat. Hist. Press; New York, pp. 70–99.

Chevallier, J. M. and Cailleux, A. (1959) Essai de reconstitution géometrique des continents primitifs, *Z. für Geomorphologie*, **3**, 257–68.

Chikishev, A. G. (1965) *Plant Indicators of Soils, Rocks and Subsurface Waters.* Consultants Bureau; New York.

Chorley, R. J. and Kennedy, B. A. (1971) *Physical Geography: a Systems Approach.* Prentice-Hall; New Jersey and London.

Clapham, A. R., Tutin, T. G. and Warburg, E. F. (1962) *Flora of the British Isles*, Cambridge University Press; Cambridge.

Clark, F. E. (1967) Bacteria in soil, in Burges, A. and Raw, F. (eds) *Soil Biology.* Academic Press; London and New York, pp. 15–49.

Clements, F. E. (1916) Plant succession: An analysis of the development of vegetation, *Carnegie Inst. Washington Publication*, No. 242, 3–4.

Clements, F. E. (1928) *Plant Succession and Indicators.* H. W. Wilson; New York.

Cloud, P. and Gibor, A. (1970) The oxygen cycle, in *The Biosphere.* Scientific American, W. H. Freeman; San Francisco, pp. 57–68.

Coates, D. R. (1983) Large-scale land subsidence, in Gardner, R. and Scoging, H. (eds) *Mega-geomorphology.* Clarendon Press; Oxford, pp. 212–34.

Colbeck, S. C. and Evans, R. J. (1973) A flow law for temperate glacier ice, *J. Glaciol.*, **12**, 71–86.

Cole, D. W. and Rapp, M. (1981) Elemental cycling in forest ecosystems, in Reichle, D. E. (ed.) *Dynamic Properties of Forest Ecosystems.* Cambridge University Press; Cambridge, pp. 241–409.

Cole, F. W. (1970) *Introduction to Meteorology.* Wiley; New York and London, pp. 112–63.

Collier, B. D., Cox, G. W., Johnson, A. W. and Miller, P. C. (1973) *Dynamic Ecology.* Prentice-Hall; New Jersey and London.

Cooke, R. U., Brunsden, D., Doornkamp, J. C. and Jones, D. K. C. (1982) *Urban Geomorphology in Dry Lands.* Oxford University Press; London and New York.

Coope, G. R. (1977) Fossil Coleopteran assemblages as sensitive indicators of climatic changes during the Devensian (last) cold stage, *Phil. Trans. Roy. Soc.*, **B280**, 313–40.

Cooper, W. S. (1926) The fundamentals of vegetational change, *Ecology*, **7**, 391–413.

Cooper, W. S. (1931) A third expedition to Glacier Bay, Alaska, *Ecology*, **12**, 61–95.

Corby, G. C. (1954) The airflow over mountains: a review of the state of current knowledge, *Quart. J. Roy. Met. Soc.*, **80**, 491–521.

Cowles, H. C. (1911) The causes of vegetative cycles, *Bot. Gaz.*, **51**, 161–83.

Crabtree, R. W. and Burt T. P. (1981) Spatial variation in solutional denudation and soil moisture over a hillslope hollow, *Earth Surface Processes and Landforms*, **6**, 319–30.

Crawford, C. and Johnston, C. (1971) Construction on permafrost, *Can. Geotech. J.*, **8**, 236–51.

Crowe, P. R. (1971) *Concepts in Climatology*. Longman; London and New York.

Cushing, D. H. (1971) Upwelling and production of fish, *Adv. Mar. Biol.*, **9**, 255–334.

Dancer, W. S., Handley, J. F. and Bradshaw, A. D. (1977) Nitrogen accumulation in kaolin mining wastes in Cornwall. 11. Forage legumes, *Plant and Soil*, **48**, 303–14.

Dansereau, P. (1957) *Biogeography: an Ecological Perspective*. Ronald; New York.

Darwin, C. (1859) *On the Origin of Species*. John Murray; London.

Davies, P. A. and Runcorn, S. K. (1980) *Mechanisms of Continental Drift and Plate Tectonics*. Academic Press; London and New York.

Davis, W. M. (1909) The geographical cycle, in Johnson, D. W. (ed.) *Geographical Essays*. Ginn & Co.; pp. 254–6.

Dean, R., Ellis, J. E., Rice, R. W. and Demeret, R. E. (1975) Nutrient removal by cattle from a short grass prairie, *J. Appl. Ecol.*, **12**, 25–9.

Defant, F. (1951) Local winds, in Malone, T. F. (ed.) *Compendium of Meteorology*. American Met. Soc.; Boston, pp. 655–72.

Derbyshire, E. (1983) On the morphology, sediments and origin of the loess plateau of Central China, in Gardner, R. and Scoging, H. (eds) *Mega-geomorphology*. Clarendon Press; Oxford, pp. 172–94.

Derbyshire, E., Gregory, K. J. and Hails, J. R. (1979) *Geomorphological Processes*. W. M. Dawson and Sons Ltd, Folkestone; Westview Press, Boulder.

Dietz, R. S. (1962) The seas' deep scattering layers, *Sci. Amer.*, **207**, 1–6.

Dokuchaev. V. V. (1886) Data on land appraisal in Nizhnii-Novgorod province, in *Collected Works*, Vol. 4, Acad. Sci., Moscow, 1950.

Donn, W. L. (1965) *Meteorology*. McGraw-Hill; London and New York, pp. 78–140.

Douglas, I. (1976) Lithology landforms, and climate, in Derbyshire, E. (ed.) *Geomorphology and Climate*. Wiley; New York and London, pp. 345–66.

Dove, H. W. (1862) *The Law of Storms* (trans. by R. H. Scott). Longman; London and New York.

Drewry, D. J. (1983) Antarctic ice sheet: aspects of current configuration and flow, in Gardner, R. and Scoging, M. (eds) *Mega-geomorphology*, Clarendon Press; Oxford, pp. 18–38.

Drozdz, A. (1967) Food preference, food digestibility and the natural supply of small rodents, in Petrusewicz, K. (ed.) *Terrestrial Ecosystems* (*Principles and Methods*), vol. 1. Panstovowe Wydownictwo; Warsaw, pp. 323–30.

Drury, W. H. and Nisbet, I. C. T. (1973) Succession, *J. Arn. Arb.*, **54**, 331–68.

Duchaufour P. (1982) *Pedology*, (trans. by J. R. Paton). Allen and Unwin; London and Boston.

Duvigneaud, P. and Denaeyer-De Smet, S. (1970) Biological cycling of minerals in temperate deciduous forests, in Reichle, D. E. (ed.) *Analysis of Temperate Forest Ecosystems*. Springer-Verlag; Berlin and New York, pp. 199–225.

Edwards, P. J. (1982) Studies of mineral cycling in a montane rain forest in New Guinea. V. Rates of cycling in throughfall and litter fall, *J. Ecol.*, **70**, 807–27.

Ehrlich, P. R., Ehrlich, A. H. and Holdren, P. R. (1977) *Ecoscience: Population, Resources, Environment*. W. H. Freeman; San Francisco.

Einstein, H. A. (1950) The bed load function for sediment transportation in open channel flows, *Tech, Bull., United States Dept. Agric.*, 1026.

Ekman, V. W. (1905) On the influence of the earth rotation on ocean currents, *Arkiv. f. Mat. Astron. och Fysik*, **2**, 11.

Ellison, W. D. (1944) Studies of raindrop erosion, *Agricultural Engineering*, **25**, 131–6.

Elton, C. (1927) *Animal Ecology*. University of Washington Press; Washington.

Embleton, C. (1979) Glacial processes, in Embleton, C. and Thornes, J. (eds) *Processes in Geomorphology*. Edward Arnold; London, pp. 272–306.

Embleton, C. (1979) Nival processes, in Embleton, C. and Thornes, J. (eds) *Processes in Geomorphology*. Edward Arnold; London, pp. 307–24.

Emmett, W. W. (1970) The hydraulics of overland flow on hillslopes, *United States Geological Survey Professional Paper*, 662-A.

Emmett, W. W. (1978) Overland flow, in Kirkby, M. J. (ed.) *Hillslope Hydrology*. Wiley; New York and London, pp. 145–76

Ernst, W. H. O. (1983) Element nutrition of two contrasted dune annuals, *J. Ecol.*, **71**, 197–209.

Etherington, J. R. (1982) *Environment and Plant Ecology*. Wiley; New York and London.

Evans, G. C. (1976) A sack of uncut diamonds: the study of ecosystems and the future resources of mankind, *J. Ecol.*, **64**, 1–39.

Fahey, B. D. (1973) An analysis of diurnal freeze-thaw and frost heave cycles in the Indian Peaks region of the Colorado Front Range, *Arct. Alp. Res.*, **5**, 269–81.

Fahey, B. D. (1983) Frost action and hydration as rock weathering mechanisms on schist: a laboratory study, *Earth Surface Processes*, **8**, 535–45.

FAO/Unesco (1974) *Soil Map of the World, 1 : 5 000 000*, vol. 1. Legend Unesco; Paris.

Farres, P. J. (1980) Soil observations on the stability of soil aggregates to raindrop impact, *Catena*, **7**, 223–31.

Fenwick, I. M. and Knapp, B. J. (1982) *Soils: Process and Response*. Duckworth; London.

Ferrell, W. (1889) *A Popular Treatise on the Winds*. Macmillan; London and New York.

Finegan, B. (1984) Forest succession, *Nature*, **312**, 109–114.

Finch, V. C. and Trewartha, G. T. (1942) *Physical Elements of Geography*. McGraw-Hill; London and New York.

Findlay, B. F. and Hirt, M. S. (1969) An urban-induced meso-circulation, *Atmos. Env.*, **3**, 537–42.

Finlayson, B. (1981) Field measurements of soil creep, *Earth Surface Processes*, **6**, 35–48.

Foster, I. D. L. (1980) Chemical yields in runoff and denudation in a small arable catchment, East Devon, England, *J. of Hydrology*, **47**, 349–68.

Fraser Darling, F. (1963) The unity of ecology, *Brit. Assoc. Adv. Sci.*, **20**, 297–306.

Freeze, R. A. and Banner, J. (1970) The mechanism of natural groundwater recharge and discharge, *Water Resources Res.*, **6**, 138–55.

French, H. M. (1975) *The Periglacial Environment*. Longman; London and New York.

Friederichs, K. (1958) A definition of ecology and some thoughts about basic concepts, *Ecology*, **39**, 154–9.

Furley, P. A. and Newey, W. W. (1982) *The Geography of the Biosphere*. Butterworths; London and Washington.

Geiger, R. (1965) *The Climate Near the Ground*. Harvard University Press; Cambridge, pp. 249–57.

Gemmel, R. P. and Goodman, G. T. (1980) The maintenance of grassland on smelter wastes in the Lower Swansea Valley. III. Zinc smelter wastes, *J. App. Ecol.*, **17**, 461–8.

Gentilli, J. (1971) Dynamics of the Australian troposphere, in Gentilli, J. (ed.) *Climates of Australia and New Zealand* (World Survey of Climatology, vol. 13). Elsevier; Amsterdam, London and New York.

Gersmehl, P. J. (1976) An alternative biogeography, *Ann. Assoc. Amer. Geogr.*, **66**, 223–41.

Gimingham, C. H. (1972) *The Ecology of Heathlands*. Chapman and Hall; London.

Gleason, H. A. (1917) The structure and development of the plant association, *Bull. Torrey Bot. Club*, **44**, 463–81.

Gleason, H. A. (1926) The individualistic concept of the plant association, *Bull. Torrey Bot. Club*, **53**, 7–26.

Goldberg, E. D. (1971) Atmospheric dust, the sedimentary cycle and man, *Comm. Earth Sci. Geophys.*, **1**, 117–32.

Goudie, A. S. (1977) *Environmental Change*. Clarendon Press; Oxford.

Goudie, A. S. (1981) *Geomorphological Techniques*. Allen and Unwin; London and Boston.

Goudie, A. S. (1983) Dust storms in space and time, *Progress in Physical Geography*, **7**, 502–30.

Gray, J. T. (1983) Nutrient use by evergreen and deciduous shrubs in southern California. I. Community nutrient cycling and nutrient-use efficiency, *J. Ecol.*, **71**, 21–42.

Gray, J. T. and Schlesinger, W. H. (1983) Nutrient use by evergreen and deciduous shrubs in southern California. II. Experimental investigations of the relationship between growth, nitrogen uptake and nitrogen availability, *J. Ecol.*, **71**, 43–56.

Gregory, K. J. and Walling, D. E. (1973) *Drainage Basin Form and Process*. Edward Arnold; London.

Gross, M. G. (1972) *Oceanography. A View of the Earth*. Prentice-Hall; New Jersey and London.

Grove, A. T. (1969) The ancient Erg of Hemsaland and similar formations on the south side of the Sahara, *Geographical J.*, **135**, 192–212.

Hack, J. T. (1957) Studies of longitudinal stream profiles in Virginia and Maryland, *United States Geological Survey Professional Paper*, 294-B.

Hadley, G. (1735) Concerning the cause of the general tradewinds, *Phil. Trans. Roy. Soc.*, **29**, 58–62.

Hall, A. D. and Fagen, R. E. (1956) Definition of system, *Gen. Systems*, **1**, 18–28.

Halley, E. (1686) An historical account of the trade winds and monsoons observable in the seas between and near the tropicks; with an attempt to assign the phisical cause of the said winds, *Phil. Trans. Roy. Soc.*, **16**, 153–68.

Hann, J. (1866) Zur Frage über den Ursprung des Föhns, *Zeit. Osterreich Ges. Met.*, **1**, 257–63.

Hanson, W. C. (1967) Caesium-137 in Alaskan lichens, caribou and eskimos, *Health Phys.*, **13**, 383–9.

Hanson, W. C. (1971) Seasonal patterns in native residents of three contrasting Alaskan villages, *Health Phys.*, **20**, 585–91.

Hardy, A. C. (1956) The Open Sea vol. 1. *The World of Plankton*. Collins; London.

Harris, C (1974) Autumn, winter and spring soil temperatures in Okstindan, Norway, *J. Glaciology*, **13**, 521–33.

Harris, C (1981) Periglacial mass-wasting: a review of research, *British Geomorphological Research*

Group Research Monograph, 4. Geobooks; Norwich.

Harrold, L. L. (1969) Evapotranspiration: a factor in the plant–soil water economy, *The Progress of Hydrology*, University of Illinois, Urbana, **2**, 694–716.

Harvey, A. M. (1975) Some aspects of the relations between channel characteristics and riffle spacing in meandering streams, *Amer. J. Sci.*, **275**, 470–8.

Harvey, D. (1969) *Explanation in Geography*. Edward Arnold; London.

Harwood, R. S. (1981) Atmospheric vorticity and divergence, in Atkinson, B. W. (ed.) *Dynamical Meteorology*. Methuen; London and New York.

Heal, O. W. H. and Maclean, S. F. (1975) Comparative productivity in ecosystems – secondary productivity, in Van Dobben, W. H. and Lowe-McConnell, R. H. (eds) *Unifying Concepts in Ecology*. W. Junk; The Hague, pp. 89–108.

Herbertson, A. J. (1901) *Outlines of Physiography*. Edward Arnold; London.

Herrera, R., Jordan, C. F., Klinge, H. and Medina, E. (1978) Amazon ecosystems: their structure and functioning with particular emphasis on nutrients, *Interciencia*, **3**, 223–32.

Hjulstrom, F. (1935) Studies of the morphological activity of rivers as illustrated by the River Fyris, *Bull. Geol. Institute Univ. Uppsala*, **25**, 221–7.

Hobbs, B. E., Means, W. D. and Williams P. F. (1976) *An Outline of Structural Geology*. Wiley; New York and London.

Hodgson, J. M. (1976) *Soil Survey Field Handbook*. Tech. Mono. No. 5, Soil Survey of Great Britain; Harpenden.

Holdridge, L. R. (1947) Determination of world plant formations from simple climatic data, *Science*, **105**, 267–368.

Holmes, R. H. and Robertson, G. W. (1958) Conversion of latent evaporation to potential evapotranspiration, *Can. J. Plant Sci.*, **38**, 164–72.

Horn, H. S. (1976) Succession, in May R. M. (ed.) *Theoretical Ecology*. Blackwell; Oxford, pp. 187–204.

Horton, R. E. (1945) Erosional development of streams and their drainage basins: hydrophysical approach to quantitative morphology, *Bull Geol. Soc. Amer.*, **56**, 275–370.

Howard, A. D. (1980) Thresholds in river regimes, in Coates, D. R. and Vitek, J. D. (eds) *Thresholds in Geomorphology*. Allen and Unwin; London and Boston, pp. 227–58.

Hudson, N. W. (1971) *Soil Conservation*. Batsford; London.

Innes, J. L. (1983) Debris flows, *Progress in Physical Geography*, **7**, 469–501.

Institute of Hydrology (1984) *Annual Report 1981–84*, Wallingford, **1**.

Isaacs, J. D. and Schwartzlose, R. A. (1975) Active animals of the deep sea floor. *Ocean Science*, Scientific American, W. H. Freeman, San Francisco, pp. 223–38.

Jeffreys, H. (1926) On the dynamics of geostrophic winds, *Quart. J. Roy. Met. Soc.*, **52**, 85–101.

Johnson, A. M. (1976) The climate of Peru, Bolivia and Ecuador, in Schwerdtfeger, W. (ed.) *Climates of Central and South America* (World Survey of Climatology, Vol. 12). Elsevier; Amsterdam, London and New York, pp. 147–218.

Johnson, F. S. (1970) The oxygen and carbon dioxide balance in the earth's atmosphere, in Singer, F. S. (ed.) *Global Effects of Environmental Pollution*. Springer-Verlag; Berlin and New York, pp. 4–11.

Jones, G. (1979) *Vegetation Productivity*, Longman; London and New York.

Jones, J. A. A. (1981) The nature of soil piping: a review of research, *British Geomorphological Research Group Research Monograph*, 3. Geobooks; Norwich.

Karpinsky, A. M. (1841) Can living plants be indicators of rocks and formations on which they grow and does their occurrence merit the particular attention of the specialist in structural geology? *Zhur. Sadovodstra*, Nos. 3 and 4.

Kendrew, W. G. (1922) *The Climate of the Continents*. Oxford University Press; London and New York.

Kennett, J. P. (1982) *Marine Geology*. Prentice-Hall; New Jersey and London, pp. 23–39, 320–94.

Kerfoot, D. E. (1974) Thermokarst features produced by man-made disturbance to the tundra terrain, in Fahey, B. D. and Thompson, R. D. (eds) *Research in Polar and Alpine Geomorphology*. Geo Abstracts; Norwich, pp. 60–72.

King, C. A. M. (1975) *Introduction to Marine Geology and Geomorphology*. Edward Arnold; London.

King, C. A. M. (1980) *Physical Geography*, Blackwell, Oxford.

King, K. (1961) Evaporation from land surfaces, *Proc. Hydrol. Symp. No. 2 Evaporation*, Queen's Printer Ottawa, 55–80.

King, L. C. (1962) *The Morphology of the Earth*. Oliver and Boyd; Edinburgh.

Kirkby, M. J. (1969) Infiltration, throughflow and overland flow, in Chorley, R. J. (ed.) *Water, Earth and Man*. Methuen; London and New York, pp. 215–17.

Kononova, M. M. (1966) *Soil Organic Matter*. Pergamon; Oxford.

Knapp, B. J. (1978) Infiltration and storage of soil

water in Kirkby, M. J. (ed.) *Hillslope Hydrology*. Wiley; New York and London, pp. 43–72.

Lachenbruch, A. (1970) Some estimates of the thermal effects of a heated pipeline in permafrost, *United Stated Geological Survey Circ.*, **632**, 1.

Lamb, H. H. (1950) Types and spells of weather around the year in the British Isles: Annual trends, seasonal structure of the year, Singularities, *Quart. J. R. Met. Soc.*, **76**, 393–438.

Lamb, H. H. (1964) Circulation of the atmosphere, in Priestley, R. *et al.* (eds) *Antarctic Research*. Butterworths; London and Washington, pp. 265–77.

Lamb, H. H., Collison, P. and Ratcliffe, R. A. S. (1973) Northern hemisphere monthly and annual mean-sea-level pressure distribution for 1951–66 and changes of pressure and temperature compared with those of 1900–39, *Met. Office Geophys. Mem.*, No. 118, HMSO; London.

Landsberg, H. E. (1981) *The Urban Climate*. Academic Press; London and New York.

Langbein, W. B. and Leopold L. B. (1966) River meanders – theory of minimum variance, *United States Geological Survey Professional Paper*, 422-H.

Leighly, J. (1949) Climatology since the year 1800, *Trans. Amer. Geophys. Un.*, **30**, 658–72.

Leith, H. F. H. and Whittaker, R. H. (eds) (1975) *Primary Productivity of the Biosphere*. Springer-Verlag; Berlin and New York.

Leopold, L. B. and Wolman, M. G. (1957) River channel patterns, braided, meandering and straight, *United States Geological Survey Professional Paper*, 282-B.

Leopold, L. B., Wolman, M. G. and Miller, J. P. (1964) *Fluvial Processes in Geomorphology*. W. H. Freeman; San Francisco.

Likens, G. E., Bormann, F. H., Pierce, R. S., Eaton, J. S. and Johnson, N. M. (1977) *Biogeochemistry of a Forested Ecosystem*. Springer-Verlag; Berlin and New York.

Likens, G. E., Bormann, F. H., Pierce, R. S. and Reiners, W. A. (1978) Recovery of a deforested ecosystem, *Science*, **199**, 492–6.

Lindeman, R. L. (1942) The trophic dynamic aspect of ecology, *Ecology*, **23**, 399–418.

Lliboutry, L. A. (1979) Local friction laws for glaciers: a critical review and new openings, *J. Glaciology*, **23**, 67–95.

Lockwood, J. G. (1974) *World Climatology, An Environmental Approach*. Edward Arnold; London.

Long, S. P. and Mason, C. F. (1983) *Saltmarsh Ecology*. Blackie; London and Glasgow.

Lowry, W. P. (1967) *Weather and Life: an Introduction to Biometeorology*. Academic Press; London and New York, pp. 129–32.

Ludlam, F. H. (1961) The hailstorm, *Weather*, **16**, 152–62.

Mabbutt, J. A. (1977) *Desert Landforms*. MIT Press; Cambridge, Mass.

Machta, L. and Hughes, S. E. (1970) Atmospheric oxygen in 1967 to 1970, *Science*, **168**, 1582–4.

Mackay, J. R. (1972) The world of underground ice, *Ann. Assoc. Amer. Geogr.*, **62**, 1–22.

McVean, D. N. and Ratcliffe, D. A. (1962) *Plant Communities of the Scottish Highlands*. HMSO; London.

Maddock, T. (1969) The behaviour of straight open channels with movable beds, *United States Geological Survey Professional Paper*, 622-A.

Manley, G. (1945) The helmwind of Crossfell, 1937–9, *Quart. J. Roy. Met. Soc.*, **71**, 197–219.

Margalef, R. (1968) *Perspectives in Ecological Theory*. University of Chicago Press; Chicago.

Marshall, J. F. and Davies, P. J. (1984) Last interglacial reef growth between modern reefs in the Southern Great Barrier Reef, *Nature*, **307**, 43–5.

Marshall, N. B. (1979) *Development in Deep Sea Biology*. Blandford Press; Poole.

Miles, J. (1979) *Vegetation Dynamics*. Chapman and Hall; London.

Miller, J. P. (1961) Solutes in small streams draining single rock types Sangre de Cristo Range, New Mexico, *United States Geological Survey Water Supply Paper*, 1535-F, Washington D. C.

Moore, P. D. (1983) Revival of the organismal heresy, *Nature*, **303**, 132–3.

Morgan, R. P. C. (1982) Splash detachment under plant covers: results and implications of a field study, *Trans. Amer. Soc. of Agric. Engineers*, **25**, 987–91.

Morner, N.-A. (1983) Sea levels, in Gardner, R. and Scoging, H. (eds) *Mega-geomorphology*. Clarendon Press; Oxford, pp. 73–92.

Mosley, M. P. (1973) Rainsplash and the convexity of badland divides, *Zeits. für Geomorph. Suppl.*, **18**, 10–25.

Mosley, M. P. (1982). The effect of a New Zealand beech forest canopy on the kinetic energy of water drops and on surface erosion, *Earth Surface Processes and Landforms*, **7**, 103–7.

Moss, C. E. (1910) The fundamental units of vegetation, *New Phytol.*, **9**, 18–53.

Murray, J. and Hjort, J. (1912) *The Depths of the Ocean*. Murray; London.

Musgrave, G. W. (1947) Quantitative evaluation of factors in water erosion – a first approximation, *J. Soil and Water Conserv.*, **2**, 133–8.

Nace, R. L. (1969) Human use of ground water, in Chorley, R. J. (ed.) *Water, Earth and Man*. Methuen; London and New York, pp. 285–94.

Newson, M. D. and Hanwell, J. D. (1982) *System-*

atic Physical Geography. Macmillan; London and New York, pp. 56–61.

Nixon, S. W. (1980) Between coastal marshes and coastal waters – a review of twenty years of speculation and research on the rôle of salt marshes in estuarine productivity and water chemistry, in Hamilton, P. and Macdonald, K. B. (eds) *Estuarine and Wetland Processes with Emphasis on Modeling*. Plenum Press; New York, pp. 437–525.

Odum, E. P. (1960) Organic production and turnover in old field succession, *Ecology*, **41**, 34–49.

Odum, E. P. (1969) The strategy of ecosystem developments, *Science*, **164**, 262–70.

Odum, E. P. (1971) *Fundamentals of Ecology*. Saunders; Philadelphia and London.

Odum, E. P. (1975) *Ecology*. Holt, Rinehart and Winston; New York.

Odum, W. E. and Heald, E. J. (1972) Trophic analysis of an estuarine mangrove community, *Bull. Marine Sci.*, **22**, 671–738.

Oke, T. R. (1978) *Boundary Layer Climates*. Methuen, London; Wiley, New York and London.

Ollier, C. D. (1975) *Weathering*. Longman; London and New York.

Ollier, C. D. (1981) *Tectonics and Landforms*. Longman; London and New York.

Oxburgh, E. R. (1974) The plain man's guide to plate tectonics, *Proc. Geol. Assoc.*, **85**, (3), 299–357.

Oxley, N. C. (1974) Suspended sediment delivery rates and the solute concentration of stream discharge in two Welsh catchments, in Gregory, K. J. and Walling, D. E. (eds) *Fluvial Processes in Instrumented Watersheds*. Institute of British Geographers, Special Publication No. 6; London, pp. 141–54.

Palmén, E. (1951) The role of atmospheric disturbances in the general circulation, *Quart. J. Roy. Met. Soc.*, **77**, 337–54.

Palmén, E. and Newton, C. W. (1969) *Atmospheric Circulation Systems. Their Structure and Physical Interpretation*. Academic Press; London and New York.

Paterson, W. B. (1981) *The Physics of Glaciers*. Pergamon; Oxford and New York.

Paul, M. A. (1983) The supraglacial landsystem, in Eyles, N. (ed.) *Glacial Geology*. Pergamon; Oxford and New York, pp. 71–90.

Pedgley, D. E. (1962) *A Course in Elementary Meteorology*. HMSO; London.

Penning-Rowsell, E. C. and Townshend, J. R. G. (1978) The influence of scale on the factors affecting stream channel slope, *Trans. Inst. Brit. Geogr.*, new series, **3**, 395–415.

Pennington, W. (1977) The Late Devensian flora and vegetation of Britain, *Phil. Trans. Roy. Soc. Lond.*, **B280**, 247–71.

Perrin, R. M. (1965) The use of drainage water analyses in soil studies, *Experimental Pedology*, Proceedings of 11th Easter School in Agricultural Science, University of Nottingham, 1964, 73–96.

Perry, A. H. and Walker, J. M. (1977) *The Ocean–Atmosphere System*. Longman; London and New York.

Petterssen, S. (1956) *Weather Analysis and Forecasting*. McGraw-Hill; London and New York.

Petterssen, S. and Smebye, S. J. (1971) On the development of extra tropical cyclones, *Quart. J. Roy. Met. Soc.*, **97**, 457–82.

Pilgrim, D. H. and Huff, D. D. (1983) Suspended sediment in rapid subsurface stormflow on a large field plot, *Earth Surface Processes and Landforms*, **8**, 451–63.

Postgate, J. R. and Hill, S. (1979) Nitrogen fixation, in Lynch, J. M. and Poole, N. J. (eds) *Microbial Ecology: A Conceptual Approach*. Blackwell; Oxford, pp. 191–213.

Raiswell, R., Brimblecombe, D., Dent, D. and Liss, P. (1980) *Environmental Chemistry*. Edward Arnold; London.

Redfield, A. C. (1958) The biological control of chemical factors in the environment, *Amer. Sci.*, **46**, 205–21.

Reichle, D. E. (ed.) (1981) *Dynamic Properties of Forest Ecosystems*. Cambridge University Press; Cambridge.

Richards, K. (1982) *Rivers Form and Process in Alluvial Channels*. Methuen; London and New York.

Richards, P. W. (1979) *The Tropical Rain Forest*. Cambridge University Press; Cambridge.

Richey, J. E. (1983) The phosphorous cycle, in Bolin, B. and Cook, R. B. (eds) *The Major Biogeochemical Cycles and their Interactions*. Wiley; New York and London, pp. 51–6.

Riley, G. A. (1956) Production and utilization of organic matter in Long Island Sound, *Bull. Bingham Oceanogr. Coll.*, **15**, 324–44.

Roberts, R. D., Marrs, R. H., Staffington, R. A. and Bradshaw, A. D. (1981) Ecosystem development on naturally colonized china clay wastes. 1. Vegetation changes and overall accumulation of organic matter and nutrients, *J. Ecol.*, **69**, 153–61.

Robinson, G. W. (1949) *Soils: Their Origin, Constitution and Classification*. Thomas Murby; London.

Rodin, L. E. and Bazilevich, N. I. (1967) *Production and Mineral Cycling of Terrestrial Vegetation*. Oliver and Boyd; London and Edinburgh.

Rossby, C.-G. (1941) The scientific basis of modern meteorology, in United States Department of Agriculture Yearbook, *Climate and Man*, 599–655.

Rossby, C.-G. (1949) On the nature of the general

circulation of the lower atmosphere, in Kuiper, G. P. (ed.) *The Atmosphere of the Earth and Planets*. University of Chicago Press; Chicago, pp. 16–48.

Russell, E. J. (1961) *The World of the Soil*, The Fontana Library, Collins; London.

Russell, E. W. (1973) *Soil Conditions and Plant Growth*. Longman; London and New York.

Rutten, M. G. (1969) *The Geology of Western Europe*. Elsevier; Amsterdam, London and New York.

Sansom, H. W. (1951) A study of cold fronts over the British Isles, *Quart. J. Roy. Met. Soc.*, **77**, 96–120.

Sawyer, J. S. (1957) Jet stream features of the earth's atmosphere, *Weather*, **12**, 333–44.

Schlesinger, W. H. (1977) Carbon balance in terrestrial detritus, *Ann. Rev. Ecol. Syst.*, **8**, 51–81.

Schumm, S. A. (1979) Geomorphic thresholds: the concept and its applications, *Trans. Inst. Brit. Geogr.*, new series, **4**, 485–515.

Schumm, S. A. and Khan, H. R. (1972) Experimental study of channel patterns, *Bull. Geol. Society of Amer.*, **83**, 1755–70.

Scorer, R. S. (1957) Vorticity, *Weather*, **12**, 72–83.

Selby, M. J. (1980) A rock mass strength classification for geomorphic purposes: with lists from Antarctica and New Zealand, *Zeit. für Geomorph.*, **24**, 31–45.

Selby, M. J. (1982) *Hillslope Materials and Processes*. Oxford University Press; London and New York.

Selleck, G. W. (1960) The climax concept, *Bot. Rev.*, **26**, 534–45.

Sellers, W. D. (1965) *Physical Climatology*. University of Chicago Press; Chicago. pp. 11–126.

Servant, J. (1975) Contribution à l'étude pedologique des sols halomorphes: l'example des sols sales du' sud et du sud-est de la France. State doctoral thesis; University of Montpellier.

Shaw, W. N. (1911) *Forecasting Weather*, Constable; London.

Shaw, W. N. (1919–31) *Manual of Meteorology* (4 vols). Cambridge University Press; Cambridge.

Shaw, W. N. (1933) *The Drama of Weather*. Cambridge University Press; Cambridge.

Shepard, F. P. (1963) *Submarine Geology*. Harper and Row; London and New York, pp. 206–399.

Shure, D. J. and Ragsdale, H. L. (1977) Patterns of primary succession on granite outcrop surfaces, *Ecology*, **58**, 993–1006.

Siegenthaler, U. and Oeschger, H. (1978) Predicting future atmospheric carbon dioxide levels, *Science*, **199**, 388–95.

Simmons, D. B. and Richardson, E. V. (1966) Resistance to flow in alluvial channels, *United States Geological Survey Professional Paper*, 422-J.

Simmons, I. G. (1979) *Biogeography: Natural and Cultural*. Edward Arnold; London.

Simpson, J. E. (1964) Sea-breeze fronts in Hampshire, *Weather*, **19**, 208–20.

Simpson, R. H. and Riehl, H. (1981) *The Hurricane and its Impact*. Blackwell; Oxford.

Skempton, A. W. (1964) The long-term stability of clay slopes, *Geotechnique*, **14**, 75–102.

Smith, D. I. and Atkinson, T. C. (1976) Process, landforms and climate in limestone regions, in Derbyshire, E. (ed.) *Geomorphology and Climate*. Wiley; New York and London, pp. 367–410.

Smith, D. I. and Newson, M. D. (1974) The dynamics of solutional and mechanical erosion in limestone catchments on the Mendip Hills, Somerset, in Gregory, K. J. and Walling, D. E. (eds) *Fluvial Processes in Instrumented Watersheds*. Institute of British Geographers, Special Publication No. 6; London, pp. 155–68.

Smith, K. (1972) *Water in Britain*. Macmillan; London and New York, pp. 56–63, 74–5.

Smith, R. A. H. and Bradshaw, A. D. (1979) The use of metal tolerant plant populations for the reclamation of metalliferous wastes, *J. Appl. Ecol.*, **16**, 595–612.

Smith, R. L. (1980) *Ecology and Field Biology*. Harper and Row; London and New York.

Smith, W. E. and Smith, A. M. (1975) *Minamata*. Holt, Rinehart and Winston; New York.

Sorokin, Y. I. (1971) On the role of bacteria in the productivity of tropical oceanic waters, *Int. Rev. ges. Hydrobiol.*, **56**, 1–48.

Spencer, H. (1899) *First Principles*. D. Appleton; New York.

Statham, I. (1977) *Earth Surface Sediment Transport*. Clarendon Press; Oxford.

Stoddart, D. R. (1965) Geography and the ecological approach – the ecosystem as a geographical principle and method, *Geography*, **50**, 242–51.

Strahler, A. N. (1975) *Physical Geography*. Wiley; New York and London.

Stuiver, M. (1978) Atmospheric carbon dioxide and carbon reservoir changes, *Science*, **199**, 253–8.

Summerhayes, V. S. and Elton, C. S. (1923) Contributions to the ecology of Spitsbergen and Bear Island, *J. Ecol.*, **11**, 214–86.

Sutcliffe, R. C. (1947) A contribution to the problem of development, *Quart. J. Roy. Met. Soc.*, **73**, 370–83.

Swaine, M. D. and Hall, J. B. (1983) Early succession on cleared forest land in Ghana, *J. Ecol.*, **71**, 601–27.

Tansley, A. G. (1935) The use and abuse of vegetational concepts and terms, *Ecology*, **16**, 284–307.

Tansley, A. G. (1939) *The British Islands and Their Vegetation*. Cambridge University Press; Cambridge.

Teal, J. M. (1962) Energy flow in the salt marsh ecosystem of Georgia, *Ecology*, **43**, 614–24.

Tedrow, J. C. F. (1958) Major genetic soils of the Arctic slope of Alaska, *J. Soil Sci.*, **9**, 33–45.

Temple, P. H. and Rapp, A. (1972) Landslides in the Mgeta area, Western Uluguru Mountains, Tanzania, *Geografiska Annaler*, **54A**, 157–94.

Terzaghi, K. (1962) Stability of steep hard slopes on hard unweathered rock. *Geotechnique*, **12**, 251–70.

Thompson, R. D. (1974) Climate and permafrost in Canada, *Weather*, **29**, 298–305.

Thorn, C. E. and Hall, K. (1980) Nivation: an Arctic–alpine comparison and reappraisal, *J. Glaciology*, **25**, 109–24.

Thornes, J. (1979) Fluvial processes, in Embleton, C. and Thornes, J. (eds) *Processes in Geomorphology*. Edward Arnold; London, pp. 213–27.

Thornthwaite, C. W. (1944) A contribution to the report of the committee on transpiration and evaporation 1943–44, *Trans. Amer. Geophys. Un.*, **25**, 686–93.

Thornthwaite, C. W. (1948) An approach towards a rational classification of climate, *Geogr. Rev.*, **38**, 55–94.

Tiagi, Y. D. and Aery, N. C. (1982) Geobotanical studies on zinc deposit areas of Zawar mines, Udaipur, *Vegetabo*, **50**, 65–70.

Townshend, J. R. G. (1970) Geology, slope form and slope processes and their relation to the occurrence of laterite, *Geogr. J.*, **136**, 392–9.

Trewartha, G. T. (1968) *An Introduction to Climate*. McGraw-Hill; London and New York.

Tricart, J. (1972) *Landforms of Humid Tropics, Forests and Savannas*. Longman; London and New York.

Trudgill, S. T. (1976) Rock weathering and climate: quantitative and experimental aspects, in Derbyshire, E. (ed.) *Geomorphology and Climate*. Wiley; New York and London, pp. 55–99.

Trudgill, S. T. (1979) *Soil and Vegetation Systems*. Oxford University Press; London and New York.

Turekian, K. K. (1976) *Oceans*. Prentice-Hall; New Jersey and London.

UNESCO/UNEP/FAO (1978) *Tropical Forest Ecosystems: a State of the Knowledge Report*. UNESCO; Paris.

United States Department of Agriculture (USDA) (1975) *Soil Taxonomy: a Basic System of Soil Classification for Making and Interpreting Soil Surveys*. Agriculture Handbook No. 436, Soil Conservation Service; Washington, DC.

Uvarov, E. B., Chapman, D. R. and Isaacs, A. (1979) *The Penguin Dictionary of Science*. Penguin Books; Harmondsworth and New York.

Van Beers, W. F. J. (1958) *The Auger Hole Method for Field Measurement of Hydraulic Conductivity*. International Institute for Land Reclamation and Improvement; Wageningen, Bulletin No. 1.

Vinogradova, N. G. (1962) Vertical zonation in the distribution of the deep sea benthic fauna in the ocean, *Deep Sea Res.*, **8**, 245–50.

Walker, D. (1970) Direction and rate in some British post-glacial hydroseres, in Walker, D. and West, R. G. (eds) *Studies in the Vegetational History of the British Isles*. Cambridge University Press; Cambridge, pp. 117–39.

Walling, D. E. (1971) Sediment dynamics of small instrumented catchments in south-east Devon, *Trans. Devonshire Assoc.*, **103**, 147–65.

Walter, H. (1936) Nahrstoffgehalt des Bodens und natürliche Waldbestrande, *Foystle. Wochenschrift Silva*, **24**, 201–13.

Ward, R. C. (1975) *Principles of Hydrology*. McGraw-Hill; London and New York.

Warren, A. (1979) Aeolian processes, in Embleton, C. and Thornes, J. (eds) *Processes in Geomorphology*. Edward Arnold; London, pp. 325–52.

Warren, V. H. (1972) Biogeochemistry in Canada, *Endeavour*, **31**, 46–8.

Webster, J. R., Warde, J. B. and Patten, B. C. (1975) Nutrient cycling and the stability of ecosystems, in Howell, F. G., Gentry. J. B. and Smith, M. H. (eds) *Mineral Cycling in Southeastern Ecosystems*. National Technical Information Science, US Department of Commerce, pp. 1–27.

White, S. E. (1976) Is frost action really only hydration shattering?, *Arctic and Alpine Research*, **8**, 1–6.

Whittaker, R. H. (1953) A consideration of the climax theory: climax as a population and pattern, *Ecol. Monogr.*, **23**, 41–78.

Whittaker, R. H. (1975) *Communities and Ecosystems*. Collier-Macmillan; London and New York.

Whittow, J. B. (1984) *The Penguin Dictionary of Physical Geography*. Penguin Books; Harmondsworth and New York.

Wiegert, R. G. (1979) Ecological processes characteristic of coastal Spartina marshes of the south-eastern U.S.A., in Jefferies, R. L. and Davy, A. J. (eds) *Ecological Processes in Coastal Environments*. Blackwell; Oxford, pp. 469–90.

Wolman, M. G. and Miller, J. P. (1960) Magnitude and frequency of forces in geomorphic processes, *J. Geology*, **68**, 54–74.

Woodwell, G. M. Whittaker, R. H., Reiners, W. A., Lukens, G. F., Delwicke, C. C. and Botkin, D. B. (1978) The biota and the world carbon budget, *Science*, **199**, 141–6.

Woodwell, G. M., Wurster, C. F. and Isaacson, P. A. (1967) DDT residues in an East Coast estuary: a case of biological concentration of a persistent pesticide, *Science*, **156**, 821–4.

Wooldridge, S. W. and Morgan, R. S. (1937) *The Physical Basis of Geography*. Longmans, Green and Co. Ltd; London and New York.

WMO (1956) *International Cloud Atlas*. World Meteorological Organisation/WMO; Geneva.

Young, A. (1974) The rate of slope retreat, in *Progress in Geomorphology*, Institute of British Geographers, Special Publication, No. 7; pp. 65–78.

Zingg, A. W. (1940) Degree and length of land slope as it affects soil loss in runoff. *Agric. Eng.*, **21**, 59–64.

Index